The Forgotten Genius of Physics, a work on Marian Smoluchowski

STUDIES IN PHILOSOPHY, CULTURE AND CONTEMPORARY SOCIETY

Edited by Bogusław Paź

VOLUME 39

Jan Grzanka

The Forgotten Genius of Physics, a work on Marian Smoluchowski

The Story of Marian Smoluchowski

Translated by Ewan Jones

Berlin - Bruxelles - Chennai - Lausanne - New York - Oxford

Bibliographic Information published by the Deutsche Nationalbibliothek
The Deutsche Nationalbibliothek lists this publication in the Deutsche Nationalbibliografie; detailed bibliographic data is available in the internet at http://dnb.d-nb.de.

Library of Congress Cataloging-in-Publication Data
A CIP catalog record for this book has been applied for at the Library of Congress.

A photograph of Marian Smoluchowski, from the Smoluchowski family collection.

ISSN 2196-0151
ISBN 978-3-631-92493-8 (Print)
E-ISBN 978-3-631-92494-5 (E-PDF)
E-ISBN 978-3-631-92790-8 (E-PUB)
DOI 10.3726/b22427

© 2025 Peter Lang Group AG, Lausanne Published by Peter Lang GmbH, Berlin, Deutschland

info@peterlang.com - www.peterlang.com

All rights reserved.

All parts of this publication are protected by copyright. Any utilization outside the strict limits of the copyright law, without the permission of the publisher, is forbidden and liable to prosecution. This applies in particular to reproductions, translations, microfilming, and storage and processing in electronic retrieval systems.

This publication has been reviewed.
www.peterlang.com

for Pola and Ela

Table of Contents

Introduction		9
Chapter I	Chance	11
Chapter II	Causality	21
Chapter III	Propaedeutics in physics	39
Chapter IV	Nobel	49
Chapter V	A dispute over atoms	57
Chapter VI	Brownian motion	67
Chapter VII	Einstein – Smoluchowski – Sutherland	91
Chapter VIII	Youth and family	105
Chapter IX	Mountaineer	129
Chapter X	Citations	167
Chapter XI	Scientist	173
Chapter XII	Aristotle	193
Chapter XIII	Philosophy	201
Chapter XIV	Beauty in physics	217
Chapter XV	Utility	223
Chapter XVI	Materialism	241
Chapter XVII	Professor	253

Chapter XVIII	The philosophy of physics	281
Chapter XIX	Children	305
Chapter XX	Commemoration	313

Annex 321
    List of Marian Smoluchowski's published works 321
    Publications issued after Smoluchowski's death 356

Conclusion 359

Vernissage – powerful, eternal nature… 363

Acknowledgements 383

List of illustrations 385

Bibliography 395

Index of Names 411

Introduction

The forgotten genius of physics, a work on Marian Smoluchowski is the story of the greatest Polish physicist, all but forgotten by his compatriots.

It seems ironic that the name Smoluchowski is quite popular in my hometown of Gdańsk, but not due to his scientific achievements or mountaineering successes. Smoluchowski had no association with Gdańsk, he was never there and nothing connects him to the city, though quite an important street is named after him in the Aniołki district. Naming the street after him was decided by something of an accident; after the war most of the street names that had German connotations were changed. It is not an average street, however, differentiated by its location, leading to the Medical University with thousands of people walking down it every day. The polish physicist's surname therefore remains in the common awareness of residents, although the majority of them think Smoluchowski was an eminent doctor working at the Gdańsk Medical Academy (the university had that status for decades after the war).

Two streets lead to the Medical University of Gdańsk, the first – for years the more important – led directly to the university's former main gate and bears the name of Maria Skłodowska-Curie. A new, bigger hospital building is currently being constructed on Marian Smoluchowski street running parallel and it is slowly becoming the more important street in the opinion of Gdańsk residents. There is a high probability that the name Smoluchowski will gain in popularity.

In post-war Poland, Maria Skłodowska-Curie was the first lady of Polish science, her personality overshadowing the achievements of many other scientists, including Smoluchowski. This state of affairs was undoubtedly influenced by the objective greatness of our countrywoman, but that was not the only reason for her dominance. The dissemination of information at the time was monopolized and the authorities decided both which names of pre-war science could enter into social consciousness and which were consigned to be forgotten. Changes started in the 1970s when talk started of the Lvov-Warsaw School of Polish philosophy started by Kazimierz Twardowski and a little later of the Lvov school of mathematics – a group of academics active in the two decades between the wars focused around Stefan Banach (1892–1945) and Hugo Steinhaus (1887–1972). In the 1980s we became acquainted with the names of Witold Gombrowicz and Czesław Miłosz. It was only in free Poland, and then not immediately, that talk started of the brilliant logician Alfred Tarski (1901–1983) and the mathematicians Stanisław Ulam and Marek Kac. But the philosopher, logician and methodologist Henryk Mehlberg (1904–1979) remains almost unknown in the country to this day. The partitions of the 19th century and several decades of the communist system in the 20th century caused many outstanding Poles to be ousted from the national memory.

Beyond a short period in the 1950s, Smoluchowski disappeared from general consciousness. The people deciding on the culture of socialist Poland were not interested in paying homage to the brilliant physicist's memory, perhaps because an attempt to make Smoluchowski a scientific icon of materialism failed. The book in the Reader's hands is intended to at least in part fill in this national lack of memory. Smoluchowski's achievements in the field of physics, described herein, have been supplemented with many biographical elements showing the extremely valuable personality of a man with broad interests. Research materials have been used presenting Smoluchowski as a pedagogue, scientist and father of a family, which were quite numerously published immediately after the physicist's sudden death. Many parts have been devoted to Marian Smoluchowski's mountaineering hobby, which he pursued with great love together with his brother Tadeusz, achieving many successes in this field and writing himself permanently into the history of European mountaineering.

Of interest to a Reader curious about the physicist's non- scientific passions may be the *Vernissage* included at the end of the book, in which more than 30 of Marian Smoluchowski's watercolours are presented. The graphic side is also an important aspect of the study – the book contains several dozen photos illustrating the life of Smoluchowski, his family and people connected with him. These will certainly help to bring the Reader closer to the scientist and the times in which he lived.

The monograph's author wanted to reach a wider readership, especially the young. They should learn of the Polish scientist, whose famous achievements – beyond physics – also gained recognition in the realms of the philosophy of science and the philosophy of nature. This work is also intended to show the unusually colourful personality of Marian Smoluchowski, a man of the renaissance with many interests to which he devoted his life.

The forgotten genius of physics, a work on Marian Smoluchowski is intended as a popular science book. Readers interested in deeper and wider analyses of physics, and especially of Smoluchowski's philosophy, are invited by the author to read *Between physics and philosophy. The philosophy of nature and physics in the writings of Marian Smoluchowski*, published in 2020 by Towarzystwo Autorów i Wydawców Prac Naukowych "Universitas". This publication uses fragments of scientific deliberations from that work.

Chapter I Chance

August 22, 1917, was a sunny day heralding the coming golden Polish autumn. Professor Marian Smoluchowski, having been rector of the Jagiellonian University for two months, had another week or two of peaceful existence ahead of him before he fell into the turmoil of preparations for the new academic year, in which he would perform his new role. This peaceful period and the fine weather tempted the scholar to take a last-minute walking tour in the Ojców region. He wanted to relax and also work on his inauguration speech. Such an expedition was no particular challenge for the athletic mountaineer, it was a sightseeing tourist hike in the surrounding hills. As a student, Smoluchowski had conquered pristine peaks in the Alps and Dolomites, and hiking trips in the eastern Carpathian or Tatra mountains, for which he still found time, kept him in good shape. In the afternoon, the professor decided to take a dip in the small Prądnik river. He did not know that a little earlier the same day an Austrian army barracks upstream (we must not forget that there was a war going on) had decided to clean out its latrine. The waste was released into the river running alongside. By a dreadful coincidence, the heavy pollution occurred just at the time the scholar was taking a swim. As a result, the attack of microbes was so serious that it caused a severe attack of dysentery. The great Polish climber, physicist Zygmunt Klemensiewicz (1886–1963) recalled the moment thus: "The yearning of the mountaineer does not die out in him, however; when far from high mountains, he visits small ones in Scotland during his studies in Glasgow, during the Lvov years he skis in the Carpathians in winter, and in the summer of 1917 Ojców, from where he got a fatal disease."[1] This is the only publication containing a subtle suggestion of the cause of Smoluchowski's dysentery.

The residents of Kraków lived in the Austro-Hungarian Empire, a country devastated by the ongoing war. Maybe the disease attacked a human body which had been malnourished for a long time, living in the constant stress of the raging conflict. Smoluchowski was an athletic man in his prime but his body was unable to defend itself and on September 5, 1917, he died at the age of just forty five. His death was determined by chance.

---

1 Z. Klemensieicz, Marian Smoluchowski, "Mountaineer. A unit of the Tourist Section of the Polish Tatra Society," Kraków, 1915–1921, p. 4 (photocopy without number).

Smoluchowski's obituary in 'Nowa Reforma' magazine

Tadeusz Godlewski on the death of Marian Smoluchowski

An accident in the common understanding is a random, unplanned event. For the last few years of his life, Smoluchowski studied the essence and role of chance, a fundamental factor in accidents. He wondered what chance was, could it be researched scientifically? He reduced the essence of chance to an incomplete knowledge of the laws functioning in nature or to an ignorance of all causes at play. The line of research the Pole set out had an influence on the emergence of a serious branch of mathematics – statistical physics, which deals with probability in research in physics.

Smoluchowski thought that by treating chance as a negation of regularity certain contradictions appear that are a source of dilemmas and, moreover, such an understanding of chance cannot be reconciled with the generally prevailing determinism. From a deterministic position, cause and effect are perceived as a constant necessity. We can talk about a lack of necessity – Smoluchowski argued – in a relative sense, in so far as necessity is not externally recognisable. Despite the evident causal relationship between cause and effect, the nature of this relationship is unknowable since the phenomenon itself is too complex, hence the impression of a seeming break with regularity[2]. Determinism demands cause and effect to be treated as events related to one another through internal relationships of necessity and so a visible lack of necessity is an apparent phenomenon resulting from the fact that part of the cause is unknowable[3]. Chance is then defined as a hidden causal relationship existing between cause and effect.

In overly complex phenomena, chance manifests as an apparent break with regularity. Understood in this way, chance was scientifically unacceptable because the probability of an event occurring can depend exclusively on the conditions acting upon it, rather than on the extent of our knowledge.[4] Smoluchowski therefore postulated removing the subject from the concept of chance, which would result in its objectification. Strict natural science is interested not in subjective statements and presumptions but in objective or mathematical probability, i.e. the relative frequency of occurrence of the random events in question. In this narrower sense, the concept of probability becomes accessible in strictly mathematical terms[5]

---

2   M. Smoluchoski, *On the Concept of Chance and the Origin of the Laws of Probability in Physics*, "Wiadomości Matematyczne" (Mathematical News), 1923, volume 27, book 2, p. 29.
3   Ibidem.
4   Ibidem.
5   Ibidem.

A similar problem was presented by the Polish physicist in his work, *Notes on the Concept of Chance in Physical Phenomena*. Smoluchowski posits the thesis that laws of nature are a subject of physics and the assumption that causality and determinism should represent an antithesis to randomness is wrong. It should be considered how chance can arise in phenomena proceeding in accordance with the immutable laws of nature. How can physics describe it with the aid of deterministic laws and is there a place for it in nature governed by those laws? If a truly unpredictable chance event, negating causal regularity, plays a certain role in physical phenomena, how can the correct course of these phenomena be predicted?[6]

Smoluchowski recognised that finding answers to such questions depends on a correct understanding of chance, which in turn can become a factor in applying it to research on probability theory. Assuming that determinism is the fundamental concept for conducting deliberations in this field, he sought to define chance by moving away from its everyday understanding and giving it the status of a scientific notion and then performing further analyses – this time in the context of generally accepted determinism. In the article *On the Concept of Chance and the Origin of the Laws of Probability in Physics*, Smoluchowski conducts an analysis of the relation of chance to stable laws of physics. To the question as to which events fall within the scope of the application of probability theory, he answers that they are usually said to be events whose occurrence depends on chance.[7]

Science studies objective regularities and scientific laws must assume the existence of a causal relationship, which does not mean that the concept of cause can be replaced by the concept of a scientific law. Scientific laws not only clarify when, where and how something happens but they also try to answer the question why as a tool to explain events occurring in nature. Probabilistic statistical forecasts can however be an additional tool for explaining events.

Smoluchowski initially specified two conditions a phenomenon must meet in order to be called accidental. They are "small cause – great effect"[8] and "different

---

6 Ibidem.
7 Ibidem.
8 Idem. *Notes on the concept of chance in physical phenomena*, commemorative book in honour of Bolesław Orzechowicz, Lviv (1916), pp. 445–458, from *The Writings of Marian Smoluchowski* collated and published by Władysław Natanson for the Polish Academy of Arts and Sciences, Vol. 3 Kraków 1928, p 448.

causes – same effects"[9]. The first condition states that a small change in the input parameters causes serious changes to the effect. A small change is understood in a relative sense – it is small compared to the possible range of changes. The second condition describes that fact that in the range of initial states there exist many configurations that lead to the same final state.

Chance plays an important role in physical phenomena and a scientific approach to it required the use of probability calculus methods. Although at the time they had not been sufficiently developed, they had to be used as researching chance through traditional means was ineffective as it went beyond the field of calculus methods.[10] The use of probability calculus persuaded the researcher that to illustrate some phenomena in physics, the language of probability theory can be successfully used as a tool to describe events, which had earlier been questioned.

In the modern period, the application of elements of probability calculus has been quite popular, especially for the requirements of common gambling. Many enlightened minds, such as the mathematician Giovanni Francesco Peverone (1509–1559) or Galileo Galilei (1564–1642) himself considered gambling's mathematical dilemmas[11]. Girolamo Cardano (1501–1576) was the first mathematician to note that scientific rules exist governing probability and the simple throw of the dice can be viewed in terms of mathematical knowledge. He proved that the possibility of getting the elusive "double six" beyond mere luck or chance could be calculated mathematically. He included his thoughts in the book *Liber de Ludo Aleae*, probably written in 1563. However, in popular opinion, Blaise Pascal (1623–1662) and Pierre de Fermat (1601–1665) are considered to be the creators of modern probability theory. In the history of science there are few stories of important discoveries as spectacular as that of the role of Pascal and Fermat, and they are the most memorable. Pascal's interest in stochastic processes was prompted by a chance encounter with the gambler Antoine Gombaud, known as "Chevalier de Méré" (1607–1684). He sought solutions to several dilemmas encountered in card and dice games which were *de facto* mathematical problems.

The classic problem of a game of chance, resolved by Pascal for Chevalier de Méré and entering the history of probability calculus, was a situation in which during a game of dice, betting on the appearance of a six once within four throws

---

9   Idem. *On the Concept of Chance and the Origin of the Laws of Probability in Physics*, op. cit. p. 31.
10 Ibidem.
11 Ibidem.

of the die, the gambler who bet against the occurrence – that a six would not appear – won more often than lost. Wanting to make the game more complicated so that players would not be aware of the system he was using, de Méré added another die and bet on a double six on the roll of two dice. He assumed that the extra die represented only six times more possibilities, so he would have to roll the two dice six more times. This reasoning proved false as after adding another die, he lost more often than he won. In his reasoning, he had used the arithmetic rule of three, which enables the calculation of a fourth value on the basis of the remaining three known values. Pascal found the solution to this problem. He asked Fermat, living in Toulouse, what he thought of his conception. He concluded his positive response with the famous statement: "I plainly see that the truth is the same at Toulouse and at Paris."[12]

Another important figure that made a huge contribution to the development of probability calculus was Pierre Simon de Laplace (1749–1827). For the needs of his deliberations, he created a figure of omniscient intellect, which later became known in science as Laplace's demon. It was a mind that at a given moment knew all the forces of nature and the current position of all the bodies making up the Universe and which would be sufficiently powerful to analyse the data and describe the movements of all the bodies in the Cosmos – from the heaviest to the lightest atoms. To such an intellect, nothing would be uncertain and it would see both the past and the future. According to this concept, "The curve described by a simple molecule of air or vapor is regulated in a manner just as certain as the planetary orbits"[13].

According to Fernando Corbalána, Laplace was not talking about some omniscient mind. The author of *The Taming of Chance* believes that the French mathematician meant the demon metaphorically. It was intended to be a scientific method of calculating the probability of an event's occurrence with the aid of probability calculus, enabling the prediction of nature's behaviours and the learning of its laws. The essence of Laplace's idea is a situation in which the demon knows the future.[14] In probability calculus, this is an extreme chance, a state in which we foresee the occurrence of a given event with 100-percent certainty. In such a situation, the essence of the problem is missed. In writing about the demon, then, was Laplace thinking

---

12 Ibidem. p. 49.
13 A A.P. Juszkiewicz, *Historia matematyki* (The History of Mathematics) Vol. 3, Warszawa 1977.
14 F. Corbalán, *Poskromienie przypadku. Teoria prawdopodobieństwa*, (The Taming of Chance) trans. K. Rejmer, Warszawa 2012, p. 54.

about calculus? It is possible, although according to Smoluchowski, "chance of that type is removed from all a priori calculation and as such can never be the basis for applying probability calculus. Because as long as we do not know the causes with sufficient precision (…), nothing at all can be foreseen regarding the outcome. But when we know, the result can be predicted with certainty so that there remains no place for probability"[15]

However, the demon's powerful mind, which could determine all the consequences of the laws of nature, can be reduced to a calculus of probability developed to know these laws of nature as well as possible.[16] However, even assuming the 100- percent efficiency of Laplace's demon, that is assuming a situation in which we managed to know in physics with absolute accuracy the momentary state of the system of all atoms and calculate its changes, then we will find that the calculus of probability maintains its value[17]

According to Smoluchowski, even absolute knowledge of the initial state of all gas particles would not cancel out the usefulness of probability calculus to describe that system and therefore the utility of the concept of probability itself. The claim can be made, therefore, that for Maxwell's demon[18] the concept of

---

15 Idem. *On the Concept of Chance and the Origin of the Laws of Probability in Physics*, op. cit. p. 33.
16 F. Corbalán, *Poskromienie przypadku* (The Taming of Chance) op. cit., p. 54.
17 M. Smoluchowski, *Uwagi o roli przypadku we fizyce*, (Observations on the role of chance in physics) Towarzystwo Filozoficzne w Krakowie, reading delivered March 1 1917, reprint: "Zagadnienia Filozoficzne w Nauce" (Philosophical Problems in Science) 2017, No 62 (special edition).
18 Smoluchowski evokes a different conception of the demon – proposed by Maxwell. Maxwell's definition of the demon was published in 1867: "if we conceive a being whose faculties are so sharpened that he can follow every molecule in its course, such a being, whose attributes are still as essentially finite as our own, would be able to do what is at present impossible to us. For we have seen that the molecules in a vessel full of air at uniform temperature are moving with velocities by no means uniform, though the mean velocity of any great number of them, arbitrarily selected, is almost exactly uniform. Now let us suppose that such a vessel is divided into two portions, A and B, by a division in which there is a small hole, and that a being, who can see the individual molecules, opens and closes this hole, so as to allow only the swifter molecules to pass from A to B, and only the swifter molecules to pass from A to B, and only the slower ones to pass from B to A. He will thus, without expenditure of work, raise the temperature of B and lower that of A, in contradiction to the second law of thermodynamics" J.C. Maxwell, Theory of Heat, trans. E. Szumilewicz, p. 338, from the service: science20.com, https://pl.wikipedia.org/wiki/Demon_Maxwella

probability calculus would be useful as a convenient tool to describe complex systems[19]. Despite the acceptance of the existence of a demon whose predictive abilities surpass those of a person, it must be assumed that he would not be able to determine all intermediate states without the help of the calculus of probability. Today, in the age of quantum physics, we know that Smoluchowski's final suggestion proved correct. It is not possible to predict the movement of particles with mathematical certainty; it will always be hampered by the uncertainty of velocity (or position). Stephen Hawking notes that even predicting a specific set of positions and velocities ceases to be possible once the existence of black holes is taken into account.[20]

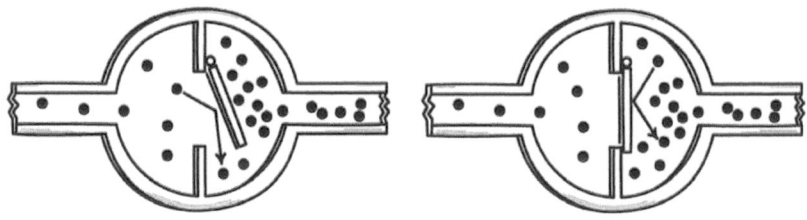

Smoluchowski's demon

Chance suited to the calculation of probability, so-called normalised chance, differs significantly from chance in the wider sense in that the effect shows a certain regularity through frequent repetition of the phenomenon regardless of the type of cause.[21] Such normalised chance enables a predicted event to be included in an empirically verifiable mathematical formula.

(access: 30.04.2020). Maxwell's demon was created for the purpose of building a perpetual motion machine. In essence, his activity is to interfere with nature at the level of cause, so he is in possession of knowledge, just like Laplace's demon.

19 M. Smoluchowski, *Uwagi o roli przypadku we fizyce*, (Observations on the role of chance in physics) op. cit.; P. Polak, *Koncepcja przypadku w pismach Mariana Smoluchowskieg* (The conception of chance in the writings of Marian Smoluchowski), w: *Krakowska filozofia przyrody w okresie międzywojennym* (The Kraków philosophy of nature in the interwar period), Vol. 3, *Smoluchowski – Natanson – others*, red. M. Heller, J. Mączka, Kraków–Tarnów 2007.

20 S.W. Hawking, *Krótkie odpowiedzi na wielkie pytania* (Brief Answers to the Big Questions), trans. M. Krośniak, War- saw 2019.

21 Idem M. Smoluchowski, *On the Concept of Chance and the Origin of the Laws of Probability in Physics*, op. cit. p. 34.

In the search for solutions, Smoluchowski gives an example illustrating a situation in which the values enabling the application of probability calculus are known.

> A shooter fires a fixed shotgun at a spinning circular disc divided into sectors and painted alternately black and white. Whether he hits a black or white sector depends on the moment he pulls the trigger. The disc can be spun so fast that certainty of firing a shot on target can be eliminated. At whatever moment the shooter pulls the trigger, the time that has elapsed from taking the decision to fire will vary within certain limits such that the probability of the shot occurring at time t is expressed by the function $\varphi(t)$ (perceptibly different from zero in the range of t to t + τ), however for this function it must be assumed that there are no unique features. If there are many revolutions of the disc in the time range τ, the influence of the individual form that the function $\varphi(t)$ could have disappears and the probability of hitting a white or black sector depends on the relative size of the fields[22] Smoluchowski disagreed with Meinong's thesis that "between a given cause and effect there exists a causal relationship but it is unknowable to us as the phenomenon is too complex. We are therefore dealing with an apparent break with regularity and chance is defined as 'a partial cause unknown to us'"[23]

He proved that ignorance of the partial cause is no obstacle to calculating probability. It is possible to calculate the effect of a "partial cause unknown to us" by invoking the law of large numbers, a rule that cannot be proven but which has shown itself to be empirically irrefutable. (…) The law of large numbers causes the irregularities wrought in the world by chance events to disappear in the overall result… our mind probably cannot reconcile itself with that in order to accept a similar principle simply because here and there its accuracy has been confirmed[24].

Probability is a mathematical concept, hence with every probabilistic theory comes the issue of its physical and operational interpretation. A frequency interpretation of probability is generally accepted. In line with this, in a set of randomly selected identical systems to which a given theory is applied, certain dynamically possible behaviours should occur with relative frequencies proportional to theoretical probabilities and this consistency is greater the more individual instances we take into consideration[25] We build probabilistic consistency on the basis of the law of large numbers, enabling physical and operational interpretation through the greatest possible use of individual instances.

---

22   Ibidem pp. 35, 36.
23   Ibidem p. 29.
24   Ibidem pp. 31–33.
25   J.J. Sławianowski, *Przyczynowość w mechanice kwantowej* (Causality in quantum mechanics) Warsaw 1969, pp. 37–38.

Chapter II Causality

Initially, Smoluchowski analysed an aspect of the general understanding of causality in which an event is considered clarified if we reduce it to causes that yield actions in a way that seems to us sufficiently known and understandable. This is not only the common way of explaining the causes of accidents occurring in everyday life, but also the typical explanation of that problem in the natural sciences. It is based on the so-called law of causality, which is the result of an unconscious belief acquired in the process of building life experience. The law, which has an almost instinctive nature, proves that every event has its cause and the same causes yield the same results. A concept of cause so defined contains (as well as expediency) anthropomorphic elements transferred from the human psyche to the lifeless external world[26].

Smoluchowski was interested in learning another dimension of the problems of causality which would enable the study of causality in scientific categories. Initially, he studied the epistemic form of causality, closer to its philosophical understanding. Later, he became interested in the ontic aspect, analysing causality from its living side, which influenced the application of methodologies enabling the de-anthropomorphisation of the concept of causality with the prospect of introducing discourse into the realm of science.

In discussing the nature of chance, Ernst Mach's functionalism[27] cannot be omitted. He proposed replacing epistemically understood causality with a

---

26 M. Smoluchowski, *Uwagi o pojęciu przypadku w zjawiskach fizycznych* (Notes on the concept of chance in physical phenomena), op. cit. pp. 167, 168.

27 This concept is interesting for at least four reasons. Firstly, Mach had a negative attitude towards the classical understanding of causality and, being a celebrated figure in the world of science, especially physics and philosophy, he had an influence on the perception and definition of the problem through the scientific circles of his time. Secondly, Smoluchowski certainly knew Mach's philosophical concepts, including those concerning causality; he was interested in his philosophy in terms of the discourse ongoing at the time on the essence of matter. Thirdly, in 1895, Ernst Mach took the chair of philosophy at the University of Vienna and often entered into disputes over kinetics and atomics with Ludwig Boltzmann, who headed the physics faculty. At this time, 1895, Smoluchowski was defending his doctorate in physics and the natural order of things and being at the university undoubtedly took direct or indirect part in the disputes between the two scientists, the subjects of which were of fundamental importance to him. Fourthly, in his scientific work, Smoluchowski paid a lot of attention to the issues of chance and causality, building the basis of probabilistic methodology in physics. It seems inconceivable in this situation for the Polish

relation called functionalism[28] in which the "causal relationship" is a sequence of successive phenomena remaining functionally related to each other. Mach understood functions in the ordinary mathematical sense, which resulted in the removal from science of the concepts of cause and effect, as unclear and ambiguous, and their replacement with the mathematical concept of a function describing the dependence of the characteristics of phenomena.

Mach's conception represented a particular continuation of Hume's critique of the principle of causality, which assumed that between events treated as cause and effect at most only a temporal sequence can be claimed. Mach went further in his reasoning and eliminated temporal consequence in the relation of causality. According to him, the functional dependence of phenomena is reciprocal and reversible and subsequent relations occurring in nature can be defined as two-way and simultaneous dependencies – hence possible to express in mathematical functions. Time in causal relationships can be disregarded because all those relationships – both spatial and temporal – are irrevocably reduced to the functional dependency of phenomena.

Ernst Mach (1838 – 1916)

---

physicist to not have had an interest in Mach's views on causality and hence it must be asked what influence Mach's philosophy had on Smoluchowski's views.

28 B.J. Gawecki, *Zagadnienie przyczynowości w fizyce* (The problem of causality in physics) Warsaw 1969, p. 179.

Mach's functionalism inspired Polish scientists to research causality. At the beginning of the 20th century there appears in scientific circles, chiefly philosophical, a range of publications devoted to the problems of causality, the calculus of probability and theories of large numbers. Smoluchowski undoubtedly knew these works since some of them concerned scholars with whom he had personal contact. An exchange of thoughts, observations and suggestions is visible in the mutual correspondence some of which is located in the Jagiellonian University.[29]

The inspirations of Polish academics like Władysław Gosiewski (1844 – 1911), Władysław M. Kozłowski, Joachim Metallmann (1889 – 1942), Władysław Heinrich (1869 – 1957) or Bolesław Gawecki, with whom Smoluchowski maintained constant contact, can be seen in his works, such as the above-cited *On the Concept of Chance and the Origin of the Laws of Probability in Physics*, as well as *Notes on the concept of chance in physical phenomena* or in the *Self-Study Handbook* (*Poradniku dla samouków*). Reference to philosophical and scientific considerations shows what a serious discourse was in progress at the time among Polish scientists, which could have inspired Smoluchowski to take up the issues of chance and causality.

Gawecki's book *The Problem of Causality in Physics* had an important position concerning causality. Discussing various conceptions of causality and chance, he conducted an analysis of Mach's functional causality and revealed sources that inspired Smoluchowski to create his own conception of chance. It cannot be ruled out that Gawecki's work influenced Smoluchowski's decision to take up research on the problem of causality.

Gawecki built his theory at least from 1913 but his first publication, titled *Causalism and functionalism in Physics*, only appeared in the *Philosophical Quarterly* (*Kwartalnik Filozoficzny*) in 1921, four years after Smoluchowski's death. However, it may have inspired the physicist. Let's look at the chronology. Smoluchowski was versed in Gawecki's conceptions thoroughly and in detail. A doctoral file stored in the Jagiellonian University archive contains a protocol[30] from a rigorosum in physics that was held on January 27, 1914, as the final pre-doctoral exam. The examiners were Władysław Natanson (1864 – 1937) and

---

29 See Smoluchowski's correspondence with Władysław Gosiewski, signature BJ Rkp. 9415 III, vol. 3, Władysław Heinrich, signature BJ Rkp. 9420 III, vol. 8 and Bolesław Gawecki, signature BJ Rkp. 9422 III.
30 Protocol of one-hour rigorosum in physics held with Gawecki on 27.01.1914 in the presence of Władysław Natanson and Marian Smoluchowski with perfect result, card 23, Jagiellonian University (UJ) Archive.

Marian Smoluchowski. On January 30, 1914, Gawecki defended a work titled *Causalism and Functionalism in Physics* and received the degree of Doctor of Philosophy at the Jagiellonian University. In his doctorate, he dealt with Mach's functionalism; Smoluchowski was the promoter[31] of Gawecki's work and promoting him proves that he knew the work itself well.

Bolesław Józef Gawecki (1889 – 1984)

It is therefore no accident that Smoluchowski published his works on causality two years later: in 1916 the essay *Notes on the concept of chance in physical*

---

31 Letter of the Dean of the Faculty of Philosophy of the Jagiellonian University, Jan Łoś, of November 18, 1913, (typescript signed facsimile) in which Professors Witold Rubczyński (1864 – 1938) and Władysław Heinrich were sent Bolesław Gawecki's doctoral dissertation along with a doctoral student's note – a request for acceptance into the rigorosum examinations. Under the dean's request appears an entry handwritten in pen: "Promotion, 14.01.1914, Promotor Prof. Marian Smoluchowski" (card 12 UJ Archive). Smoluchowski was not the promoter in the modern-day sense. As Gawecki writes in his CV, the doctorate was taught by Prof. Heinrich. Smoluchowski gave promotion *sub auspiciis Imperatoris*, which means he had to study Gawecki's doctoral thesis thoroughly. In his 1913 book, Kazimierz Kumaniecki writes: "Pursuant to the regulation of August 28, 1888, *sub auspiciis Imperatoris* promotion may be considered only for a philosophy candidate who has not only passed the necessary examinations, but also demonstrated that his dissertation far exceeds the ordinary measure in terms of scientific value," K. Kumaniecki *Zbiórnajważniejszych przepisów uniwersyteckich* (Collection of the most important university regulations), Kraków, 1913, p. 100.

phenomena and in 1917 the text *Subject, task, method and the division of physics*[32]. In 1918, after his death, an article appeared in German: *Über den Begriff des Zufalls und den Ursprung der Wahrscheinlichkeitsgesetze in der Physik* (On the concept of chance and the origin of the laws of probability in physics)[33].

Władysław M. Kozłowski, a philosopher close to Smoluchowski intellectually, wrote in 1906: "It should be remembered above all that knowledge strives firstly to remove qualities which, being absolutely different, cannot be reconciled. This can be done in two ways: 1. The ontological involves accepting, possibly without a qualitative foundation, the states which serve to explain the quality. 2. The mathematical consists of reducing qualitative differences to quantitative ones"[34]. Smoluchowski proposed differentiating between approaches in the research of quantitative and qualitative phenomena.[35]

Kozłowski was a pragmatist and, although his understanding of causality was more of a philosophical nature, his remarks became an inspiration for the philosophising physicist. The statement: "Causality in the world of phenomena, as a result of the ontological shaping of our fundamental concepts of it, naturally assumes the nature of action, whatever critical-philosophical concept we have of it,"[36] – was for Smoluchowski a suggestion to rethink the phenomenon of causality in terms of an analysis of the understanding of chance. The most important change in the perception of causality was an evolution in the understanding of the role of chance appearing in cause and effect. Smoluchowski defined chance as "a certain specific species of causal references."[37]

During the period of Gawecki's work on causality, Kozłowski was valued in Polish science as a philosopher specialising in research on causality. Gawecki knew Kozłowski's research; he wrote several books about him. In 1906, Kozłowski

---

32 *Przedmiot, zadanie, metoda oraz podział fizyki* (Subject, task, method and the division of physics) in *The writings of Marian Smoluchowski*, vol 3, op. cit.
33 "Naturwissenschaften" (Natural Sciences) 1918, vol. 6 (17), s. 253–263
34 W.M. Kozłowski, *Przyczynowość jako podstawowe pojęcie przyrodoznawstwa* (Causality as a fundamental concept of the natural sciences) Warsaw, 1906, p. 29.
35 M. Smoluchowski, *Uwagi o roli przypadku we fizyce*, (Notes on the role of chance in physics) op. cit., p. 38.
36 Idem *O fluktuacjach termodynamicznych i ruchach Browna* (On thermodynamic fluctuations and Brownian motion) "Prace Matematyczno-Fizyczne" (Mathematical-Physical Works) 1924, t. 25, s. 187–263, cit. after *The Writings of Marian Smoluchowski* collated and published by Władysław Natanson for the Polish Academy of Arts and Sciences, vol 2, Kraków 1927. P. 299
37 Idem *On the Concept of Chance and the Origin of the Laws of Probability in Physics*, op. cit. p. 37

published a small study, *Causality as a fundamental concept of the natural sciences*. In it, he proved that "the notion of dependence is broader than the notion of causality, which is one of the forms of dependence (causal dependence); than the notion of function, which is dependence according to a quantitative-defining law. Therefore, we should add to the concept of dependence a more *precisely* defined characteristic in order to get causality. We find this characteristic by considering one of the problems that runs through the whole development of the philosophical concept of causality: the problem of the temporal succession of cause and effect"[38].

Kozłowski's considerations on Mach's conception are visible in Gawecki's *The problem of causality in physics*. Kozłowski points out that Mach wants to eradicate the notion of causality and replace it with two principles: the principle of continuity and the principle of differentiation. He states that:

> Temporal and spatial relations are useless and lead to confusion in the concept of causality. Time and space, as scientific concepts, are "abstract" auxiliary hypotheses. The "concept" of time arises through changes and has significance only as long as changes exist. Temperature changes with time – that means it is dependent on the angle of the Earth's rotation. Therefore, spatial and temporal relationships are ultimately reduced to the interdependence of phenomena, which replaces all basic relations coming from outside (time, space and causality). Therefore, in place of a "metaphysical" dependence between cause and effect, there will eventually be a "purely logical" relationship between "the conceptual elements determining the fact"[39].

Władysław Mieczysław Kozłowski (1858–1935)

---

38 W.M. Kozłowski *Przyczynowość jako podstawowe pojęcie przyrodoznawstwa* (Causality as a fundamental concept of the natural sciences) op. cit. p.37
39 Ibidem pp. 16–17

Kozłowski analyses the schema of a specific example proposed by Mach. Breaking away from the temporal and spatial form of phenomena, we realise the relation of causal to functional dependence in the following way: we have a clockwork mechanism in which all the cogs intermesh and there is no hindrance to movement in either direction. By turning any cog left or right we can set the whole mechanism in motion in the appropriate direction. But if each cog has catches (like those on a watch with a mainspring) allowing movement only in one direction, and not allowing it in the other, such a mechanism will be able to be moved only in one direction and moving any cog will set in motion only those following it (in the direction of the designated movement), the cogs preceding it remaining static.

The first mechanism, as Kozłowski indicates, presents functional dependence, the second a causal relationship. The first joins the moving parts in such a way that they are all dependent on each other and this dependence manifests in any direction; the second creates a one-way fixed chain (the advent of a cause elicits an effect) and all the subsequent ones (not in time but in the direction determined by irreversibility) but nothing changes in the previous one, there is no dependence here from effect to cause i.e. in reverse (in the opposite direction).

> We can now answer the question (…) whether the mathematic form of a law of nature expresses everything contained in that law. Here of course we do not mean that functional dependence expresses only a quantitative relationship while laws of nature express a relationship between the qualitative contents at play; that substitution of quality with quantity is justified and does not impede every specific interpretation of the law. Mathematical symbols are used in this case like nominative numbers, indicating not only how many but what. What interests us is whether beyond the form of a function itself something implicit lies hidden without which this form would lose all connection with reality and therefore all non-mathematical meaning. Such an assumption exists and represents an indispensable component of all mathematical formulae expressing laws of nature: this is that the dependence relationship exists here only in the direction from certain parameters to others, it [is – ed.] not two-way, as the mathematical form of the law would lead us to suppose.[40]

It follows from Kozłowski's further reasoning that every mathematical function, when used to express a law of nature, contains the implied stipulation of a one-way dependency of parameters – "one-way dependence is the nature of causality. Therefore the previous theorem means that every mathematical function, insofar as it expresses physical relations, contains an implied causal theorem. Hence it

---

40 Ibidem pp. 21–22

follows that not only can functional dependence not replace causality but that it is through it that it acquires meaning in application to real-life phenomena."[41]

Kozłowski's thought guided Gawecki's understanding; he also disagreed with Mach's thesis, though he justified his position more forcefully, asserting that the proposed reform consisting of the elimination of the temporal corollary and introducing invertible functions cannot be applied to all types of physical relationships.[42] "Not all relationships found in physics can be interpreted only as mathematical functions: apart from Mach's function, there are also essentially irreversible functions, which can be called "physical" functions where there is an element of time, differentiating one of the terms, when B follows A but A cannot follow B. This happens in a situation in which the order in which the states of a phenomenon occur after each other is constant, determined by nature itself"[43].

Gawecki made a distinction between causalism, operating through physical functions (understood as a certain dependence relationship between the quantities characterising the parts of a given physical phenomenon, following each other in time in a certain defined order) and functionalism utilising reversible functions, which Mach gave exclusivity in science.[44]

The irreversibility of succession in this group of phenomena cannot be eliminated. This is not some intellectual invention but stems from the real world we study directly, and this directness is the source of the causal account of phenomena, known as causalism. "Not all relationships found in physics in a mathematical form can be interpreted simply as mathematical functions" Gawecki stated.[45]

The main source of error, Gawecki writes, was Mach's prejudice against the causal method as against a metaphysical superstition, which resulted in extreme mathematisation of the relation. Rather than replacing a pre-scientific concept of cause with a scientific concept, Mach entirely ruled out causality, claiming that functionalism is fully sufficient.

In reversible phenomena, "cause" and "effect" do not express any content justifying their application in science.

> To take account of reversible phenomena, the concept of function is sufficient in the mathematical sense. However, in the case of irreversible phenomena, there is no way to eliminate the temporal element; hence it seems expedient to apply to these phenomena

---

41   Ibidem p. 23
42   B.J. Gawecki, *Zagadnienie przyczynowości w fizyce*, (The problem of causality in physics) op. cit., p. 105
43   Ibidem
44   Ibidem, p. 106
45   Ibidem, pp. 105 – 106

a causal (...) method operating through temporally interpreted "physical functions". Indicating a certain defined direction of the permanent succession of real phenomena, it can be treated as a causal clarification of these phenomena. The general determination of natural phenomena should not be prejudged: after all, we are talking here about the nature of a causal law for physics rather than about the principle of causality. The assumption that every phenomenon is related in a defined way to other phenomena applies both to reversible and irreversible phenomena.[46]

Gawecki's and Kozłowski's above-presented deliberations on Mach's functionalism were a source of inspiration to Slomuchowski in building a conception of chance. Gawecki's reasoning lead him to a particular way of perceiving causality. Seeing the validity in separating the causal and functional understanding of causality, Smoluchowski understood that these two conceptions need not be mutually exclusive but, conversely, can complement one another, being understood dually in a way in which there is room for both causalism and functionality. The breakthrough here is the differentiation of perceiving causality epistemically and ontically (cognitive and living), where in the first instance the participation of the time factor is permissible, and in the second it is superfluous.

Proof of the existence of a causal relationship between the energy state of particles, leading to fluctuating collisions, and perceptible movements in rubber resin was persuasive and obvious to Smoluchowski. Since the year 1900, he had been convinced of the molecular-kinetic nature of Brownian motion[47] and an epistemic recognition of cause and effect was to him an explanation of the event occurring. According to the formula of Bunge's causality principle, the effect is permanently and necessarily correlated with the cause, because it is generated by the cause. In this same way, the Polish physicist applied the causality principle to the phenomenon of Brownian motion.[48]

Smoluchowski treated the proof of the causality of Brownian motion as an argument for epistemic causalism. The study of this phenomenon was a search for an epistemic cause (according to Gawecki – 'causal' causality), of the vibrations

---

46   Ibidem pp. 183 – 184
47   M. Smoluchowski, *O fluktuacjach termodynamicznych i ruchach Browna* (On thermodynamic fluctuations and Brownian motion), op. cit. p. 299
48   In the theory of diffusion, Smoluchowski connects the macroscopic phenomenon of viscosity to the microscopic concept of the mean free path of a molecule. In his works devoted to the problem of Brownian motion he wrote explicitly that he was striving to clarify the internal mechanism of diffusion and relate it to the phenomena of molecular movements. In his view, the macroscopic phenomenon of diffusion is a manifestation of molecular movement or density fluctuations.

of molecular particles resulting in the transfer over time, through collisions, of the energy of solvent particles to molecules in suspension. Using statistics, he built a mathematical apparatus enabling the scientific study of the causes of Brownian motion, creating a research method that involved collating statistics on the shifts achieved by particles over specified times.[49]

Moving away from philosophical considerations, he conducted mathematical proof clarifying the origin of Brownian motion. Despite a critical view of epistemic causality, he did not abstain from its application in scientific research. Accepting positivistic criticism, he adopted for research purposes the principle of causality "cleansed of obscure human and metaphysical impurities, as the quintessence of all experiences and observations, all of which confirm the unchanging regularity of nature"[50]. He emphasised its value in the form he outlined, arguing " we also acknowledge as absolutely right those who hold the assumption of causality in this form to be a cardinal condition of thinking about nature"[51]. The case presented illustrates a causal and epistemic understanding of the cause-effect relationships in irreversible physical phenomena. A reading of philosophically understood causality was not limited to learning its epistemic nature, because by applying a research method involving the collation of statistics on the movements of particles in solution, Smoluchowski introduced mathematical relations into causality, enabling the calculation of the effect by a scientific method. Presented this way, the issue did not pose a methodological problem due to the mathematical tools applied. At the same time, he started to analyse another aspect of chance – the ontic one, i.e. the living manifestation of the effect's coming into being. He established how ontic causality should be defined and how to understand chance in order to be able to introduce a

---

49 Smoluchowski understood that if kinetic molecular theory, operating through statistical methods, predicts the possible existence of processes involving a departure from the course of phenomena normal in the macro-world, the researcher's task is to theoretically analyse the conditions in which it would be fair to expect them to come about and to predict and reproduce quantitatively a hypothetical picture of their course. One of the phenomena that particularly lends itself to this purpose, namely Brownian motion, had already been indicated by Helmhotz, but physics owes to Smoluchowski the development of a mathematical method enabling this motion to be described and studied in detail.
50 M. Smoluchowski, *Uwagi o roli przypadku we fizyce*, (Notes on the role of chance in physics) op. cit., p. 24
51 Ibidem, p. 25

scientific methodology enabling calculations to be performed. The mathematical tool to achieve this goal was to be the application of the calculus of probability.

In the essay *On the Concept of Chance and the Origin of the Laws of Probability in Physics,* Smoluchowski analyses the problem of chance in specific examples. He examines typical cases occurring in unstable (changeable) states. Grasping the ontic aspect of causality is an important step towards scientific research of a certain type of phenomenon in nature. We gain a conception of the notion of causality enabling the application of new mathematical methods to describe events, such as statistical calculation and the calculus of probability. Taking a deterministic standpoint, we treat cause and effect as a permanent necessity connected by the internal relationships linking partial events. An ontic account of causality enables a description of the physical effects occurring with the aid of probability calculus. This stems from the possibility of defining the mathematical relations connecting partial events to the effect. Smoluchowski's idea of an ontic understanding of causality can be seen in a particular approach to Mach's conception since his line of reasoning made the Polish physicist aware of the possibility of disregarding time in causal relationships and including causality in mathematical functions, however with the difference of not analysing causality in its epistemic aspect but studying its ontic nature. That was the intention of another perception of causality, which was related to a shift in considerations of causality from the philosophical to the methodological space of science, especially of physics. Interpretation of the essence of chance became the tool enabling this intention.

Smoluchowski arrived at the essence of causality's ontic nature through analysing chance as a special type of causal relation. In this approach, a given event depends on chance, which is a function dependent on a variable causing the event. This cause satisfies a certain law of probability and is primary in relation to the effect, for which we note the constancy of the law of probability. The occurrence or non-occurrence of an event depends on a very small change in the variable constituting its cause. The key issue is the occurrence of the event itself in terms of cause and effect, disregarding the factor provoking the relation, while focusing on the statistical possibility of its occurrence.

This philosophical *résumé* of various functions of chance brings us closer to its nature but does not touch on the function of chance that interested Smoluchowski, which would enable the application of probability theory to calculate the probability of its occurrence. A chance lending itself to calculation of the resultant effect through probability calculus is dependent only on the appropriate frequency of a situation's occurrence and its repeatability. The effect does not depend on the probability of an earlier chance event occurring but only on the sequence of further chance events. We can consider a situation safe when we are dealing with relative frequencies falling within the scope of the law of large numbers and the essence of chance is reduced to the law of probability bringing about the cause of an event. A situation cited by Gosiewski refers to the research of Andrey Andreyevich Markov (1856–1922), and specifically to the so-called Markov property. Gosiewski's conclusion is of great importance to an understanding of Smoluchowski's conception in the context of the possibility of mathematically determining the occurrence of a chance event with the aid of probability calculus. "Chance lending itself to a calculation of probability," Smoluchowski wrote, "differs from chance in the broader sense due to an essential and characteristic property whereby the effect shows a certain regularity through frequent repetition of the phenomenon, regardless of the type of cause"[52].

An important change in the understanding until this point is the shift of attention from the cause to the occurrence itself of an event defined as a chance event which takes place through the cause-effect relation. In this situation, consideration of the element provoking the causality relation is not important to the researcher, i.e. researching (in the Aristotelian sense) the cause, and the study of a direct relation between cause and effect is not paramount. We study the effect itself and in fact the statistical possibility of its occurrence. We focus on the possibility of calculating the ontic side of the effect of the causality under examination.

The subtle difference causing a shift of emphasis from perceiving the occurrence of a causal event to the effect achieved makes it possible to calculate this result using the tool that is the calculus of probability. Its use enables the event to be verified mathematically and scientifically. Redefining the problem enables chance to be located within the realm of physics. Expressing the occurrence of an event through the use of a mathematical formula was of tremendous importance as it enabled a scientific definition of the problem under consideration. This became possible as a result of shifting the emphasis of the understanding of causality from a discourse seeking the cause of a given effect to deliberations concerning the possibility of a chance occurrence.

Chance is that aspect of an event's occurrence which we can put in mathematical terms. Let's analyse the example of the breakdown of the element radium, which will illustrate the essence of Smoluchowski's concept. The half-life of the radium 226

---

52 Idem *O pojęciu przypadku i pochodzeniu praw fizyki opartych na prawdopodobieństwie* (On the Concept of Chance and the Origin of the Laws of Probability in Physics), op. cit. p. 34

isotope is 1,599 years. Over this period, the semi-decay of any specified quantity of radium occurs. There is no way to determine or know which atoms undergo decay, over what time or due to what cause. However, this is no obstacle to determining over what time half the isotope atoms will decay. In this case, the most important problem is not of an epistemic, philosophical nature, because we are not asking what the cause of the decay is, but are focusing on the statistical aspect of the occurrence of chance, and hence on the possibility of calculating it. The ontic fact of decay becomes the most important thing, as a result of which we achieve the effect in the form of a different number of radium 226 atoms. This causality is obviously of a philosophically causal nature but due to the omission of the element of knowing the cause, we are working in the physical realm of this event's ontic occurrence.

We are dealing with a similar situation in analysing an electron changing orbit in an atom. The passage of the electron from any one orbit to another is an event occurring with no visible cause. Because no reason has been found for the movement of an electron at a given moment from one shell to another, it cannot be determined when an electron will transition from a state of higher energy to a state of lower energy. Here we are dealing with chance events for which no known cause exists. An electron does not move at a given moment from one level to another due to some specific known cause. From a statistical point of view, the lower level is more desirable than the higher one, so sooner or later the electron will jump, though we cannot predict when this will happen[53].

The claim that there exists no cause in the random events mentioned is a too far-reaching conclusion and it could be agreed that there is no cause, but only in a given frame of reference. The cause is unknown to us, which does not mean it does not exist. In Smoluchowski's theory, cause is not analysed not because it is assumed to not exist but because in a statistical account of the effect its existence or non-existence has no real significance. Ignorance of a phenomenon's cause does not affect an estimation of the probability of its occurrence.

In his book devoted to the philosophical problems of modern physics, Werner Heisenberg treats objective probability as a measure of potential related to Aristotelian potency. However, he emphasises that in statements of physics, apart from objective probability, there also exists probability of a subjective nature, and argues that probabilistic statements in quantum physics contain their own "mixture" of objective and subjective probability[54].

---

53 A. Lemańska, *Determinizm* (Determinism), https://www.kul.pl/files/57/encyklopedia/leman- ska_determinizm.pdf (dostęp: 1.11.2021).
54 S. Amsterdamski, Z. Augustynek, W. Mejbaum, *Prawo, konieczność, prawdopodobieństwo*, (Law, necessity, probability) Warsaw 1964, p. 75.

Smoluchowski's proposition offers an extremely important and effective mathematical tool subjecting new areas of reality to empirical research. The dilemma in modern physics of whether indeterminism is ontological indeterminism or whether there also exists some deeper level of reality according to which quantum phenomena are determined, is one of the main problems for physicists and philosophers, provoking discussions which still seem far from unambiguous resolution[55].

Marian Smoluchowski during the Lviv years; the picture contains the dedication: "Jadziuli's beloved brother Marian". Most probably, the picture was intended for the sister of Smoluchowski's wife – Zofia

An undoubted merit of Smoluchowski was the distinction between the philosophical and physical understanding of causality, chance and probability. Mach's functionalism, while referring to mathematical functions, was philosophically postulative and hence quite severely biased. The scientific aims and the methodologies of the two academics focused on different aspirations. For Smoluchowski,

---

55   A. Lemańska, *Determinizm* (Determinism), op. cit.

who saw the possibility of applying probability calculus to physics, the phenomenological approach proved a dead end in researching the problem. He clarified the difference in the physicist's and philosopher's approach to the problem in the following way:

> The various philosophical analyses of the concept of probability provide no great clarification in this regard. In general, the philosopher is usually concerned here with something entirely different to the physicist. The philosopher (...) is not in the habit of considering the question more closely as to the nature of objective facts that form the basis of probable propositions. In contrast, the strict natural sciences are not interested in subjective statements and presumptions, justified or not, but are interested in objective or "mathematical" probability, i.e. the relative frequency of specified chance events occurring.[56]

The characteristic Smoluchowski outlines of the physicist's and philosopher's differing approaches in perceiving the occurrence of events does not represent a declaration of a radical separation of the strict sciences from philosophical concepts.

Smoluchowski distinguishes between different understandings of chance depending on the means of its perception; conceptual chaos and difficulties arising from studying the fundamentals of probability calculus are caused by the varying perspectives of the representatives of the three different fields in which the problem is considered: mathematics, physics and philosophy[57]. Smoluchowski's point is to look at the problem of chance from a meta-scientific position, enabling a departure from colloquial language and defining a different way of perceiving the same problem. The mathematician[58] concerns himself with the formal side of probability calculus, calculating the probability of a complex phenomenon on the basis of component, elementary phenomena. The physicist, in defining what he considers chance, is usually guided by some intuition and verifies the consequences of probability calculus experimentally. The philosopher, however, is interested in the subject's psychological aspect or

---

56 Idem *O pojęciu przypadku i pochodzeniu praw fizyki opartych na prawdopodobieństwie* (On the Concept of Chance and the Origin of the Laws of Probability in Physics), op. cit. p. 30
57 M. Stawarz, *Punkt wyjścia filozoficznych rozważań Mariana Smoluchowskiego na temat przypadku i prawdopodobieństwa*, (The starting point of Marian Smoluchowski's deliberations on chance and probability) "Semina Scientiarum" (The Seeds of Science) 2008, nr 7, p. 84.
58 An extract of Smoluchowski's handwriting concerning deliberations on probability, found by Małgorzata Dziekan, is similar in content to the essay, *Notes on the role of Chance in physics*.

considers the problem of how to fit probability into a system of formal logic. However, he avoids the issue of what objective conditions external phenomena must be subjected to in order to explain the application of the concepts of chance in probability.[59]

In terms of the mentioned conceptual chaos resulting from an attempt to move away from the common understanding of the issue and seek a scientific approach to the problem, the perception of chance, according to Smoluchowski, is determined by the position taken. Epistemically guided philosophy is not interested in the practical use of the concepts of chance and probability. The philosophical approach to probability theory is epistemological in nature. The philosopher is interested chiefly in the problem of fitting probability into the system of formal logic, hence shifting considerations of the nature of chance onto the ontological plane changes that perception, bringing the study of chance into the realm of science and enabling the practical application of probability calculus.

The evolution presented above of introducing chance and the calculus of probability into the realm of science, and particularly physics, was initiated by Smoluchowski, who is sometimes called the pioneer of statistical physics. This fact is not widely known and the few statements by researchers appreciating his contribution to the development of the field only highlight the excessively modest mention in literature of the Polish physicist's achievements in applying probability theory to physics.

Joachim Metallmann wrote: "I would see Smoluchowski's merit in that in place of the sterile notion of subjective chance, as a negation of determinism, he simply introduced (…) statistical regularity"[60]. In turn, Ł. Storczak highlighted that: "It has transpired that Smoluchowski's deep investigations were necessary in order to finally establish that statistical regularity is an entirely new type of regularity, strictly determined by physical conditions"[61].

Mark Kac emphasised: "Smoluchowski certainly did not realise that he had started to write a new chapter of statistical physics, known today as stochastic processes. (…) The novelty and originality of Smoluchowski's approach lies in his bold replacement of an impossibly difficult dynamic problem (…)

---

59 M. Smoluchowski, *Notes on the role of chance in physics*, handwritten manuscript, Jagiellonian University signature 9398 IV, k. 3.
60 J. Metallmann, *Zagadnienie przypadku*, (The problem of Chance) "Przegląd Współczesny" (Modern review) 1933 year XII, vol. XLIV, p. 93–94.
61 Ł.I. Storczak, *Diskussija o prirodie fiziczeskogo znania*, (Discussions on the nature of physical knowledge) "Woprosy fiłosofii" (Questions of philosophy) 1948, no. 1.

with a relatively straightforward stochastic process"[62] and the brilliant mathematician Stanisław Ulam, in analysing the contribution of physicists to the creation of mathematics, wrote: "In some more specific actions of mathematics – for instance the calculus of probability – physicists such as Einstein and Smoluchowski opened up new fields even for mathematicians"[63].

Stanisław Loria, in summing up this activity of Smoluchowski, stated:

entering the realm of statistical physics, Smoluchowski found problems most suited to the character of his mentality and theoretical talents. A. Sommerfeld expressed the same opinion in a concise but highly accurate characterisation: "Für Smoluchowski war die Statistik Lebensluft" (Statistics was life for Smoluchowski). Importantly, since then, the productivity of his work has increased immeasurably; the horizon of theoretical generalisations has broadened, now revealing to him the relationships between phenomena, hitherto seemingly distant from one another; the understanding of the deeper philosophical meaning of the achieved and achievable results of knowledge has matured[64].

As written in the introduction – chance is a random, unplanned event. Gawecki's doctorate prompted Smoluchowski to take up the study of the essence of chance for several years. Chance also had it that in August 1917, in a place where in the whirlwind of the war devastating Europe there seemed to be an oasis of safety and calm, Smoluchowski's life came to an abrupt end. Chance acted in just the way the Polish scholar had characterised it. Smoluchowski widowed a thirty-year-old wife and orphaned a fifteen-year-old daughter, seven- year-old son, and also Polish science and physics.

---

62   M. Kac, *Marian Smoluchowski and the Evolution of Statistical Physics*, w: S. Chandrasekhar, M. Kac, R. Smoluchowski, *Polish Men of Science. Marian Smoluchowski. His Life and Scientific Work*, ed. R.S. Ingarden, Warszawa 1986, p. 17.
63   S.M. Ulam, *Przygody matematyka*, (Adventures of a mathematician) trans. A. Górnicka, Warsaw 1996, p. 322
64   S. Loria, *Marian Smoluchowski i jego dzieło* (1872–1917), (Marian Smoluchowski and his work (1872–1917)) "Postępy Fizyki" (Advances in Physics) 1953, vol. 4, book. 1, p. 12.

Chapter III Propaedeutics in physics

Fourteen months after Smoluchowski's death, World War I ended, Poland regained independence and the distinguished researcher was absent just when he would have been particularly needed – in a period when the structures of Polish science were being built. The professor had a great deal of experience, gained at universities in Vienna, Paris, Glasgow and Berlin. He had lectured in Münster, Cambridge and Göttingen and also worked at the famous Cavendish Laboratory.

He had headed faculties at the Lviv and Jagiellonian universities. In 1916 and 1917, as dean of the Philosophy Faculty, he had gone to Vienna to win funding for research, and travelled to Germany and Austria to scientific institutions in Leoben, Frieburg and many others[65]. He was also very interested in the level of Polish science and the quality of science teaching at secondary schools.

The experience gained at European universities bore fruit in ideas both in the field of the organisation of studies, as confirmed by the essay, *The Organisation and activity of physics facilities*, in which Smoluchowski considers the idea of creating scientific institutions unrelated to university centres:

> That is why the creation of facilities unrelated to any higher school, devoted exclusively to research, without additional didactic aims, is so important to the progress of science. Such facilities have long existed in the fields of astronomy, meteorology, geophysics, but in physics they are a relative novelty. An outstanding initiative towards these pursuits was the creation of the "Kaiser Wilhelm Forschungs Institut" in Dahlem near Berlin, which is intended to fulfil the programme outlined above, although so far this programme covers only certain fields (physical chemistry), among those that fall within the scope of general physics. In other fields, similar activities have been developed for many years by the science departments of state institutions: The Physikalisch- Technische Reichsanstalt in Berlin, the Bureau des Poids et Mesures in Paris, the National Physical Laboratory in Teddington near London and the Bureau of Standards in Washington. They have already played an outstanding role in science; they would undoubtedly be even higher than that if the undeniable truth had been remembered during their founding that the activity of such an institution depends primarily on the abilities of its staff, with the facility's equipment coming only in second place; in other words, that people, not the facility, create science[66].

---

65 S.M. Ulam, *Marian Smoluchowski and the Theory of Probabilities in Physics*, Los Alamos, New Mexico 1956, p. 478.

66 M. Smoluchowski, *Organizacja i działalność zakładów fizycznych* (The organisation and activity of physics facilities) "Polish science, its needs, organisation and development," Rocznik Kasy Pomocy dla Osób pracujących na polu naukowym, im. Dra

Today, Smoluchowski's comments may seem banal, but at the time, in a Poland being reborn, they were revealing. Such people with authority and knowledge were needed then in Polish science.

Smoluchowski was a pioneer in practising the propaedeutics of science. He was very committed to the popularisation of the sciences among young people in academies and high schools. He devoted a great deal of time to lectures for teachers, travelled with lessons to conferences and meetings and wrote articles for pedagogic publications. He was deeply involved in raising the level of student education in the natural sciences and was aware of the enormous deficit left by the period of Poland's partitions.

The entrance to the former Cavendish Laboratory building

Józefa Mianowskiego"(Yearbook of the Józef Mianowski Aid Fund for People working in the scientific field), vol I, Warsaw, 1918, pp. 19–25 cit. after *Pisma Mariana Smoluchowskiego* (The Writings of Marian Smoluchowski) vol 3, op. cit. p. 236.

Tadeusz Godlewski (1878–1921), author of *Marian Smoluchowski, his life and scientific activity*, wrote:

> Smoluchowski played the most active role in work on the development of natural knowledge, for the whole of his life and in every way available to him: As a researcher and scholar – through his works and publications and by participating in scientific events; as a professor – through original and impeccably prepared lectures (large parts of these lectures were issued as lithographs by the audience, with some of the script corrected by him), by running seminars and workshops, guiding independent work. (…) He always took lively and active part in scientific meetings of the Copernicus Society of Naturalists, of which he was president in 1906 and 1907. In 1909, a physicists' group was established at this society at his initiative, tasked with cooperation and discussion of the latest scientific work, of which Smoluchowski was the founder, and until the end of his time in Lviv he was the president and the most active of all the members. In Kraków, as the director of the Institute, he knew how to focus around himself a group of young staff members, who were drawn by his unique individuality but who, like all his students without exception, and like everyone else who came into close contact with him through work, soon learnt not only to admire him with all their soul, but to love him with all their heart. Unfortunately, first war and then later death permanently severed this collaboration[67].

Describing the current state of Polish society's scientific knowledge in *Mathematical News*, Smoluchowski did not limit himself to listing the negative aspects of "social anaemia" but also outlined the direction it should move in to catch up with more developed societies:

> Enthusiasm for truth, fanatical strivings for integrity and truth – this is the ethical bedrock appropriate to these [natural – J.G.] sciences and reinforcing them. They are doing battle against bluff and cliché, the diseases that afflict our society and distort our literary language. We must admit that our traditional classical upbringing is in large measure the cause of the social anaemia well known to us and the tendency towards bureaucratic formalism; it can, at best, evoke in more noble individuals a love of passive aesthetic contemplation, or a literary nostalgia for times long gone. We do not need that! We have dozed in lethargy through many decades during which the world was forging ahead at a frantic pace[68].

---

67 T. Godlewski, *Marian Smoluchowski. Jego życie i działalność naukowa*, (Marian Smoluchowski. His life and scientific activity) "Wiadomości Matematyczne" (Mathematical News) 1919, vol. 23, books. 1–3, pp. 26–27.
68 Ibidem p. 26

Marian Smoluchowski during a stay at the University of Humboldt (picture taken in Berlin probably in 1897)

This statement represents an exception to the scholar's position since Smoluchowski very rarely referred to social or national issues.

It is worth remembering that the rational teaching of physics was very important to Smoluchowski, as he underscored in May 1917 at a meeting of the Association of Higher School Teachers in Kraków in the paper *On the importance of the pure sciences for general education in secondary school*. He sought with truly fiery energy to show what enormous social and national importance an increase in, and the rational teaching of, science in secondary schools would have.

> The natural sciences – Godlewski summarises Smoluchowski's thinking – (...) shape and develop the ability to observe, develop perceptiveness and independence of judgement; but, in order to fulfil this goal, they must be taught not *ex cathedra*, but acquired first-hand by a student in the laboratory. They lay the groundwork for scientific understanding, both inductive and deductive, they teach logical rigour, stimulate philosophical reflection. But they also shape the student's ethical side, as, recognising only the laws of nature, they have always been a remedy to and antidote against blind faith in authority, against slavish mental servitude[69].

---

69 Ibidem pp. 25–26

These were not merely general remarks. Smoluchowski was critical of the high-school textbook widely used in the Kingdom, which contained dogmatic definitions of no didactic or scientific value. The sciences did not start with a description and analysis of physical phenomena but by forcing at best debatable definitions into the child's mind. He considered strange the question: "What is a body?" to which the answer was: "A body is any amount of a material made up of molecules"; he considered inaccurate the definition of physical change,

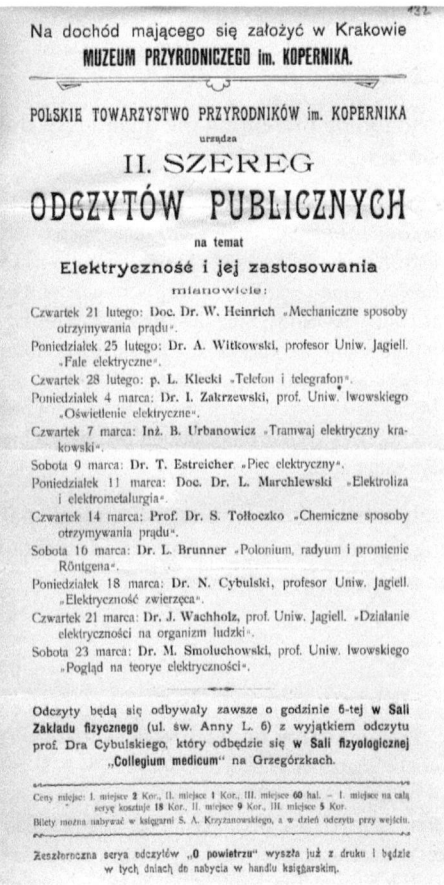

Smoluchowski regularly took part in lectures on the propaedeutics of physics

according to which it was "such a change through which the nature of molecules does not alter"[70].

Smoluchowski advocated greater emphasis being put on the study of scientific subjects, the value of which to the process of shaping young minds he considered underestimated:

> In the secondary school system, the main principle should not be the slogan: "science for science," but rather: "science for life." I think among things that the mathematics teaching programme must be adapted to practical needs; that, in defining the dimension of that study, we must treat it mainly as a tool. (…) It is also understood that, in line with this view of the role of the pure sciences, I consider some extension of the time devoted to the study of mathematics essential[71].

He justified this conviction by presenting the qualities of the pure sciences that are lacking in the humanities.

> How much higher is mathematics, whose laws are not externally imposed dogma but stem from internal necessity, from inexorable logic, such that anyone who cares to think can judge their veracity. It is the best school of pure logic and as such is a truly beneficial factor in general education regardless of the usefulness of its content. (…) There is another important matter related to this: the relationship of the pure sciences, and especially physics, to philosophy. I confess that I consider a philosophical grounding a cardinal condition of general education. (…) I regard the propaedeutics of philosophy [especially – J.G.] highly as a subject, which should feature not only in middle school but in all secondary schools, of course on condition that that subject be very skilfully taught, which no easy matter at all[72].

The crux of these deliberations is declarations summing up the analyses he conducted and indicating the subjects in which instruction should not be neglected at the level of secondary and university education.

---

70   M. Smoluchowski, *Dzisiejszy stan teorii atomistycznej*, (The current state of atomic theory) a lecture given on courses for teachers in Lviv, March 12, 1913, reprinted in "Kosmos" (Cosmos) magazine, 1918, vol. 38, pp. 355–373. cit. after *The writings of Marian Smoluchowski*, vol. 3, op. cit. p. 61.
71   Idem *Znaczenie nauk ścisłych w wykształceniu ogólnym* (The importance of the pure sciences in general education), speech given during a congress of the members of the Association of Higher School Teachers, May 27, 1917, "Muzeum", vol. XXXII, June 1917, pp. 286 – 294, cit. after *The Writings of Marian Smoluchowski*, vol. 3, op. cit. pp. 125 – 126.
72   Ibidem pp. 128 – 129.

I think however – Smoluchowski states – that every sensible, unprejudiced person agrees that school should prepare us: for life, for real, modern day life, not that of 2,000 years ago, or that invented by poets and writers. (...) It is high time for us to accept that we live in the 20th century; that we forge the weapon with which we fight in current times, i.e. schooling in the pure sciences, familiarity with the laws of nature, technical skills, economic astuteness. We need people with a modern outlook but with a certain life idealism; people loving science and capable of positive work, in line with the socio-national obligation. Such people can only be prepared through education adapted to the life demands of the present day, taking account of the sciences in broad scope and with a goal-oriented method"[73].

This does not mean the Polish scholar did not value education in the humanities. He saw how important is, "the study of the Motherland's history, and also especially the study of the native language and literature, which generally, next to mathematical and natural studies, I consider the most important school subject"[74]. He himself, as Goetel mentions, was schooled in literature.

Smoluchowski's love and talent for literature was also reflected strongly in his scientific work and lectures. His lecture, popular or strictly scientific, was a master work, polished in every respect. Let us listen to what one of his students, Władysław Żłobicki, writes about it (...). "He was a first-rate literary talent, as a result of which he gave his popular-scientific essays an extremely alluring form. His talks were often listened to by laymen unfamiliar with the mysteries of physical science and they nonetheless felt satisfaction, like a child feels at the sight of precious stones, though he does not know their true worth"[75].

An extremely important work of Marian Smoluchowski was the impressively proportioned *Self- Study Handbook,* covering various aspects of education. The deceptively-titled book represents a summary of knowledge contemporary to Smoluchowski and the general comments on the subject of teaching are undoubtedly and extraordinary work.

On several dozen pages of the *Self-Study Handbook,* Smoluchowski analyses a hundred and fifty years of the development of science, focusing chiefly on physics. The considerations start with an analysis of the essence of experiments, going on to investigate the importance of methodological differences in

---

73   Ibidem, pp. 124 and 131.
74   Ibidem, p. 130
75   W. Goetel, *Ze wspomnień osobistych o Maryanie Smoluchowskim,* (From personal recollections of Marian Smoluchowski) "Kosmos" (Cosmos) 1917, no 5–12, p. 225.

the experiments conducted. Chance always played an important role in scientific discovery, though it would not be of such significance had it not awakened in the experimenters an observational sense which enabled a perception of seemingly imperceptible or insignificant phenomena.

The very extensive introduction contains many philosophical deliberations outlining the scholar's worldview. Smoluchowski worked on the *Handbook* for several years, surely at the expense of his scientific work, which proves the importance he attached to the cultivation of science. He was one of those great academics who not only contributed to the emergence of a proper body of teaching materials but also introduced them to the daily practice of teaching. Zofia Gołąb-Meyer described the particular singularity of this outstanding work:

> Smoluchowski's uniqueness among his peers lies in his extraordinary devotion to a just cause and the creation of a unique work of great practical utility. In the Handbook, Smoluchowski meticulously discussed more than 500 textbooks for all levels of teaching, for general schools, popular science books, to university course books. It is safe to say that further interwar generations of Polish physics teachers, directly or indirectly, learnt, and taught their students, drawing on "handbook" knowledge. Because the *Handbook* was written in Polish, it did not have its due influence on European and global education[76].

Smoluchowski's death was a shock to European and world science. Letters of condolence came to Kraków from the greatest universities. "His name will always be associated with the budding atomic theory," wrote famous German physicist Arnold Sommerfeld[77]. Einstein added: "Everyone who knew Smoluchowski personally valued him not only as an enthusiastic scholar but also because he was a noble, sensitive and magnanimous person… Fate prematurely drew a line under his fruitful research and teaching work, however let us continue to respect and remember his work and life"[78]. Tadeusz Godlewski found beautiful words:

---

[76] Z. Gołąb-Meyer, *Poglądy Mariana Smoluchowskiego na nauczanie fizyki z perspektywy stulecia* (Marian Smoluchowski's views on the teaching of physics from a century's perspective), in: *Marian Smoluchowski – od teorii atomistycznej do fizyki współczesnej* (Marian Smoluchowski – from atomic theory to modern physics), ed. A Strzałkowski, Kraków 2003, pp. 45 – 46.

[77] Cit. after: A. Wróblewski, *Marian Smoluchowski: Polak, który stworzył nową gałąź fizyki* (Marian Smoluchowski: A Pole who created a new branch of physics), Interia – historia, 20 June 2017, https://historia.interia.pl/ aktualnosci/news-marian-smoluchowski-polak-ktory-stworzyl-nowa-galaz- fizyki,nId,2407763 (access: 13.02.2020)

[78] Ibidem

What an enormous loss world science has suffered by his death! When a working man dies at the end of a normal human life, despite the sadness at what has passed, an inevitable necessity must be seen in what has to happen. But Smoluchowski died in the 46th year of life, he died in full mental fortitude, at a time when he was approaching the zenith of activity that he had developed through his work and abilities. When one looks at the development of his work, which bore ever more wonderful fruit every year, everyone must be overcome by unrestrained regret that he was not allowed to go further in life and to work more at that pace. And what a loss the whole of Poland has suffered by his death! If the evil forces of the world have tried for a century and a half in every possible way to undermine our existence, to hamper our development and cut down our life, Smoluchowski's work is the strongest confirmation that those efforts were not always fruitful. Because if they cry out to the world today that there is no Poland, that Poland is dead, then the fruits of his work are the loudest protest against that, they are the highest, the most beautiful negation of that because a nation that produces great, creative people is most fully alive. But if that is why Smoluchowski's life was one of the most precious national treasures, then it is why his death is the greatest national loss and calamity. His work's achievements will remain eternal in science, always living, never forgotten, and will be an immortal pride and glory of Polish science. That he left us so early; that he abandoned his beloved and noble work at a time when he could have done so much more; that he had to leave us in such difficult times when we need above all truly brave and great people; that a person cannot be found who could replace him on those uplands on which he stood; therefore his death will stay in our memory as a great harm that has befallen us and we will always feel his absence as something nothing can rectify, in every sense of the expression an irreparable loss and truly, for the culture of the whole nation, the most severe misfortune[79].

Władysław Natanson, speaking in the name of The Kraków Academy of Learning, stated:

> In full strength, in the blossoming of creativity, one of the most wonderful minds that we have boasted in our Republic of Science, Marian Smoluchowski, leaves us… he reaped a rich crop of discoveries, generously sowed thoughts the fruits of which will fall to the coming generations. He adored science with a pure and hot love and gave the strenuous labour of his whole life to it without reckoning or measure. He left us his example and his works[80].

---

[79] T. Godlewski, *Marian Smoluchowski*, op. cit., pp. 29–30; see also: *Marian Smoluchowski (1872–1917). Fizyk, taternik – romantyk nauki*, (Marian Smoluchowski: Physicist, mountaineer – romantic of science), op. cit., p. 19.

[80] Cit. after: A. Teske, *Marian Smoluchowski: życie i twórczość* (Marian Smoluchowski: life and works) Warsaw 1955, p. 268.

In his memorial memoires he wrote that everyone who had met Smoluchowski directly was impressed by his personality and emphasised: "I would like to recall here the grace of His life, the courtly softness of heart, combined with the refinement of kindness. I would like to recreate the strange charm of his person, remember how reserved he was, how humble, how sweetly shy, how full he was of a pure, almost unwitting joy"[81].

But in the context of the above considerations it can also be seen that Smoluchowski's premature death deprived the reborn Poland of a personality who could have contributed through his knowledge and engagement to the formation of institutions of science, influencing the curricula of secondary schools and universities. There was an opportunity to create a strong scientific centre of physics on a par with the Polish mathematical school that emerged around Stefan Banach and Hugo Steinhaus in interwar Lviv. At the university in Lviv, another powerhouse of Polish science – Kazimierz Twardowski – was laying the foundations of a world-renowned centre of logical- philosophical thought, known later as the Lwów-Warsaw school. Stanisław Ulam saw the opportunities that would have presented themselves to Polish physics in November 1918 under the direction of Smoluchowski. He was working then as a junior professor outside the great centres of scientific activity. It transpired that a person of exceptionally high abilities could reach the forefront of European thought in the field of physics even though the environment they worked in was isolated and did not have a long scientific tradition. It seems that pioneering works in a relatively new field (such as statistical mechanics was) had opened the way to world science after making contact with other researchers. A similar situation appeared in Poland just after the end of the First World war when a small group of mathematicians, thanks to their enthusiastic work, managed to create a new school of mathematics in such fields as set theory, topology and the foundations of mathematics despite the lack of a pre-war mathematical tradition in the country[82].

The Polish physicist's premature death also put an end to the chance of a Pole winning the first Nobel Prize in the field of the pure sciences.

---

81 Cit. after: A. Strzałkowski, *Wprowadzenie* (Introduction) in: *Marian Smoluchowski (1872–1917). Fizyk, taternik – romantyk nauki (Marian Smoluchowski 1872 – 1917)*, (Marian Smoluchowski: Physicist, mountaineer – romantic of science), op. cit., p.13.

82 S. Ulam, *Marian Smoluchowski and the Theory of Probabilities in Physics*, op. cit., p. 479.

Chapter IV Nobel

Smoluchowski's authority as a scientist would have grown exponentially in the memory of future generations had the physicist lived to see a Nobel Prize, which – as the history of the prize shows – in the course of the next few years was highly probable.

Smoluchowski would have had the chance to receive the prize several times, although he was never nominated in his lifetime. At least three Nobel laureates – Richard Zsigmondy (1865–1929; Nobel Prize in Chemistry in 1925), Jean Baptiste Perrin (1870–1942; Nobel Prize in Physics 1926), and Theodor Svedberg (1884–1971; Nobel Prize in Chemistry 1926) – were awarded the prize for work in which they directly or indirectly used Smoluchowski's research. The basic reason that Smoluchowski never received it was that he died earlier and that prize is not awarded posthumously.

In 1925, Smoluchowski could have got the Nobel in chemistry together with Richard Zsigmondy, the professor from Graz, "for proving the heterogeneous nature of colloidal solutions and for the methods used which have laid the foundation of modern colloid chemistry"[83].

Richard Zsigmondy (1865–1929)

---

83 Description on the Nobel Prize website https://www.nobelprize.org/prizes/chemistry/1925/zsigmondy/facts/ (access: 25.11.2021).

Zsigmondy's work was experimental in nature and was closely related to Smoluchowski's theoretical achievements. The Polish physicist's role in this research was huge. Being an experimenter, Zsigmondy needed a theoretical basis for his research, and got it from Smoluchowski.

During the Nobel lecture he gave in Stockholm, he said: "This induced me to ask the theoretical physicist M. von Smoluchowski to derive an experimentally verifiable formula by which the presence of spheres of attraction could be deduced from the speed of coagulation; von Smoluchowski willingly agreed to my suggestion and gave, in addition, a complete theory of coagulation on a mathematical basis."[84]

The following year, as many as two Nobel Prizes were directly related to the great Pole. Frenchman Jean Baptiste Perrin received the Nobel in physics for his experimental work on Brownian motion, confirming the correctness of the Einstein- Smoluchowski molecular theory, while the chemistry prize went to Theodor Svedberg from Uppsala, whose experimental work on solutions was concurrent with Smoluchowski's theoretical deliberations related to them[85].

Theodor Svedberg (1884–1971)

---

84 R.A. Zsigmondy, *Properties of Colloids*, Nobel Lecture, Chemistry 1922–1941, Singapore, New Jersey, London, Hong Kong 1999, p. 53.
85 See B. Cichocki, Nobel Prize "Delta" 1997, no 12

Svedberg received the Nobel "for his work on disperse systems"[86]. Disperse systems are dispersed colloidal systems, composed of at least two phases of which at least one is a finely divided material dispersed in a continuous second phase, known as a dispersion medium.

Together with his collaborators, he published a series of works devoted to the issue of fluctuation. In the introduction to the first article, titled *A new method of testing the validity of the Boyle-Gay-Lussac law for colloidal solutions*,[87] he wrote that to interpret the results of the measurements obtained, a theory was needed that would could relate the observed quantities to the concepts of the kinetic theory of matter. Svedberg found it in two of Smoluchowski's works. They were two articles, of which the first, – *Über Unregelmässigkeiten in der Verteilung von Gasmolekülen und deren Einfluss auf Entropie und Zustandsgleichung* (On irregularities in the distribution of gas molecules and their effect on entropy and the equation of state)[88] – was included in a commemorative publication of February 20, 1904, marking Ludwig Boltzmann's 60th birthday. In it, Smoluchowski formulated a theory of fluctuations in the density of particles in a gas and considered the influence of fluctuation in the equation of state. In the second text, titled *Molecular-Kinetic Theory of the Opalescence of Gases in the Critical State and a Few Related Phenomena*[89], dating from 1907, he applied his theory to the phenomena of the opalescence of gas in the neighbourhood of a critical state and to

---

86 Description on the Nobel Prize website https://www.nobelprize.org/prizes/chemistry/1926/svedberg/facts/ (acess: 25.11.2021).

87 T. Svedberg, K. Inouye, *Eine neue Methode zur Prüfung der Gültigkeit des Boyle-Gay-Lussacschen Gesetzes für kolloide Lösungen*, (A new method to test the validity of the Boyle-Gay-Lussac law for colloidal solutions) "Zeitschrift für Physikalische Chemie-Stöchiometrie und Verwandtschaftslehre" (Journal of Physical Chemistry-Stoichiometry and Related Teaching) 1911, issue. 77, pp. 145–190.

88 M. Smoluchowski, *Über Unregelmäßigkeiten in der Verteilung von Gasmolekülen und deren Einfluß auf Entropie und Zustandsgleichung* (On irregularities in the distribution of gas molecules and their effect on entropy and the equation of state), in: *Festschrift Ludwig Boltzmann gewidmet zum sechzigsten Geburtstage* (Commemorative publication dedicated to Ludwig Boltzmann on the occasion of his sixtieth birthday) Leipzig 1904, pp. 626–641.

89 Idem, *Teoria kinetyczna opalescencji w stanie krytycznym oraz innych zjawisk pokrewnych* (Molecular-Kinetic Theory of the Opalescence of Gases in the Critical State and a Few Related Phenomena), Rozprawy Wydziału Matematyczno- Przyrodniczego Akademii Umiejętności (Papers of the Mathematical-Natural Department of the Academy of Learning), Kraków, 1907, vol. XLVII, Series A, pp. 179–199.

explaining the phenomenon of blue skies and other related effects in which the influence of molecular fluctuations is revealed on the macroscopic scale[90].

The collaboration between Smoluchowski and Svedberg lasted seven years – from 1907 to 1914. During this period, the two held an intensive scientific correspondence. Starting in 1908, Smoluchowski cited and discussed, sometimes critically, Svedberg's results in all his work on Brownian motion and fluctuations. In his work on these problems, Svedberg also mentioned Smoluchowski's publications and letters[91].

During the Nobel Prize awards ceremony on May 19, 1927, Svedberg presented in his speech the results of his research conducted in 1923–1926 (a few years after Smoluchowski's death). In describing them, he had no opportunity to recall his collaboration with Smoluchowski. It was highlighted by Professor Henrik Gustaf Söderbaum, secretary of the Royal Swedish Academy of Sciences. Outlining Svedberg's merits, he said:

> As we have recently heard, Einstein evolved a theory for this so-called Brownian movement which was then developed to a high degree by the now late Smoluchowski... If we now consider a very small volume fraction, the result is that, as Smoluchowski has calculated in detail, the number of particles present simultaneously within this volume can change from one moment to another. Svedberg and his collaborators have been able to confirm this extremely interesting conclusion that a "few- molecular" system having definite limits within a large volume of a material with a definite mean temperature may contain a varying number of particles, partly by counting the colloidal particles, partly in the case of solutions of radioactive substances by counting the number of so-called scintillations[92].

The great French physicist Jean Baptiste Perrin received an honorary Nobel Prize in 1926 "for his work on the discontinuous structure of matter, and especially for his discovery of sedimentation equilibrium"[93]. This concerned research which confirmed the Einstein-Smoluchowski theory experimentally.

---

90 See B. Średniawa, *Rola współpracy Mariana Smoluchowskiego i Teodora Svedber-ga w prowadzonych w pierwszych latach XX wieku badaniach ruchów Browna i fluktuacji* (The role of the collaboration of Marian Smoluchowski and Theodor Svedberg conducted in the first years of the 20th century in research on Brownian motion and fluctuations), "Postępy Fizyki" (Progress in Physics) 1991, vol. 42, b. 4, p. 441.
91 Ibidem, p. 451
92 Ibidem, p. 452
93 Description on the Nobel Prize website, https://www.nobelprize.org/prizes/physics/1926/perrin/facts/ (access: 25.11.2021).

Jean Baptiste Perrin (1870–1942)

In 1913, Perrin published *Atoms*[94]. In it, he described the scope of research as well as the conclusions and thoughts stemming from the experiments he conducted, intended *de facto* to disprove the Einstein-Smoluchowski theory. He cites in it many times Smoluchowski's research work in the field of Brownian motion, fluctuation and opalescence, which were the theoretical complement to work by Einstein published earlier. Separately, Perrin devoted the whole of the fifth chapter, titled *Fluctuations. Smoluchowski's theory*[95], to Smoluchowski's fluctuations. In it, he objectively describes the theoretical results Smoluchowski achieved, which he used in his experiments:

> No longer confining himself to the case of rarefied substances, Smoluchowski succeeded a little later (…) in calculating the mean density fluctuation of any liquid whatever, and proved that, even with condensed fluids, the fluctuations should become noticeable in spaces visible under the microscope when the fluid is near the critical state. He thus succeeded in explaining the enigmatic opalescence which is always shown by fluids in the neighbourhood of the critical state. This opalescence, which is absolutely stable, indicates a permanent condition of fine grained heterogeneity in the fluid.[96]

Perrin repeatedly refers to Smoluchowski's research, thereby proving the importance of his theoretical deliberations and the scholar's contribution to proving the atomic makeup of matter.

---

94   J. Perrin, *Les Atomes* (Atoms), Paris 1913.
95   Idem, *Atoms,* New York 1916, p. 134
96   Ibidem, p. 135

The Polish physicist's research used in the work of the above-mentioned scientists is not the only example of the scholar's scientific achievements that could have yielded him a Nobel Prize. A discovery that created a near paradigm shift in physics was the development of the foundations of the field of stochastic processes and the introduction of probability theory to the pure sciences. This was an achievement worthy not only of the Nobel Prize, due to which Smoluchowski's name was permanently inscribed into the history of science. He was a pioneer of statistical physics. Mark Kac notes: "Smoluchowski may not have been aware of it but he begun writing a new chapter of Statistical Physics, which in our times goes by the name of Stochastic Processes. (…) The novelty and originality of the Smoluchowski approach lie in his bold replacement of an impossibly difficult dynamical problem (…) by a relatively simple stochastic process."[97]

The Polish physicist also provided the first correct equation relating ζ potential to measurable quantities, i.e. electrophoretic mobility, potential flow and pressure difference in electro- osmosis. These are known in world literature as the Smoluchowski equations[98], which also qualified him for the Nobel Prize. Many years later, Indian astrophysicist, mathematician and Nobel laureate Subrahmanyan Chandrasekhar (1910–1995) wrote that only the Polish scientist's untimely death caused him to never receive the important distinction.

In 1973, Chandrasekhar was awarded the Marian Smoluchowski Medal of the Polish Physical Society in recognition of his contribution to stochastic methods in physics and astrophysics and in particular for an article published in the "*Przegląd Fizyki Współczesnej*" (Modern Physics Review) in 1943 which highlighted Smoluchowski's contribution to that work. In his speech during the awards ceremony, Chandrasekhar noted that the Nobel Prize in Chemistry that Richard Zsigmondy received in 1925 and Svedberg a year later, was awarded for research aimed at experimentally confirming Smoluchowski's theoretical predictions concerning colloidal and dispersion systems. Had Smoluchowski lived, he would certainly have received the prize himself.[99]

---

[97] M. Kac, *Marian Smoluchowski and the Evolution of Statistical Physics*, op. cit., p. 17.
[98] See M. Kozmulski, *Potencjał ζ i równanie Smoluchowskiego* (ζ Potential and the Smoluchowski equation) "PAUza Akademicka. Tygodnik PAU" (Academic PAUza, the Weekly of the Polish Academy of Artsand Sciences), 2017, No 380–381, p. 7.
[99] B. Duplantier, *Brownian Motion, "Diverse and Undulating"*, in: *Einstein, 1905– 2005. Poincaré Seminar 2005*, eds. T. Damour, O. Darrigol, B. Duplantier, V. Rivasseau, Progress in Mathematical Physics book series (PMP, vol. 47), Basel 2006, p. 251–252.

Despite the lack of a Nobel Prize and Smoluchowski's premature death, he stood the chance of having his name permanently inscribed in every physics textbook. His greatest success was discovering the essence of Brownian motion as fundamental proof of the kinetic-atomic makeup of matter and ending the dispute raging in the second half of the 19th century on the atomic makeup of matter. Unfortunately, he did not make use of that opportunity, restraining himself from publishing his research, and consequently Albert Einstein is widely regarded as having discovered the essence of Brownian motion.

Chapter V A dispute over atoms

Since the times of ancient Greece, atomic science had remained in the realm of speculative philosophy, elements of a scientific atomic theory emerging only in the 19th century. The start of these changes occurred in 1805, when John Dalton (1766–1844) explained the simple numerical rules observed in the formation of chemical compounds through the combination of unchanging atoms into certain groups, molecules and particles[100]. Dalton proved that every chemical compound contains the same amounts by mass of the same elements and this can be easily explained if we assume an atomic makeup of matter.

At the end of the 19th century, phenomenological views on the structure of matter dominated the beliefs of European physicists. Atomic theory was considered to be obsolete, an unscientific fantasy destined to be forgotten. The rejection of this theory by most of the scientific community at the end of the 19th century seemed to be permanent. Marian Smoluchowski outlined in a few words the general intellectual mood of the era of atomic theory's fall. There existed, in his view, very important factual arguments at the time against atomic theory and dissuading from that course. However, scholars involved in the pure sciences, tough people determined in their pursuits, did not succumb to the general mood[101].

Those doubting the legitimacy of atomic theory cited as their chief argument the inconsistency of everyday observations with the theory's assumptions. They were convinced that if the indefinite and statistical nature of phenomena were to dominate in nature, then both these features must be applicable to observable phenomena in the macroscopic world and should consequently be perceptible in everyday life. In the observable world, there is no way to confirm characteristics of nature that would prove the accidental or random status of matter. It could even be said that explaining physical events with the aid of a mathematical tool such as the calculus of probability is contrary to everyday life experience and so any theory based on such mathematics is mere intellectual speculation. Another hurdle hampering acceptance of atomic theory (mentioned earlier) was the reversibility of natural phenomena assumed in kinematics. The dispute between energeticists, representing a phenomenological perception of matter,

---

100 See. M. Smoluchowski, *Ewolucja teorii atomistycznej* (The evolution of atomic theory) in: *The Writings of Marian Smoluchowski*, vol. 3, op. cit., p. 17.
101 Ibidem, p. 63

and advocates of the conception of atomic theory, was not superficial; it concerned both an understanding of the essence of science and paradigms forming the foundation of scientific thought. The approach to science had to undergo a change, which, Smoluchowski claimed, should be based on experiment as well as on the theoretical aspect – on mathematics, with particular emphasis on the developing calculus of probability.

A fact representative of the situation at the time was that when in 1895 Ernst Mach was named professor of philosophy at the University of Vienna, where Boltzmann had worked since 1893, taking the floor during a lively academic debate on the values of atomic theory, he cut the discussion short with one statement: "I don't believe that atoms exist." In subsequent academic disputes, Mach asked Boltzmann several times: *"Eines haben Siegesehen?"* (Have you seen one [an atom – J.G.])[102].

Smoluchowski noted a situation characteristic of the climate in science at the time, when following the publication of Boltzmann's work *Vorlesungen über Gastheorie (Lectures on gas theory)* in a German scientific journal, the following statement appeared: "Kinetic theory, as we know, is as flawed as various mechanical theories of gravity, in particular it wrongly understands the principle of the conservation of energy; however, if someone really wants to learn about it, let them take up Boltzmann's work"[103].

The emergence of key evidence for kinetic-atomic theory was initiated by Robert Brown (1773–1858), a Scottish botanist researching in 1827 microscopically fine particles floating in a liquid and making small, irregular movements visible with the use of a high-magnification microscope. He called them molecular movement[104]. Brown was not the first person to have made such observations. In 1784, a Dutch doctor, Jan Ingen-Housz (1730–1799) had also noticed the effect. However, Brown was the first to conduct systematic[105] research on small pollen grains suspended in liquid. He became convinced that the irregular oscillating movements were made by tiny particles of organic or inorganic substances. Smoluchowski describes this fact in the following way:

---

102 See J. Berstein, *Einstein and the Existence of Atoms*, "American Journal of Physics" 2006, vol. 74, no 10, p. 864

103 M. Smoluchowski, *Dzisiejszy stan teorii atomistycznej* (The current state of atomic theory) op. cit., p. 61.

104 Idem, *Ewolucja teorii atomistycznej* (The evolution of atomic theory), op. cit. p. 21.

105 See J. Berstein, *Einstein and the Existence of Atoms*, op. cit. p..865.

The name "molecular movement" comes from the English botanist Brown, who in 1827 noticed through microscopic research that tiny pollen grains, suspended in liquid, vibrate and make irregular movements reminiscent of the movements of a swarm of mosquitoes or an army of ants; upon closer research, he became convinced that this is a general phenomenon exhibited by any small particles of an organic or inorganic substance when they are in such conditions. This same phenomenon had actually already been noticed by other scholars before Brown, like Needham 1750, von Gleichen 1764, but they did not take the matter up more closely. A slightly different idea to our current view is associated with the name Brown, however, as by molecule, a concept not as clearly crystallised at the time as today, he meant only tiny moving particles; the name meant just the observed fact with no thought as to its explanation[106].

No thorough research was conducted on the phenomenon for a long time although every naturalist observed them in the course of their microscopic work because the movement itself intrigued researchers. Various ideas constantly appeared to explain the essence of this phenomenon but there was no work focused on understanding the movements' cause. It transpired that the correct explanation of the nature of the phenomenon was a key argument for accepting kinetic-atomic theory. The contradictions accompanying the observations of various authors, as well as the complexity and inconsistency of the theoretical explanations contributed, according to Smoluchowski, to the phenomenon being widely ignored. When any research was sought to be conducted, contradictions appeared during its course, which discouraged their continuation. An example may be the big differences achieved in the observed velocities of particles. Smoluchowski writes:

> The real core of the issue, however, remained untouched: whether the observed phenomenon corresponds quantitatively with the requirements of atomic-kinetic theory. In this regard, there seemed to be a fundamental contradiction. The velocity of particles, given for example by F. Exner, stands at more or less 0.0003 cm/sec (for rubber particles of 0.001 mm diameter in water).
> Meanwhile, accepting, in line with kinetic theory, that the average energy of liquid molecules must equal the kinetic energy of these particles, they would obtain a theoretical velocity of 0.4 cm/sec, so over a thousand times greater[107].

The discrepancies were extremely large and inexplicable. Apart from laboratory problems, Smoluchowski mentions the climate prevailing among academics to whom the notion of a molecular-kinetic composition of matter was alien:

---

106 M. Smoluchowski, *O fluktuacjach termodynamicznych i ruchach Browna* (On thermodynamic fluctuations and Brownian motion), op. cit., p. 297.
107 Ibidem, p. 298.

Contradictions between the observations of various authors as well as the diversity and inconsistency of the theoretical explanations were surely the reason why the phenomenon was generally ignored, such that we find no mention of it in any of the extensive works or physics textbooks (with the one exception of Lehmann's "Molekularphysik" (Molecular Physics)) until recent years. The foreignness to the general scientific community of the notion of its molecular-kinetic essence is evidenced by the characteristic fact that when, in the last decade of the last century, the school of energeticists and phenomenologists (Mach, Ostwald, Zermelo etc.) decried atomic-kinetic views as naïve, unscientific beliefs, there was nobody, even among proponents of this view, who would highlight Brownian motion as obvious evidence of thermal molecular motion[108].

Scientists inclined towards atomic theory lacked the courage to publically air their views, as Smoluchowski saw for himself. The hypotheses emerging around Brownian motion represent, according to the Polish researcher, who describes their history in the work *On thermodynamic fluctuations and Brownian motion*, an extremely interesting chapter in the history of physics. Their history proves, he claims, that a known and often-observed phenomenon was forgotten and found no acceptance or interest on the part of official science. Ultimately, it shows how a true theory provided the key to understanding and more closely researching the phenomenon of Brownian motion[109].

The strength of the influence of physicists from the school of energetics on the intellectual attitudes in physicists' circles is evidenced by two examples illustrating the scale of tensions arising in the academic world. The first of these concerns Ludwig Boltzmann and his conception of the theory of gases. It was particularly fiercely fought over, which may have deepened Boltzmann's depression, which may have been one of the possible causes of his suicide. The Austrian physicist's theories had been treated by scientific detractors as a compromising manifestation of speculation. They looked for contradictions in his arguments, seeking inconsistencies between experience and kinetic theory.

The second example concerns Marian Smoluchowski, who, after developing equations to describe Brownian motion, hesitated to publish them. The delay in publication was due, as he said himself, to the author's caution as he awaited more precise experimental results, but also to fear of the reaction of a scientific community disinclined towards atomic hypotheses. Smoluchowski wrote:

> There prevailed in science an excessively critical current, it could be called: cowardly sober. It is not easy to clip the wings of the human mind but he who could not keep himself from speculations at least held back from declaring them publically. I remember

---

108 Ibidem
109 Ibidem, p. 279

how I myself for a long time hesitated and procrastinated over announcing my contributions to kinetic theory. This current primarily turned against the most powerful theory that science had produced to date, i.e. atomic theory.[110]

However, despite resistance, kinetic theory systematically developed and the logical consequence was that Boltzmann's hypothesis on the "reversibility of phenomena" had to be accepted. Ice in a glass of water, according to Boltzmann and in line with kinetic theory, can cool by itself and simultaneously warm up the water in the glass but this is such an improbable phenomenon that in practice it is ignored.[111] This was a hypothesis most physicists found impossible to accept.

Ludwig Boltzmann (1844 – 1906)

The period from 1900, when Smoluchowski became acquainted with Franz Serafin Exner's article *Notiz zu Brown's Molecularbewegung* (*Notes on Brownian molecular motion*), until 1903, when he finished his research on Brownian motion, culminating in the development of a hypothesis and building proof in the form of mathematical models describing its nature, was a time that shaped his views in the field of atomic-kinetic theory.

---

110 Idem, *Dzisiejszy stan teorii atomistycznej*, (The current state of atomic theory) op. cit., p. 62.
111 Idem, *Ewolucja teorii atomistycznej* (The evolution of atomic theory), op. cit. pp. 19 – 21.

Carl von Nägeli (1817–1891)

Wilhelm Ostwald (1853–1932)

It is no accident that the terms "kinetics" and "atomistics" are used interchangeably in this paper as Smoluchowski used the names of the two theories in the following way:

Perhaps I may take the opportunity to explain parenthetically why I use these names to some extent as synonyms, although the correct meanings of the terms: "atomic" and "kinetic" theory are different. As we know, the former atomic theory of Dalton has been enriched since the times of the formulation of the principle of conservation of energy, by the additional supposition that atoms and molecules are in constant motion, that heat is a store of kinetic energy and the measure of the kinetic energy of this internal motion is what we call temperature. Since the time of Robert Mayer and Helmholtz, therefore, atomic theory has fused with kinetic theory into a unified whole[112].

The work of Einstein and Smoluchowski became a turning point in research on Brownian motion. The results achieved proved the important role of randomness in nature.

Smoluchowski proved that the random collisions of molecules are a cause of the bizarre zig-zag movements that the French physicist Louis George Gouy (1854 – 1926) compared to the movements of ants around an anthill. He demonstrated that the space occupied by a molecule among the whole collection of other molecules is random and determined by chance circumstances. Hence it cannot be assumed that the molecules of a gas or liquid are distributed regularly and evenly. Theodor Svedberg, the Swedish Nobel laureate who confirmed experimentally conclusions on random unevenness, achieved this by counting ultra-microscopic particles in colloidal solutions.

The hypothesis of Einstein and Smoluchowski defined Brownian motion as chaotic shifts of colloidal particles suspended in a liquid or gas, caused by collisions with liquid particles; the more intensive, the less the liquid's viscosity, the smaller the size of the molecules of the solution and the higher the temperature of the liquid. The random wandering of emulsion particles is caused by their bombardment by water molecules, which are much smaller, numerous and fast-moving. The bombardment of particle by molecules is on average the same from all sides. However, if the particle is small enough, it happens that the number of molecules colliding with it from one side will be different at some point (greater or smaller) than the number of molecules striking from the other side. As a result, the particle receives a stronger impulse every so often on the side marked by the impact of a larger (at a given moment) group of molecules. The described movements of particles, according to Smoluchowski, occur in the suspension not so much through the bombardment made by the liquid particles as through the fluctuations occurring in the density of particles in the direct vicinity of the suspension particle.

---

112 Idem, *Dzisiejszy stan teorii atomistycznej*, (The current state of atomic theory) op. cit., p. 62.

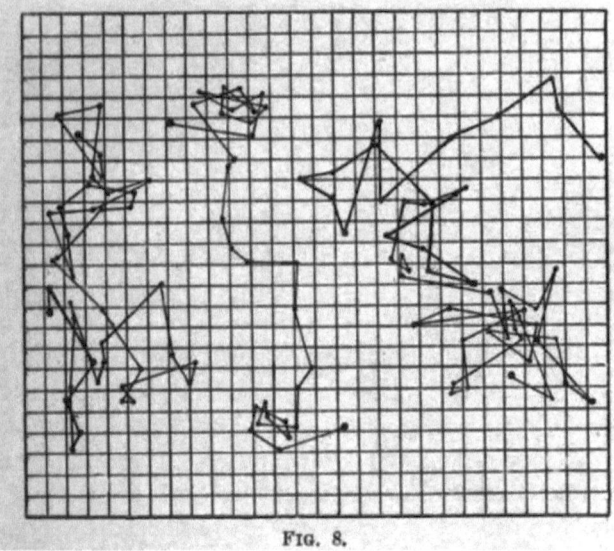

Brownian motion according to the observations of Perrin in 1908. The drawing presents successive positions (every 30 seconds) of three molecules of 0.53 μm diameter.

Smoluchowski raised this issue in response to Swiss botanist Carl Wilhelm von Nägeli, who argued that the size of a solution's molecules relative to the size of suspension particles is so small that they are in no position to effectively cause the movements of the suspension molecules. The effect of one collision with a molecule is extremely small, just $\Delta v \sim 10^{-3}$ μm/s. In a 1906 article, Smoluchowski explained this problem – Brownian motion is generated not by individual particles but by the occurrence of self-generated fluctuations of solution molecules. He mentions these in a paper from 1904, calling them "systems of a swarming nature"[113].

---

113 Idem, *Zarys kinetycznej teorii ruchów Browna i roztworów mętnych*, (An outline of the kinetic theory of Brownian motion and cloudy solutions), "Rozprawy Wydziału Matematyczno-Przyrodniczego Akademii Umiejętności" (Dissertations of the Faculty of Mathematics and Natural Sciences of the Academy of Learning, vol. 6, section A, Kraków 1906; op. cit. after *The Writings of Marian Smoluchowski commissioned by the Polish Academy of Arts and Sciences collated and published by Władysław Natanson*, vol. 1, Kraków 1924, pp. 495 – 497.

In another of von Nägeli's objections he claimed that solution particles, striking suspension particles from all sides simultaneously, cannot cause oscillations as they aresimultaneous collisions. Smoluchowski's response reveals in- depth consideration of the problem. That the solution particles collide with suspension particles 10^{20} times per second from various sides does not mean that a suspension particle should remain motionless as this is the same error of understanding as if a person playing a game of chance (for instance rolling dice) believed that he will never bear a greater loss or ever achieve a greater gain than the stake for one throw. We well know, states Smoluchowski, that good and bad luck do not usually entirely balance out, that the longer the game lasts, the greater the average sum either won or lost. The researcher points out that the difference in collisions, positive or negative, can run to 10^8 or 10^{10} collisions per second[114].

Without getting into the mathematical structures, it is worth emphasising that the essential idea on which the theory of Brownian motion rests is the assumption that suspension particles (at a certain temperature) behave to a certain extent the same a gas molecules and the average kinetic energy of translational motion performed by the centre of mass of such a particle is equal to the average energy of the translational motion that a gas particle possesses at that temperature[115]. This corresponds to the basic assumption of the kinetic theory of gases and is a direct consequence of the Maxwell principle. The thermodynamic principle of energy equipartition assumes that the available energy a molecule (for example gas) possesses is evenly distributed in all its possible uses (the so-called degrees of freedom) – regardless of whether that is a degree of freedom related to the vibrations of a particle, rotational energy or translational motion.

---

114   Ibidem
115   See Idem, *O fluktuacjach termodynamicznych i ruchach Browna* (On thermodynamic fluctuations and Brownian motion), op. cit., p. 300.

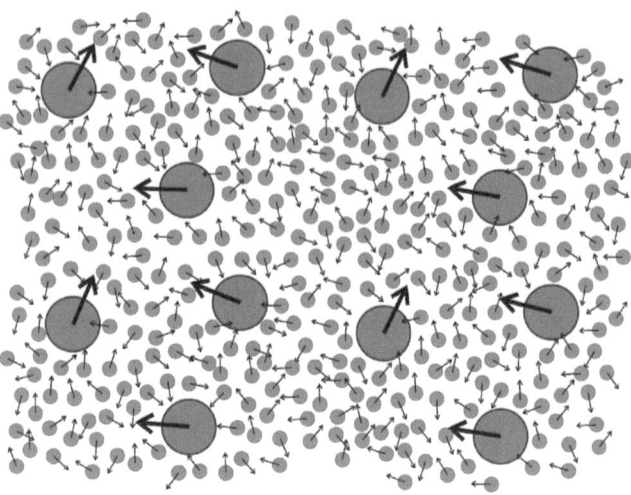

Diagram of the formation of Brownian motion

Much later, in *The Conclusion to the Self-Study Handbook*, Smoluchowski summarised this breakthrough moment in physics. He stated that kinetic theory, having awakened from a temporary lethargy, showed new vitality, not only in predicting a range of previously unknown phenomena (heat transfer in rarefied gases, the transpiration of rarefied gases, radiometric forces) but also resolving the dispute with thermodynamics. He drew attention to the fact that in Brownian motion, taking place in microscopically small particles suspended in liquids or gases, we have a visible example of particle movements[116].

---

116 M. Smoluchowski, *Poradnik dla samouków: wskazówki metodyczne dla studiujących poszczególne nauki. Fizyka, Geofizyka, Meteorologia*, (The Self-Study Handbook: methodical tips for those studying particular sciences. Physics, Geophysics, Meteorology), vol. I and II, Warsaw 1917, pp. 344 – 345.

Chapter VI Brownian motion

Research on Brownian motion, the fluctuations of gases and opalescence, effectively changed the understanding of the designations of theoretical terms functioning in 19th-century science and philosophy, which fundamentally contributed to the triumph of kinematic-atomistic theory. However, the discovery was accompanied from the outset by an ambiguity regarding the prolegomena of recognising the phenomenon.

For a hundred years, a narrative has been built in physics in which the discovery of the essence of Brownian motion is attributed to Einstein. There are many examples confirming this fact; for Perrin the discoverer was Einstein, as he wrote himself in *Atoms*. Initially, Smoluchowski was mentioned along with Einstein as the co-creator of the kinetic theory of Brownian motion. In his Nobel Lecture, given on December 11, 1926, he noted: "It is due to Einstein and Smoluchowski that we have a kinetic theory of the Brownian movement which lends itself to verification"[117]. In awarding the Nobel Prize to Svedberg on May 19, 1927, Professor H.G. Söderbaum, Secretary of the Royal Swedish Academy of Sciences, said: "As we have recently heard, Einstein evolved a theory for this so-called Brownian movement which was then developed to a high degree by the now late Smoluchowski"[118].

The scientific community quite quickly forgot the Polish physicist's contribution to the discovery, however. In the multiply-revised work on the history of physics – *Kulturgeschichte der Physik* (*A Cultural History of Physics*) by Károly Simonyi – only Einstein's name appears. In *The Collected Papers of Albert Einstein* from 1989, John Stachel also attributes the discovery to Einstein[119]. In the book *A Brief History of Time*, Hawking discusses how proof for the atomic makeup of matter came about, but mentions only Einstein's publications[120]. The same is

---

117 J.B. Perrin's Nobel Lecture, given on December 11, 1926, https://www.nobelprize.org/prizes/physics/1926/perrin/lecture/ (access: 3.02.2022).
118 Speech by H. G. Söderbaum, Secretary of the Royal Swedish Academy of Sciences, given during the ceremony to award the Nobel Prize to Theodor Svedberg on May 19, 1927, https://sandwalk.blogspot.com/2007/10/nobel-laureate-svedberg.html (access: 3.02.2022).
119 *The Collected Papers of Albert Einstein. Vol. 2: The Swiss Years: Writings, 1900–1909*, eds. J. Stachel, D.C. Cassidy, J. Renn, R. Schulmann, New Jersey 1989, p. 206.
120 S.W. Hawking, *A brief History of Time. From the big bang to black holes*, trans. P. Amsterdamski, Warsaw 1990, p. 67

true in Percy Williams Bridgman's *The Nature of Thermodynamics*[121]. The abovementioned titles are just a few examples of the innumerable quantity of works the tone of which is identical.

It has occasionally happened that researchers – like for instance Indian astrophysicist Subrahmanyan Chandrasekhar in his work *Stochastic Problems in Physics and Astronomy*[122] – have admitted: "It is somewhat disappointing that the more recent discussions of the laws of thermodynamics [e.g., P. Bridgman etc.] contain no relevant references to the investigations of Boltzmann and Smoluchowski. The absence of references, particularly to Smoluchowski, is to be deplored since no one has contributed so much as Smoluchowski to a real clarification of the fundamental issues involved"[123].

Over time, a story emerged to reinforce this general trend, according to which, between March 18 and December 19, 1905, Albert Einstein wrote five papers that changed the nature of physics. They concerned the theory of relativity, the photoelectric effect and the theory of Brownian motion. Physicist John Stachel has noted that Einstein's publication of these papers has caused 1905 to be considered in the general discourse of physicists a "wonder year" analogous to the Newtonian year of 1666, dubbed the *annus mirabilis*. In 1905, Einstein, like Newton, published research that laid the foundations of a revolution in 20th-century physics[124]. Roger Penrose claims that the publication of Einstein's first five papers started the fourth revolution in science, manifesting in the way nature is perceived. Generally, legends do not have to be fully in line with the truth; they live their own lives. Newton did not make all his discoveries in 1666; in the case of Einstein, a question mark needs to be placed next to the papers analysing Brownian motion.

That scholar's later achievements in physics made him an almost monumental figure in science. This represents an addition hurdle to any attempt at discussion as it is hard to argue with legends. Albert Einstein, unfortunately, rarely cited the work of other authors in his papers, although he used many of their results. Despite the general attribution of the discovery of Brownian motion to Einstein, there is considerable evidence that Marian Smoluchowski and William

---

121 S. Loria, *Marian Smoluchowski i jego dzieło* (Marian Smoluchowski and his work), op. cit. P. 7
122 S. Chandrasekhar, *Stochastic Problems in Physics and Astronomy*, "Reviews of Modern Physics" 1943, vol. 15, no. 1, pp. 1–89.
123 Ibidem.
124 A. Einstein, *5 prac, które zmieniły oblicze fizyki* (Einstein's Miraculous Year: Five Papers That Changed the Face of Physics) trans. P. Amsterdamski, Warsaw 2005, p. 15.

Sutherland were at least co-contributors to this spectacular breakthrough in science.

Smoluchowski took up the problem from 1900. As he writes in *On thermodynamic fluctuations and Brownian motion*, he had been convinced of their molecular-kinetic essence since 1900[125]. He obtained results confirming his theories in 1903, three years prior to the publication date, as he described in *An outline of the kinetic theory of Brownian motion and cloudy solutions*: "several years ago I developed for this phenomenon a kinetic theory, which seemed to me most probable; I have not yet published the results, wanting to check them further against more precise experimental measurements"[126].

The aggressive positions of Mach and Ostwald contributed to the postponement of Smoluchowski's decision to publish *Zur kinetischen Theorie der Brownischen Bewegung und der Suspensionen*[127] (On the Kinetic theory of the Brownian Molecular Motion and of Suspensions), which was a summary of research on Brownian motion which came out in 1906. The paper was published in the prestigious German science journal *Annalen der Physik* (Annals of Physics) – the same publication in which Einstein's papers appeared, but a year later. That publication of Einstein's inclined Smoluchowski towards publishing the results of his own research.

As is often the case, there were at least a few reasons for Smoluchowski's hesitation to publish, and today we can only speculate as to which played the most important role. It is hard to judge whether the failure to publish research results immediately after they were obtained was due to fear of an excessively aggressive response from academic circles or whether Smoluchowski had to come to terms with how historic his equation was. To some extent, as he writes himself, he expected experimental support for his theory, which would have enabled him to be sure of its validity. Time has shown how wrong the Polish researcher was. Firstly, for experimental confirmation of the theory of Brownian motion to be achieved, it would have to have been sufficiently well known in scientific publications for experimental physicists to take it up. Smoluchowski repeatedly stated the methodological thesis that all experiments acquire scientific sense when they

---

125 M. Smoluchowski, *On thermodynamic fluctuations and Brownian motion*, op. cit. P. 299.

126 Idem, *Zarys kinetycznej teorii ruchów Browna i roztworów mętnych* (An outline of a kinetic theory of Brownian motion and cloudy solutions), op. cit. p. 490.

127 Idem, *Zur kinetischen Theorie der Brownschen Molekularbewegung und der Suspensionen*, (On the Kinetic Theory of the Brownian Motion and of Suspensions) "Annalen der Physik" (Annals in Physics) 1906, vol. 21, pp. 756–780.

are preceded by solid theoretical preparation, so he should have published his theory. Secondly, as it transpired, proving the validity of the theory of Brownian motion was no simple matter. That process took several years, over a dozen scientists participated in it, and it only really ended with Perrin's publication *Les preuves de la réalité moléculaire* (Evidence for molecular reality) in 1911, six years after Einstein's work was published.

549

5. *Über die von der molekularkinetischen Theorie der Wärme geforderte Bewegung von in ruhenden Flüssigkeiten suspendierten Teilchen;*
von A. Einstein.

In dieser Arbeit soll gezeigt werden, daß nach der molekularkinetischen Theorie der Wärme in Flüssigkeiten suspendierte Körper von mikroskopisch sichtbarer Größe infolge der Molekularbewegung der Wärme Bewegungen von solcher Größe ausführen müssen, daß diese Bewegungen leicht mit dem Mikroskop nachgewiesen werden können. Es ist möglich, daß die hier zu behandelnden Bewegungen mit der sogenannten „Brownschen Molekularbewegung" identisch sind; die mir erreichbaren Angaben über letztere sind jedoch so ungenau, daß ich mir hierüber kein Urteil bilden konnte.

Wenn sich die hier zu behandelnde Bewegung samt den für sie zu erwartenden Gesetzmäßigkeiten wirklich beobachten läßt, so ist die klassische Thermodynamik schon für mikroskopisch unterscheidbare Räume nicht mehr als genau gültig anzusehen und es ist dann eine exakte Bestimmung der wahren Atomgröße möglich. Erwiese sich umgekehrt die Voraussage dieser Bewegung als unzutreffend, so wäre damit ein schwerwiegendes Argument gegen die molekularkinetische Auffassung der Wärme gegeben.

§ 1. *Über den suspendierten Teilchen zuzuschreibenden osmotischen Druck.*

Im Teilvolumen V^* einer Flüssigkeit vom Gesamtvolumen V seien z-Gramm-Moleküle eines Nichtelektrolyten gelöst. Ist das Volumen V^* durch eine für das Lösungsmittel, nicht aber für die gelöste Substanz durchlässige Wand vom reinen Lösungs-

A page from Einstein's paper On the movement of small particles suspended in stationary liquids required by the molecular- kinetic theory of heat (1905)

756

4. Zur kinetischen Theorie der Brownschen Molekularbewegung und der Suspensionen; von M. von Smoluchowski.

[Bearbeitet nach einer am 9. Juli 1906 der Krakauer Akademie vorgelegten und demnächst in dem Bullet. Int. Crac. erscheinenden Abhandlung.]

§ 1. Die viel umstrittene Frage nach dem Wesen der von dem Botaniker Robert Brown 1827 entdeckten Bewegungserscheinungen, welche an mikroskopisch kleinen, in Flüssigkeiten suspendierten Teilchen auftreten, ist neuerdings durch zwei theoretische Arbeiten von Einstein[1]) wieder in Anregung gebracht worden. Die Ergebnisse derselben stimmen nun vollkommen mit einigen Resultaten überein, welche ich vor mehreren Jahren in Verfolgung eines ganz verschiedenen Gedankenganges erhalten hatte, und welche ich seither als gewichtiges Argument für die kinetische Natur dieses Phänomens ansehe. Obwohl es mir bisher nicht möglich war, eine experimentelle Prüfung der Konsequenzen dieser Anschauungsweise vorzunehmen, was ich ursprünglich zu tun beabsichtigte, habe ich mich doch entschlossen, jene Überlegungen nunmehr zu veröffentlichen, da ich damit zur Klärung der Ansichten über diesen interessanten Gegenstand beizutragen hoffe, insbesondere da mir meine Methode direkter, einfacher und darum vielleicht auch überzeugender zu sein scheint als jene Einsteins.

Dem Mangel einer direkten experimentellen Verifikation suche ich teilweise wenigstens durch eine zusammenfassende Übersicht der bisher bekannten Versuchsresultate abzuhelfen, welche im Verein mit einer kritischen Analyse der verschiedenen Erklärungsversuche deutliche Hinweise darauf zu geben scheint, daß das Brownsche Phänomen in der Tat mit den theoretisch vorauszusehenden Molekularbewegungen identisch ist. Den Schluß bilden einige Bemerkungen über die Suspensionen

---

1) A. Einstein, Ann. d. Phys. 17. p. 549. 1905; 19. p. 371. 1906.

Extract from Smoluchowski's publication An outline of a kinetic theory of Brownian motion and cloudy solutions (1906)

Albert Einstein (1879–1955)

Another scientist who made a contribution to the discovery of Brownian motion was a physicist from the University of Sydney, William Sutherland. Reflections on the Australian researcher's work can be found in Bertrand Duplantier's work titled *Brownian Motion, "Diverse and Undulating"*[128]. Duplantier attributes at least co-discovery of the essence of Brownian motion to Sutherland. He states that in 1905 Einstein and Sutherland independently developed a quantitative theory of Brownian motion that enabled Perrin to establish the precise value of the Avogadro number (which the scholar did in famous experiments in 1908–1909). He believes that Einstein and Sutherland achieved success where many others had failed and that Marian Smoluchowski simultaneously performed an analysis according to a different, more probabilistic, schema of understanding (German: *Gedankenweg*), which led him to similar conclusions[129].

Duplantier saw Smoluchowski's significant contribution to developing a theory of Brownian motion, which is not widely known, however he was not very familiar with all the Polish physicist's works and maybe that is why he did not name Smoluchowski among the creators of the theory of Brownian motion. Connecting the process of Smoluchowski's and Sutherland's research to the end result in the form of Einstein's papers shows that new circumstances concerning this discovery are still being established. We will probably never know the full facts.

Sutherland was a well- known physicist among his peers – he and Josiah Willard Gibbs (1839–1903) – were the only physicists from outside Europe to

---

128 B. Duplantier, *Brownian Motion, "Diverse and Undulating"*, op. cit., pp. 201– 293.
129 Ibidem, p. 218.

be invited to contribute to the Boltzmann Festschrift (commemorative book) in 1904[130].

Not having access to laboratory facilities, his research was limited to theoretical work, chiefly in the field of molecular dynamics. He assumed that the particles that make up matter exert an attractive force on each other in addition to gravity. Although his theory contrasted with Boltzmann's and those of other physicists contemporary to him, who adopted a purely kinematic view, it is currently widely accepted. Modern texts usually refer to the "Sutherland model" and describe the force in terms of the "Sutherland potential."

In his work, Duplantier frequently refers to the Polish physicist, emphasising his achievement. He believes that the name Smoluchowski is closely connected with the theory of diffusion and with Brownian motion. He cites the opinion of Mark Kac, who believes that Smoluchowski's proof that the notion of a game of chance forms the foundation of our understanding of physical phenomena was a real intellectual feat, adding also that "We are indebted to him for his original and bold introduction of the calculus of probability in statistical physics, and he deserves a place beside the great names of Maxwell, Boltzmann, and Gibbs."[131]. However, he attributes the discovery of Brownian motion only to Einstein and Sutherland, though he assigns an important, but complementary, role to Smoluchowski.

During a congress of the Australasian Association for the Advancement of Science in January 1904 in Dunedin, New Zealand, Sutherland submitted two papers of which one – *The Measurement of Large Molecular Masses* – concerned Brownian motion[132]. In it, he proposed the so-called quantitative theory of Brownian motion, based on the principles of statistical mechanics. In the introduction to his arguments, Sutherland notes:

> Determining the large molecular masses of such substances as the physicist works with presents a problem almost insurmountable using the chemist's usual methods. Vapour density is unachievable and the molecular reduction of a solvent's freezing point produced in its massive molecules is so small that it can be seen as an experimental error in the solvent's solidification temperature in its purest possible obtainable state. The proposed method is based on the measurement of the coefficient of substance diffusion through a solvent. The only difficulty related to the measurement of the diffusion

---

130 Ibidem, p. 221.
131 Ibidem, p. 239.
132 W. Sutherland, *A Dynamical Theory of Diffusion for Non-Electrolytes and the Molecular Mass of Albumin*, "Philosophical Magazine and Journal of Science" 1905, series 6, vol. 9, pp. 117–121.

velocity of a substance of large molecular mass is that the measurement experiment must be extended by a time inversely proportional to a given velocity. Thomas Graham (1805–1869), a pioneer of diffusion research, measured the diffusion velocity of protein in water. Due to this, being able to establish the dynamic dependency between the diffusion velocity of a substance and the size of it molecules, it is possible to measure the molecular mass of substances such as protein and the products of its more or less complete breakdown[133].

This paper represents the culmination of several years' work by Sutherland on the problem of Brownian motion.

There is an importance episode connected with Smoluchowski's discovery of the essence of Brownian motion. In September 1903, he wrote an article, *On irregularities in the distribution of gas moleccules and their influence on entropy and the equation of state*, published in 1904 in a commemorative book celebrating Boltzmann's sixtieth birthday. In it he wrote about observations of density heterogeneity (density fluctuations). The book came out a year before the publications of Einstein, who – as John Stachel demonstrates – read the article before writing his own. Stachel, Director of the Center for Einstein Studies, points out the visible influence of Smoluchowski's paper of 1904 on Einstein's 1905 paper on Brownian motion. He shows that Einstein, as a paid reviewer, received the Boltzmann commemorative book and reviewed three other papers from it for a supplement to the "*Annalen der Physik*" (Annals in Physics) journal (devoted to reviews of papers). Hence the almost certain supposition that he read the Pole's work[134]. Godlewski writes about Smoluchowski's article:

> In a wonderful paper printed in a celebratory volume devoted to Boltzmann, he demonstrates the possibility of a new kinetic approach to the phenomena observed in gases. When in the ordinary kinetic theory we understand the deviations in certain velocities from the mean velocity by taking into account the laws of distribution, he takes into consideration the inequalities in the spatial distribution of molecules and therefore density deviations from the mean, that is the continual thickening and thinning. With the aid of the calculus of probability, he calculates the average positive or negative deviation of density from the norm within a given gas element and shows that it is dependent upon the number of molecules and that, generally, the greater (the number), the lower the volume considered. In this approach, the appropriately and kinetically derived

---

133   Ibidem, p. 117.
134   Similarly, J. Stachel writes in: *Einstein on Brownian Motion* in *The Collected Papers of Albert Einstein*, op. cit. p. 216: In 1904 he published an article on density fluctuations in gases that has several features in common with his later work, as well as with Einstein's work on Brownian motion. Einstein may have read this paper, which appeared in Meyer, S. 1904.

expression for the value of entropy also changes, to which the probability logarithm of the corresponding constellation and density must also apply[135].

That Eisntein became acquainted earlier with Smoluchowski's paper, as well as ambiguities that appear while reading the works, prompts a comparison of four key articles – two by Smoluchowski from July 1903[136] and 1906[137], with two of Einstein's from May[138] and December 1904[139].

Certainly Einstein's May work differs radically from December's, despite being only seven months apart. In *The case of Brownian Motion*, Roberto Maiocchi writes that at first glance, the May article has little in common with the phenomenon discovered by Brown. It does not even contain mentions of all the earlier research on the phenomenon. "That these movements may then coincide with the Brownian motion – claims Maiocchi – is certainly an idea present in Einstein's thinking, but the explanation of Brownian motion does not appear to be his basic objective at all[140].

The first article was inspired by an innovative thought, but concerning diffusion. It was most probably a continuation of Einstein's doctoral thesis, *Eine Neue Bestimmung der Molekűldimensionen* (On a new determination of molecular dimensions), which he completed on April 30, 1905.

---

135 T. Godlewski, *Marian Smoluchowski*, op. cit., p. 6; see also: *Marian Smoluchowski (1872-1917). Fizyk, taternik – romantyk nauki*, (Marian Smoluchowski (1872-1917). Physicist, mountaineer, romantic of science) op. cit., pp. 4–5.
136 M. Smoluchowski, *O nieregularnościach w rozkładzie cząsteczek gazu i ich wpływie na entropię i na równanie stanu*, (On irregularities in the distribution of gas molecules and their influence on entropy and the equation of state), trans. B.J. Gawecki, Warsaw 1956; first printed as: *Über Unregelmäßigkeiten in der Verteilung von Gasmolekülen und deren Einfluß auf Entropie und Zustandsgleichung*, op. cit.
137 Idem, *Zur kinetischen Theorie der Brownschen Molekularbewegung und der Suspensionen*, (On the kinetic theory of Brownian motion and suspensions), op. cit. pp. 756–780.
138 A. Einstein, *Über die von der molekularkinetischen Theorie der Wärme geforderte Bewegung von in ruhenden Flüssigkeiten suspendierten Teilchen*, (On the Movement of Small Particles Suspended in Stationary Liquids Required by the Molecular-Kinetic Theory of Heat) "Annalen der Physik" (Annals in Physics) 1905, vol. 322, No. 8, pp. 549–560.
139 Idem, *Zur Theorie der Brownschen Bewegung* (On the theory of Brownian motion) "Annalen der Physik" (Annals in Physics), 1906, vol. 324, No. 2, pp. 371–381.
140 R. Maiocchi, *The Case of Brownian Motion*, "The British Journal for the History of Science" 1990, vol. 23, no. 3, p. 263.

In the text *Über die von der molekularkinetischen Theorieder Wärme geforderte Bewegung von in ruhenden Flüssigkeitensuspendierten Teilchen* (On the movement of small particles suspended in a stationary liquid demanded by the molecular-kinetic theory of heat), Einstein describes the diffusion process of suspended particles subjected to a constant force, which are not initially evenly distributed in a liquid, assuming that the cause of such movement was molecular mixing[141]. The problem of Brownian motion, as he writes himself, was not well known to him, even in terms of the validity of the molecular- kinetic understanding of heat[142]. Einstein was not as familiar with the issue as Smoluchowski although he was interested in the subject, as evidenced by discussions and correspondence with Michele Besso[143]. There is a visible lack of certainty in the conclusions drawn.

> It is possible that the motions to be discussed here are identical with the so-called "Brownian molecular motion", Einstein writes, however, the data available to me on the latter are so imprecise that I could not form a definite opinion on this matter. If it is really possible to observe the motion to be discussed here, along with the laws it is expected to obey, then classical thermodynamics can no longer be viewed as strictly valid (…) for microscopically distinguishable spaces, and an exact determination of the real size of atoms becomes possible. Conversely, if the prediction of this motion were to be proved wrong, this fact would provide a weighty argument against the molecular-kinetic conception of heat[144].

The duality of this situation has been noted by many authors. Einstein's reasoning was quite convoluted and proved faulty in some respects. Starting out from an attempt to calculate the way in which the viscosity coefficient of a liquid changes as a result of adding particles to the solution, he arrived at a model for a formula for the diffusion coefficient. Through this formula, he verified the value of the Avogadro number[145]. Einstein's December publication is interesting

---

141 Ibidem, pp. 264–265.
142 A. Einstein, *Über die von der molekularkinetischen Theorie der Wärme geforderte Bewegung von in ruhenden Flüssigkeiten suspendierten Teilchen* (On the movement of small particles suspended in a stationary liquid demanded by the molecular-kinetic theory of heat) op. cit., p. 549.
143 Michele Besso (1873–1955) was an engineer, a close friend of Einstein from his time working at the Federal Polytechnic Institute in Zurich and later at the Patent Office in Bern.
144 A. Einstein, *Über die von der molekularkinetischen Theorie der Wärme geforderte Bewegung von in ruhenden Flüssigkeiten suspendierten Teilchen*, op. cit. p. 549.
145 R. Maiocchi, *The Case of Brownian Motion*, op. cit., p. 266.

as it constitutes the culmination of seven months of consideration and – importantly – a preliminary confirmation of the theory's validity by some in the scientific community. In December he lost his uncertainty, left the issue of diffusion and focused on the more important matter from which he had initially distanced himself. The second article dispelled all the doubts that had arisen in the first, hence his explanation:

> The present paper shall supplement my above-mentioned paper in several points. We will derive here not only the translatory, but also the rotational motion of suspended particles for the simplest special case when the particles have a spherical shape. We will also establish the shortest observation times for which the result given in the paper is still valid[146].

Einstein starts with thermodynamics, using the concept of osmotic pressure and Boltzmann's famous theorem. From the general theory of fluctuations around the state of equilibrium, he calculates the mean value of the Brownian shift of a particle over the course of one second. Smoluchowski takes the opposite path – he starts by solving the stochastic problem and by tracing the intricate movement of a particle, and seeks an answer to the question about the probability with which a particle moving in straight lines of equal length, backwards and forwards, will find itself, after a certain number of steps, as a given point[147]. Smoluchowski's work was also more diligent, better justified, and the approach to the problem more astute.

John Stachel named three characteristic elements of Einstein's approach to his work, which will be helpful in an attempt to compare the achievements of the two scientists:

> (1) he based his analysis on the osmotic pressure rather than on the equipartition theorem; (2) he identified the mean square displacements of suspended particles rather than their velocities as suitable observable quantities; and (3) he simultaneously applied the molecular theory of heat and the macroscopic theory of dissipation to the same phenomenon, rather than restricting each of these conceptual tools to a single scale, molecular or macroscopic[148].

An unexplained issue, raised by Smoluchowski, was the problem of Einstein's application of Stokes' law in his calculations: "If the dimensions of sphere M

---

146 A. Einstein, *On the Theory of Brownian Motion*, op. cit., p. 371
147 S. Loria, *Marian Smoluchowski i jego dzieło* (Marian Smoluchowski and his work), op. cit., p. 19.
148 *The Collected Papers of Albert Einstein*, op. cit., p. 210.

are large in comparison with the free path of surrounding particles, we can use Stokes' regular formula to calculate the resistance"[149], however, "Einstein did not take account at all of (…) particles so small that they do not obey Stokes' formula[150].

Smoluchowski proves that Einstein's application of Stokes' formula results in his omission of small particles, not subject to the formula, from which he concludes that in the research on Brownian motion this important element remained beyond the investigation. According to Maiocchi, Einstein not only did not omit them, but applied Stokes' formula to them. Hence, in his view, a criticism arises of Einstein's theory and Perrin's experiments, which have one common attribute – the application of Stokes' law to ultramicroscopic particles. The use of this law was equally fundamental to Einstein's work and to Perrin's reasoning. Victor Henri posed a specific question: "Can we really apply Stokes' law to the displacement of granules of 1 μ diameter in water? It may be that this law is not applicable to such small granules". Moreover, he identified the illegitimate use of Stokes' law as the probable cause of the observed disagreement between Einstein's formula and his own data[151].

The application of Stokes' law to microscopic particles in Einstein's and Perrin's work represents an extension to the use of the law far beyond the recognised boundaries of its applicability. Thus an object was studied, having little in common with the object originally intended to be studied, and was ultimately subjected to a law that does not apply to it. The result of such an approach was the following sequence of actions:

> First of all, Stokes' law could only be considered as being proved experimentally for particles of some millimetres in radius, and to apply it to orders of magnitude of one micron was an extremely rash move. Secondly, the law presupposes a sphere which moves in a continuous medium, whereas in the kinetic model the surrounding medium is discontinuous and does not operate on the sphere with continuous force but with irregular collisions. Thirdly, the velocity with which the particle moves in the fluid appears in the law but the research on Brownian motion had by now made it clear that the velocities which can be measured (…) are not at all the velocities of displacement of the particles[152].

---

149   M. Smoluchowski, *Zarys kinetycznej teorii ruchów Browna i roztworów mętnych* (An outline of the kinetic theory of Brownian motion and cloudy solutions), op. cit., p. 505.
150   Ibidem, p. 506.
151   R. Maiocchi, *The Case of Brownian Motion*, op. cit., p. 277.
152   Ibidem, p. 277–278.

Smoluchowski expands upon the resultant inconsistencies:

> It was understood that the numbers Exner considered the measure of particle velocities in no way correspond to the velocity of actual movement; they are chance shifts resulting from the geometric composition of a great number of small deflections possessing all possible orientations in space (…) the displacement achieved over a unit of time must be a unit of a significantly lower order than the actual velocity of movement, occurring in an immeasurably intricate, zig-zag path[153].

The above deliberations are in opposition to the writings of Sutherland, who in 1902 analysed in his works on ionisations, ionic velocities and the dimensions of atoms, the calculation of the size of ions based on Stokes' law[154].

Einstein and Besso, discussing in 1903 the theory of dissociation, which demands the assumption of molecular aggregates combined with water, opened the way for a straightforward calculation of the size of ions in a solution based on hydrodynamic considerations. Einstein came upon the idea of determining it with the help of classical hydrodynamics, as he wrote in a letter to Besso on March 17, 1903, proposing something that, according to Duplantier, seems to be merely a calculation that Sutherland had made earlier. Einstein writes:

> Have you already calculated the absolute size of ions on the assumption that these are spheres and are large enough that the equation of the hydrodynamics of viscous fluids are applicable?
> Given that we know the absolute charge of the electron, this should be quite an easy matter. I would have done it myself, but I lack the literature and time. You could also take advantage of diffusion in order to obtain results with neutral salt molecules in solution[155].

As the editors of the collected works of Einstein note, this extract is particularly important as both key elements of the method of determining molecular sizes – the earlier-mentioned theories of thermodynamics and diffusion – related to hydrodynamics, probably only include Stokes' law[156]. Do Sutherland's deliberations of 1902 bear only a chance resemblance to Einstein's method?

---

153 M. Smoluchowski, *O fluktuacjach termodynamicznych i ruchach Browna*, (On thermodynamic fluctuations and Brownian motion) op. cit. p. 299.
154 W. Sutherland, *Ionization, Ionic Velocities, and Atomic Sizes*, "The London, Edinburgh, and Dublin Philosophical Magazine and Journal of Science" 1902, series 6, vol. 4, pp. 625–645.
155 A. Einstein, M. Besso, *Correspondence 1903–1955*, transl., notes and introduction P. Speziali, Paris 1979.
156 *The Collected Papers of Albert Einstein*, op. cit., pp. 170–182.

Equally striking is that an earlier letter, from February 11- 17, 1903, this time written by Besso to Einstein, clearly shows that the two scientists had discussed Sutherland's work. The letter is made up of two parts – the first concerns experimental data related to the dissociation of bi-ionic molecules, while the second concerns what Besso calls "Sutherland's hypothesis". He states that the "ionic hydrates" theory, as he calls it, temporarily rescues this hypothesis with regard to Ostwald's dilution law. Besso also discusses the role of imperfect semi-permeable membranes as a possible experimental test of Sutherland's hypothesis. In the French edition of the two scientists' correspondence, Pierre Speziali pointed out that in this letter Besso discussed further work by Sutherland, titled *Causes of Osmotic Pressure and of the Simplicity of the Laws of Dilute Solution*[157]. After reading the letters from 1903 one cannot help but wonder whether Besso and Einstein were also familiar with Sutherland's theses of 1902 on ionic sizes. In that case, the Australian scientist's suggestion to use Stokes' hydrodynamic law to determine the size of molecules would have been a direct inspiration for Einstein's dissertation and subsequent work on Brownian motion![158]. Sutherland had assumed the existence of atoms, and tackled a practical question; the measurement of large molecular masses. He was interested in these masses because of their role in the chemical analysis of organic substances. While that is what everyone now uses the Sutherland-Einstein equation for, it was perhaps not of such widespread interest at the time. However, we have just seen from the Einstein-Besso correspondence how extremely important Sutherland's idea was of determining the sizes of ions or molecules by means of classical hydrodynamics[159].

We have here two different narrations of the use of Stokes' law in research on Brownian motion, written from two different approaches to the problem. Smoluchowski conducted an analysis of Brownian motion from the perspective of the microcosm, hence in his opinion the use of Stokes' law for granules of 1 μ diameter was unjustified. Sutherland, and later Einstein, in considering the problem of diffusion, approached the problem from the perspective of the macrocosm, analysing particles on the assumption that they are spheres large enough that hydrodynamic equations for viscous liquids apply in their case and hence the use of Stokes' law seemed to them fully justified. In a 1952 paper, Loria

---

157 W. Sutherland, *Causes of Osmotic Pressure and of the Simplicity of the Laws of Dilute Solutions*, "The London, Edinburgh, and Dublin Philosophical Magazine and Journal of Science" 1897, series 5, vol. 44, p. 271.
158 B. Duplantier, *Brownian Motion*, op. cit., pp. 219–220.
159 Ibidem, p. 221.

indicates that "more important – as the later development of the theory turned out – was the advantage of Smoluchowski's method, due to its *par excellence* microphysical nature[160].

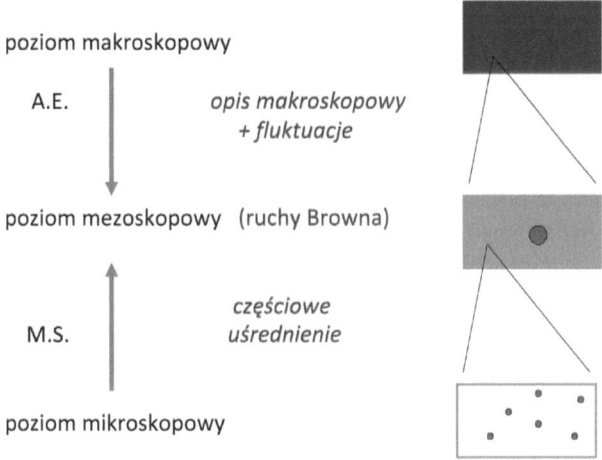

A diagram by Prof. Bogdan Cichocki showing the different perspective in Einstein's and Smoluchowski's perception of Brownian motion

Einstein's main assumption is the legitimacy of using classical hydrodynamics to calculate the influence of the molecules of dissolved substances, treated as rigid spheres, on the viscosity of the solvent in a diluted solution. His method lends itself well to determining the size of solute molecules, which are large in comparison to the molecules of the solvent, so he applied it to the dissolution of sugar molecules. In 1905, Sutherland published a description of a method for determining the masses of large molecules, which coincides in part with Einstein's method. Both use the molecular theory of diffusion developed by Walther Nernst (1864–1941)[161], based on Jacobus "Henry" van 't Hoff's analogy between

---

160 S. Loria, *Marian Smoluchowski i jego dzieło* (Marian Smoluchowski and his work) op. cit., p. 22.
161 W. Nernst, *Stöchiometrie Verwandtschaftslehre* (Stoichiometry relationship theory) "Zeitschrift für Physikalische Chemie" (Journal of Physical Chemistry) 1888, vol. 2, pp. 613–639.

solutions and gases, and Stokes' hydrodynamic friction law[162]. The derivation of equipartition represents the technically difficult part of Einstein's dissertation. It rests on the assumption that the motion of the fluid can be described by the hydrodynamical equations for stationary flow of an incompressible homogeneous fluid, even in the presence of solute molecules; that the inertia of these molecules can be neglected; that they do not interact, and can be treated as rigid spheres moving in the liquid without slipping, under the sole influence of hydrodynamical stress. As with Sutherland, it requires the identification of the force on a single large molecule, which appears in Stokes' law, with the apparent force due to the osmotic pressure[163].

Einstein was in possession of two theories about particles in a fluid. The first was Stokes' hydrodynamic theory, based on the hypothesis that the fluid is a continuous medium which adheres to a large solid surface moving through it, without any turbulence, and where the molecular agitation does not seem to play any role. The second was van 't Hoff's osmotic theory, based on the hypothesis that a particle in solution is similar to any other fluid molecule, and therefore is subjected to the same laws of molecular agitation. Duplantier underscores that Einstein's insight was necessary, as well as his profound knowledge of statistical mechanics, to understand and prove that the two points of view were simultaneously valid for particles as big as Brownian particles[164].

At the very start of his May article, Einstein refers to Brownian motion[165], while in the December text he writes about the movement of particles suspended in a fluid as proposed by the molecular theory of heat[166].

However, the mathematical arguments presented in the May article concern mainly the relationship between the diffusion coefficient and temperature, while the December paper is written differently. The text *A New Determination of Molecular Dimensions,* created in 1904 and 1905, directed Einstein's attention to the problem of diffusion and he focused on that problem in the May article.

---

162 B. Duplantier, *Brownian Motion*, op. cit., p. 223.
163 Ibidem.
164 Ibidem, p. 226.
165 A. Einstein, *Über die von der molekularkinetischen Theorie der Wärme geforderte Bewegung von in ruhenden Flüssigkeiten suspendierten Teilchen*, (On the Motion of Small Particles Suspended in a Stationary Liquid, as Required by the Molecular Kinetic Theory of Heat) op. cit., pp. 549–560.
166 Idem, *Zur Theorieder Brownschen Bewegung* (On the theory of Brownian motion), op. cit. pp. 371–381.

Stachel captures the essence of that article in the following way – Einstein derived the diffusion equation from an analysis of the time-dependence of the particle distribution, calculated from the probability distribution for displacements. This derivation is based on his crucial insight into the role of Brownian motion as the microscopic process responsible for diffusion on a macroscopic scale[167]. Einstein's December article is entirely devoted to the analysis and calculation of Brownian motion. Moving into a different area of deliberation, Smoluchowski writes:

> I will not enter into a discussion here on the very ingenious reasoning with the aid of which Einstein arrived at his formulae, however I believe that both methods used by him rely on indirect reasoning (e.g. neither transferring the laws of osmotic pressure to particles M suspended in a fluid and calculating the velocity at which they diffuse through the liquid, nor applying Boltzmann's theorem (on the influence of potential forces on the statistical composition of mechanical systems) to the non- potential force which is the resistance experienced by particles M in the movement through the medium, is entirely beyond reproach), which does not seem entirely persuasive. In each case, concurrence with the direct method used here, which better explains the mechanism of the whole phenomenon, should be seen as a desirable confirmation of both means of calculation[168].

According to Einstein, so-called osmotic pressure is caused by pressure exerted by the solute, separated from the pure solvent by a membrane, i.e. by acting on a wall that is permeable for the solution but not for the substance dissolved in the solution. We therefore have osmotic pressure, which must act on the solution in order to prevent the flow of the solvent with the solute through the semipermeable membrane that separates solutions of various concentrations. The cause of osmotic pressure is the difference in concentrations of chemical compounds in solutions on either side of the membrane and the system's desire to equalise them. This difference can also be defined as the degree of disorder of electrolyte molecules, which is a determinant of the degree of entropy of an electrolyte in solution. The entropy is greater the higher the degree of disorder of the electrolyte's molecules.

The osmotic pressure described by Einstein is measurable in a situation in which the mass is not distributed evenly and the pressure's value appears on the membrane separating the electrolyte from the pure solution. According

---

167 *The Collected Papers of Albert Einstein*, op. cit., p. 212.
168 M. Smoluchowski, *Zarys kinetycznej teorii ruchów Browna i roztworów mętnych*, (An outline of the kinetic theory of Brownian motion and cloudy solutions) op. cit., p. 506.

to Einstein, it appears on the boundary of the partial volume V^*, when a situation arises that electrolyte molecules are dissolved in the partial volume V^* of a fluid of total volume V despite there being no membrane. Smoluchowski called this 'artificial osmotic pressure', which incidentally is a particular case of internal pressure arising in a situation where mass is not distributed evenly. In the paper *On thermodynamic fluctuations and Brownian motion*, Smoluchowski shows that the osmotic pressure formulae applied work well in the description of thermodynamic fluctuations that occur in a solution with uneven distribution of particles, hence applying artificial osmotic pressure to the calculation of diffusion is justified, but for the calculation of Brownian motion it is an indirect and problematic tool.

The nature and direction of the changes occurring in the course of spontaneous processes in an isolated thermodynamic system intended for the calculation of the effect of diffusion action is similar to calculations of the fictional action of osmotic pressure. Both processes take place, in line with the second law of thermodynamics, if the thermodynamic system moves from one state of equilibrium to another. Smoluchowski comments on Einstein's idea thus:

> According to Einstein, a conclusion of the existence of diffusion phenomena can be taken as a starting point, linking it with the laws of osmotic pressure, as a result of which we get a duality of views on the phenomenon of diffusion since on one hand we explain it, quite rightly, as the result of microscopic Brownian motion and on the other we consider it a result of "osmotic pressure", which in this case is more of a fictitious macroscopic notion. This assumption enables a calculation of the velocity of diffusion movement if we rely on the generally accepted hypothesis that diffusion can be reduced to the fictitious action of osmotic pressure, which shifts particles from places of greater concentration towards places of lower concentration[169].

We may have a situation in which molecules find themselves in one half and the other remains entirely empty. However, the number of such cases of extreme density is so negligible that we do not need to take them into account here[170]. This is the situation on the basis of which Einstein builds his essential argument in his May article. Therefore, Smoluchowski's thesis, that research on Brownian

---

169 Idem, *O nieregularnościach w rozkładzie cząsteczek gazu i ich wpływie na entropię i na równanie stanu* (On irregularities in the distribution of gas molecules and their influence on entropy and the equation of state) op. cit., p. 309–310.

170 Idem, *Über Unregelmäßigkeiten in der Verteilung von Gasmolekülen und deren Einfluß auf Entropie und Zustandsgleichung*, (On irregularities in the distribution of gas molecules and their influence on entropy and the equation of state) op. cit., p. 637.

motion in terms of osmotic pressure is an indirect conception and not entirely convincing, is justified. The physicist Robert Alicki notes:

> However, the question arises as to why Einstein made such an identification, which is not obvious. Osmotic pressure occurs when an area of a solution is bordered by a semipermeable barrier, which is not the case in the problem of diffusion. It can therefore be speculated that Smoluchowski's idea of 1903 helped Einstein adopt that assumption. Smoluchowski showed that local fluctuations in the density of gas reduce entropy (now described as the "microscopic function of entropy" 'S') and also increase the momentary free energy. As Smoluchowski writes in paragraph 5," The importance of this "microscopic" entropy function lies in the fact that gas can "by itself" perform the work $T(S_0 - S')$..." and later he writes about Maxwell's demon[171]. In other words, the fluctuations themselves, without the need to introduce a semi- permeable barrier, can generate forces (of the osmotic pressure type) that move a Brownian particle against the force of resistance, performing work. Later, Smoluchowski withdrew from this and today we also do not consider that work, but the image itself of a random force generated by fluctuations remains[172].

Another issue, to which Smoluchowski devotes a great deal of attention in a 1904 article, is entropy. In § 3 he writes:

> Some interest may be aroused by a variation in the usual notion of entropy, based on the molecular structure of gas discussed here. If we applied a 'macroscopic' formula for entropy (…) (per unit of mass) to calculate the total entropy of a unit of mass from the entropy of individual parts of its volume (…) we would have to take into account that the relative number of such parts of the volume where the density ϱ is increased or decreased (…). The 'microscopic' entropy of individual parts of the volume is therefore higher or lower than the normal value in an even density distribution; but the mean value is lower and at the same time depends significantly on the dimensions used to calculate the parts of the volume (consisting of a number of particles v)[173].

---

171 "Similarly, Maxwell's 'demon' would break the second law of thermodynamics though in our case it would not pay attention to velocity but to the density of molecular swarms. Maybe instead of that – as one of my friends noted – an equally perfect one-way valve could be applied (…) it would not even have to be unusually small to create appreciable densities. In order to achieve this result, an additional contribution – admittedly not taken account of in more detail here – would be differences in velocity andtemporal irregularities," ibidem, p. 61

172 Robert Alicki – professor of physics, lecturer at the University of Gdańsk. The quoted statement was sent to the author via e- mail on February 4, 2019.

173 M. Smoluchowski, *O nieregularnościach w rozkładzie cząsteczek gazu i ich wpływie na entropię i na równanie stanu*, (On irregularities in the distribution of gas molecules and their influence on entropy and the equation of state), op. cit., pp. 58– 59.

Let's compare Smoluchowski's text with an extract from Einstein's article in which he develops his conception as follows:

> Let z gram-molecules of a nonelectrolyte be dissolved in the partial volume V^* of a liquid of total volume V. If the volume V^* is separated from the pure solvent by a wall that is permeable to the solvent but not to the dissolved substance, then this wall is subjected to the so-called osmotic pressure (…) But if instead of the dissolved substance, the partial volume V^* of the liquid contains small suspended bodies that likewise cannot pass through the solvent-permeable wall, then according to the classical theory of thermodynamics we should not expect-at least if we neglect the force of gravity, which does not interest us here-that a force be exerted on the wall; because according to the customary conception, the "free energy" of the system does not seem to depend on the position of the wall and of the suspended bodies, but only on the total masses and properties of the suspended substance, the liquid, and the wall, as well as on the pressure and temperature. To be sure, the energy and entropy of the interfaces (capillary forces) should also be considered in the calculation of the free energy; but we can disregard them since the changes in the position of the wall and the suspended bodies considered here shall proceed without changes in the size and condition of the contact surfaces[174].

Let's compare three issues described in different language but relating to the same problems in Smoluchowski's 1903 article and Einstein's of May 1904. The first is the noticeable relationship between entropy size and osmotic pressure size. The orientation of the problem of diffusion on entropy or osmotic pressure depends on the research tools used, but in both cases the results indicate the degree of the electrolyte's disorder.

The second issue concerns the research methods proposed by the two scientists. Smoluchowski wants to study entropy by isolating individual volume parts from the total volume. Einstein proposes studying osmotic pressure by isolating the partial volume V^* from total volume V. The scholars' ideas of separating small volumes are similar; in both cases a partial volume is separated from the solution's volume and studied in relation to the overall volume.

The third problem is the concordance of the two physicists' thoughts on the subject of the forces at work in a solution. Smoluchowski writes that "the internal cohesion forces cancel each other out everywhere and create only a constant

---

174 A. Einstein, *Über die von der molekularkinetischen Theorie der Wärme gefor- derte Bewegung von in ruhenden Flüssigkeiten suspendierten Teilchen*, (On the Motion of Small Particles Suspended in a Stationary Liquid, as Required by the Molecular Kinetic Theory of Heat) op. cit., pp. 549–550.

'internal pressure', calculated as if the mass was evenly distributed"[175], and Einstein that "a force be exerted on the wall; because according to the customary conception, the 'free energy' of the system does not seem to depend on the position of the wall and of the suspended bodies, but only on the total masses and properties of the suspended substance, the liquid, and the wall"[176].

How does Smoluchowski's "internal pressure" differ in the above case from Einstein's "free energy"? Is the application of osmotic pressure not similar to the application of entropy? The similarity of the ideas and terms is difficult to underestimate. The important difference is that Smoluchowski's text appeared a year before Einstein's, who could have read Smoluchowski's paper straight after its publication.

Let us note that in the case of research on entropy we are dealing with stochastic processes, while in the case of research on osmotic pressure the processes are statistical.

Approaching the problem of diffusion through researching osmotic pressure is to narrow the description of a specific situation of the phenomenon compared to a description of diffusion by means of entropy, which is fuller and more universal.

Einstein's idea of using osmotic pressure and indirectly van 't Hoff's equation as well as other equations determining standard enthalpy and entropy in his proof was an assumption that enabled the formulation of a conception of diffusion and Brownian motion. The source of the idea lies in Smoluchowski's suggestion, as evidenced by Einstein's original intention to arrive at the average value of a particle's Brownian shifts in one second from the general theory of fluctuation around a state of equilibrium, using the concept of osmotic pressure and Boltzmann's famous theorem. This is contained in Smoluchowski's proposals in the paper *On irregularities in the distribution of gas molecules and their influence on entropy and the equation of state*, that the application of formulae for osmotic pressure was indispensable in the calculation of diffusion but not entirely justified in calculating Brownian motion. In his May 1905 paper, Einstein writes:

---

175 M. Smoluchowski, *O nieregularnościach w rozkładzie cząsteczek gazu i ich wpływie na entropię i na równanie stanu*, (On irregularities in the distribution of gas molecules and their influence on entropy and the equation of state), op. cit., p. 66.

176 A. Einstein, *Über die von der molekularkinetischen Theorie der Wärme gefor- derte Bewegung von in ruhenden Flüssigkeiten suspendierten Teilchen*, (On the Motion of Small Particles Suspended in a Stationary Liquid, as Required by the Molecular Kinetic Theory of Heat) op. cit., pp. 549–550.

But from the standpoint of the molecular-kinetic theory of heat we are led to a different conception. According to this theory, a dissolved molecule differs from a suspended body in size alone, and it is difficult to see why suspended bodies should not produce the same osmotic pressure as an equal number of dissolved molecules. We will have to assume that the suspended bodies perform an irregular, even though very slow, motion in the liquid due to the liquid's molecular motion; if prevented by the wall from leaving the volume V^*, they will exert forces upon the wall exactly as dissolved molecules do. Thus, if n suspended bodies are present in the volume V^*, i.e., $n/V = v$ in the unit volume, and if the separation between neighbouring bodies is sufficiently large, there will correspond to them an osmotic pressure p.

Smoluchowski writes: "If we were to apply a 'macroscopic' formula of entropy to calculate the total entropy of a unit of mass from the entropy of individual parts of the volume, we would have to take into account that the relative number of such parts of the volume, where the density is the number of such parts v by volume in which the density ϱ, is increased or decreased[177].

Perhaps Einstein saw the author's intention, transposing the contents of his argument from gases to liquids; the idea was not new to him, he had encountered it in van 't Hoff's analogy between solutions and gases. He also made certain assumptions about Brownian motion, not being entirely convinced of its legitimacy. Smoluchowski underscores the similarity of the action in the two media: "Deliberations concerning Brownian motion were based on the analogy of particles suspended in a liquid medium with gas particles, an analogy which is expressed quantitatively in the fact that the kinetic energy of the translational motion must be the same in both cases[178].

In the introduction to his paper *On irregularities in the distribution of gas molecules and their influence on entropy and the equation of state,* Smoluchowski calls for research into the influence of unevenness in the local distribution of gas molecules on deviations from the mean values of velocities of individual molecular gas particles. However, even for microscopically small sizes it would be imperceptible under the microscope. It would be possible if "swarming systems" were created from individual molecules[179]. The microscopic entropy of

---

177 M. Smoluchowski, *O nieregularnościach w rozkładzie cząsteczek gazu i ich wpływie na entropię i na równanie stanu*, (On irregularities in the distribution of gas molecules and their influence on entropy and the equation of state), op. cit., p. 60.
178 Idem, *O fluktuacjach termodynamicznych i ruchach Browna* (On thermodynamic fluctuations and Brownian motion) op. cit., p. 328.
179 In the case of liquids, Smoluchowski calls these systems fluctuations.

individual parts of the volume is sometimes greater and sometimes smaller than the normal value in even density distribution[180].

Smoluchowski's accurate interpretation of the mechanism causing the movement of molecules and the problem's omission by Einstein, who does not describe the mechanism causing Brownian motion, suggests a difference that separates the two scientists in terms of the research and experiments they conducted. According to Smoluchowski, the movements of particles cannot be cause by individual particles of water or other solvent due to too great a difference between their size and the dimensions of the particles. He is convinced that single molecules must create swarms of particles or fluctuations which could effectively cause the particles' movements. Einstein does not mention this important condition of Brownian motion.

> I have found in Einstein's formulae some of my own results and his end result agrees completely with mine although we used an entirely different method. Hence, I offer my reasoning, especially as my method seems to me transparent and therefore more persuasive than Einstein's method, which is not free from criticism. I add to that the discussion of other theories and the factual material accumulated by earlier researchers, which I believe is highly persuasive of a kinetic interpretation of these phenomena. At the end of my paper I include a few comments on the so- called colloidal suspensions theory related to this subject[181].

For research on the size of particle shifts in Brownian motion, the position is important of French physicist Paul Langevin (1872–1946), who in 1908 presented the third – after Einstein and Smoluchowski – version of a formula for the mean squared displacement of a Brownian particle. This formula was experimentally verified quite quickly. At the same time, it was the simplest derivation of a formula for the squared displacement of molecules (in Langevin's own words it was "infinitely simpler" than Einstein's). In part I, Langevin refers to two of Einstein's papers in which the latter derives the function Δ_x^2 given in Langevin's equation. Langevin's own analysis in part II also generates the equation $\Delta\_x^2 = RT/N1/3\pi\mu a \tau$! Smoluchowski, using different methods, obtained for Δ_x^2 the same expression of the same form as $\Delta\_x^2 = RT/N1/3\pi\mu a \ \tau$! but which differs from it by the coefficient of 64/17. Does Smoluchowski's theory predict a

---

180 M. Smoluchowski, *O nieregularnościach w rozkładzie cząsteczek gazu i ich wpływie na entropię i na równanie stanu*, (On irregularities in the distribution of gas molecules and their influence on entropy and the equation of state), op. cit., p. 59.

181 Idem, *Zarys kinetycznej teorii ruchów Browna i roztworów mętnych* (An outline of the kinetic theory of Brownian motion and cloudy solutions), dz. cyt., s. 490–491.

value Δ_x^2 larger by a factor of 64/27, or smaller by a factor of 27/64 than that predicted by the Einstein/Langevin formula of $\Delta\_x^2 = RT/N1/3\pi\mu a\ \tau$!? The natural reading is that Smoluchowski's predictions are larger than Einstein's and Langevin's by a factor of 64/27. Indeed, an inspection of Smoluchowski's paper confirms this interpretation[182].

---

[182] D.S. Lemons, A. Gythiel, *Paul Langevin's 1908 Paper "On the Theory of Brownian Motion"*, "American Journal of Physics" 1997, vol. 65, no. 11, November.

Chapter VII Einstein – Smoluchowski – Sutherland

It is worth emphasising at the outset that in literature on the subject there are no references to Smoluchowski's numerous comments on Einstein's publications. The issue does not appear in academic or popular-science works, though the Polish physicist had a range of doubts that he presented in detail. He highlighted the different starting point for theoretical deliberations; he wrote that Einstein's solutions were more abstract in nature, resulting from his general research in the field of statistical mechanics, he was also the first to provide a formula to determine the shifts of particles suspended in a liquid, but left the question open as to whether the theoretically described phenomenon corresponds to what was known as Brownian motion[183]. Smoluchowski referred to his discovery many times, comparing it with Einstein's. A note of regret is discernible in his argument that having such an advanced theory, properly worked out mathematic formulae and conclusions, he hesitated to publish them, all the more so as he was aware of his discovery's importance.

The considerations presented above, as well as the comparison of texts, are speculative in nature but they reveal quite significant ambiguities, especially concerning who was responsible for this key discovery for early 20th-century physics. In summary, it can be seen that from Smoluchowski's various comments there emerges a picture of Einstein's conceptions and calculations which are not very clear or entirely consistent. Einstein approached the problem chiefly from a mathematical perspective and did not fully explain the essence of the discovery. The Polish physicist pointed out that "Einstein's deliberations are of a more abstract nature as they stem from his general research in the field of statistical mechanics"[184], adding "but for now he has left the question open whether the theoretically foreseen phenomenon corresponds to what was known as Brownian motion"[185]. This is without doubt a significant comment. In the article

---

[183] M. Smoluchowski, *Zarys kinetycznej teorii ruchów Browna i roztworów mętnych*, (An outline of a kinetic theory of Brownian motion and cloudy solutions), op. cit. pp. 490– 491

[184] Idem, *O fluktuacjach termodynamicznych i ruchach Browna*, (On thermodynamic fluctuations and Brownian motion), op. cit., p. 299.

[185] Ibidem

On thermodynamic fluctuations and Brownian motion, Smoluchowski writes straightforwardly:

> The author of this paper has been convinced of the molecular-kinetic essence of Brownian motion since 1900, when he learnt of its existence from Exner's work and from that time has gradually sought to resolve that theory, first by considering the probable arrangement of particles, then the phenomena of movement caused by a combination of free paths, and finally the mechanism itself of the movements of particles caused by the collisions of surrounding molecules[186].

Einstein's paper *On the Movement of Small Particles Suspended in Stationary Liquids Required by the Molecular- Kinetic Theory of Heat*[187], was submitted for publication on May 11, 1905, and on June 30, his most important work, *On the Electrodynamics of Moving Bodies*[188], was published. Einstein changed our perception of reality, our understanding of the fundamental paradigms of time and space, and simultaneously during this same period developed the basic proof for the atomic composition of matter, although he never specifically dealt with it.

The very title of Smoluchowski's paper *On irregularities in the distribution of gas molecules and their influence on entropy and the equation of state* must have intrigued a physicist of Einstein's intellect and knowledge. He encourages people to become acquainted with it and Einstein probably read the Polish physicist's text, a consequence of which was preparation of the paper *On the Movement of Small Particles Suspended in Stationary Liquids Required by the Molecular- Kinetic Theory of Heat,* the contents of which imply associations with Smoluchowski's concepts. The fact that the two articles describe phenomena occurring in different media is of no matter as Smoluchowski writes about the analogy of Brownian motion and gaseous and liquid media in § 17, titled *Approximate direct calculation of shifts in a liquid medium*:

> The essential concept on which the theory of Brownian motion rests is the assumption that suspension particles (of a certain temperature) behave to a certain degree analogously to [gas – J.G.] molecules, namely that the mean kinetic energy of translational movement made through the medium of the particle's centre of mass equals the mean energy of translational movement of a gas molecule at that temperature. (...) This

---

186 Ibidem
187 A. Einstein, *Über die von der molekularkinetischen Theorie der Wärme geforderte Bewegung von in ruhenden Flüssigkeiten suspendierten Teilchen* (On the Movement of Small Particles Suspended in Stationary Liquids Required by the Molecular-Kinetic Theory of Heat) op. cit., p. 549–560.
188 Idem, *Zur Elektrodynamik bewegter Körper* (On the Electrodynamics of Moving Bodies), "Annalen der Physik" (Annals of Physics), 1905, No. 17, pp. 891–921.

assumption corresponds to the basic assumption of the kinetic theory of gases and is a direct consequence of Maxwell's principle of energy equipartition[189].

An analysis of Smoluchowski's and Einstein's articles at least substantiates the hypothesis presented earlier that the key proof for the theory of Brownian motion, based on osmotic pressure and described in the May article, had its origins in Smoluchowski's publication.

Einstein, as Stachel writes, published a total of four main papers between 1905 and 1908 on the subject of Brownian motion in liquids: *On the Movement of Small Particles Suspended in Stationary Liquids Required by the Molecular-Kinetic Theory of Heat* (1905), *On the Theory of Brownian Motion* (1906–1907) as *Theoretische Bemerkungen über die Brownsche Bewegung* and in 1908 under the title *Elementare Theorie der Brownschen Bewegung (Elementary Theory of Brownian Motion)* as well as a synopsis for his own lecture at the Bern Society of Natural Sciences (1907).

During this period, three articles also appeared on related subjects. The first, originally published as a dissertation in 1905, concerned the determination of molecular sizes. The other two (1907 and 1908) contain proof for the existence of Brownian motion and present its measurement. It is also worth mentioning Einstein's remark on the measurement of Brownian motion following a speech he gave at a meeting in Salzburg in 1909 organised by the Society of German Natural Scientists and Physicians, and other statements made in the course of the discussion[190]. There is no reference in these articles to any notes or thoughts prior to 1905. In *The Collected Paper of Albert Einstein*, in documents dating from 1902–1905, there are no significant notes on the problem of Brownian motion. The only paper close to the subject is *A New Determination of Molecular Dimensions*, but in it Einstein does not address the issue of Brownian motion.

Some scientific publications state that in 1905–1906, two theorists – Albert Einstein and Marian Smoluchowski – independently of one another and following different paths, explained the essence of Brownian motion and proved the validity of atomic theory. This appears in various publications from the early 20th century but in the light of the considerations presented here, an essential doubt arises as to the correctness of such a claim. An analysis of Smoluchowski's paper *On irregularities in the distribution of gas molecules and their influence on entropy and the equation of state* and Einstein's *On the Movement of Small*

---

189   Idem, *O fluktuacjach termodynamicznych i ruchach Browna*, (On thermodynamic fluctuations and Brownian motion), op. cit., pp

190   *The Collected Papers of Albert Einstein*, op. cit., p. 206.

Particles Suspended in Stationary Liquids Required by the Molecular-Kinetic Theory of Heat, suggests that the Polish physicist's contribution to Einstein's work is highly probable.

The theoretical formulae devised by Smoluchowski and Einstein were confirmed by the careful measurements of two physicists – Svedberg and Perrin. As Smoluchowski notes, Perrin's measurements in particular were made very accurately[191].

The University of Vienna, which was a strong scientific centre in the 1890s, became weakened. In turn, Paris became stronger, headed by the Sorbonne, and it was to there that the main focus of discussion on physics and chemistry shifted, with the inseparable participation of German universities, as in the 19th century. In 1899, Smoluchowski started working at the University of Lviv, which with the shift of discussion to Paris distanced him from the main current of lively scientific discussion. These changes meant that he participated less actively in the ongoing discussion of scientists working on the theory of Brownian motion.

From 1905, discussion on the subject centred mostly on French and German scientists, with the participation of Swedes, primarily Svedberg, and scholars from the Cavendish Laboratory. Unlike Einstein, Smoluchowski's voice went largely unheard in the discussion many academics were engaged in at the time. Einstein constantly took part in solving the problems that arose in the experimental proof of the atomic composition of matter. Over the course of the first five to seven years after the two scientists' publications, Einstein's position as the main, and sometimes only, person responsible for the discovery became firmly established in the minds of scientists. Smoluchowski took part more seldom in these discussions, while Einstein assisted, commented, and inspired the experimental research of scholars investigating Brownian motion.

---

191 M. Smoluchowski, *O fluktuacjach termodynamicznych i ruchach Browna*, (On thermodynamic fluctuations and Brownian motion), op. cit., pp. 299–300

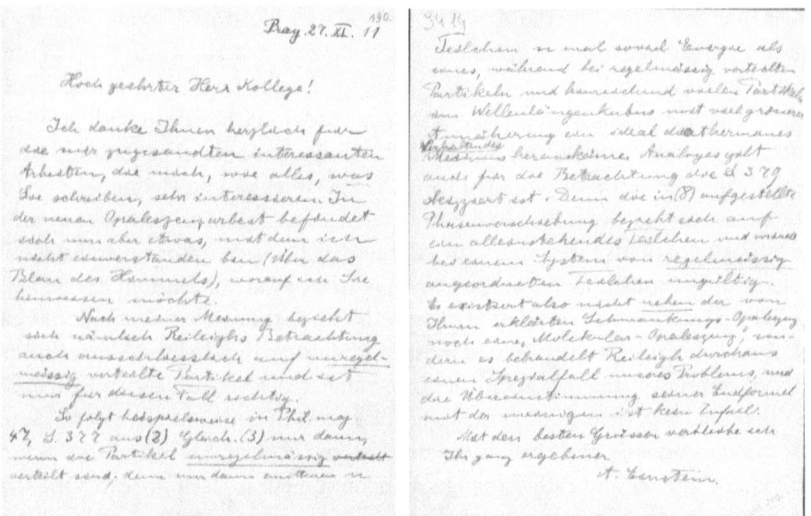

Einstein's letter to Smoluchowski

In time, Smoluchowski naturally ceased to be associated with co-creating the theory of Brownian motion. He worked far away, in Lviv, and from 1912 at a small provincial university in Kraków, and his part in the discovery waned in the memory of those who knew about it earlier and did not figure in the awareness of scientists who were just entering that field of research. Those who do not take part in the discussion cannot be right and from a certain point Smoluchowski ceased to participate in the ongoing academic discourse. The fact that someone read out one of his articles on his behalf, as the physicist Paul Langevin did at one conference, changed nothing. Einstein's ongoing scientific activity cemented the conviction of who should be considered responsible for the discovery.

A similar view – concerning Sutherland – appeared in the work of scientific historian Roderick W. Home[192]. He notes that in the English-speaking world, in which '*Philosophical Magazine*' was one of the leading titles in the field, few people were involved in German-style theoretical physics, due to which the Australian's name was not widely associated with research on diffusion and viscosity. There is much evidence that experimentally-oriented British physicists

---

192 R.W. Home, *Sutherland, William (1859–1911)*, in: *Australian Dictionary of Biography*, vol. 12, http://adb.anu.edu.au/biography/sutherland-william-8719 (access: 12.01.2022).

failed to appreciate Sutherland's work. This is well illustrated by an obituary in *'Nature'* magazine, which emphasises that his work was known to the scientific world but was so full of generalisations that assessing its value was too hard a task[193]. These works attracted a large group of readers as they dealt with problems that Einstein also worked on. Sutherland indulged in bold speculations, which came down to extensive comparisons with experimental knowledge.

So it was that in Great Britain the Australian scientist had no audience aware of the importance of his research on the relationship between diffusion and viscosity in the way that some readers on the continent were aware of Einstein. He was also not helped by the way he presented his arguments, as they were full of intricate calculations concerning the molecular mass of albumin, which he could have omitted without compromising the quality of his research. However, he was firmly focused on the problem of determining the molecular mass of large molecules and clearly saw the dependence of diffusion viscosity as an incidental result on the way to achieving a greater aim rather than as an achievement in itself. It seems the time has finally come for the community of physicists to recognise Sutherland's achievement and christen the famous relation anew with a double name[194].

The subtleties of the intricacies of Smoluchowski's and Sutherland's contribution to research on Brownian motion were not decisive but certainly contributed to them not currently being widely associated with that scientific success.

William Sutherland (1859–1911) at the age of 20.

---

193 B. Duplantier, *Brownian Motion*, op. cit., pp. 221–222.
194 Ibidem.

Of course, the relation of diffusion and viscosity is generally known as the Einstein relation, not the Einstein-Sutherland relation. I think this happened because at the start of the 20th century, theoretical physics was for the most part a German discipline. Although the theory had already been devised, it was not initially published, however it was taken up by continental researchers who had read Einstein's papers and also noticed that a similar theory had been published in an article titled *A Dynamical Theory of Diffusion for Non-Electrolytes and the Molecular Mass of Albumin*[195].

The facts presented above do not provide sufficiently concrete arguments to explain the events that occurred in the period between Smoluchowski's earlier-developed theses and Einstein's later theories, and also do not explain Sutherland's influence on the discovery of Brownian motion. They represent conjectures, each of which is unconvincing individually, but cannot be underestimated when viewed as a whole.

The fact that Einstein never confirmed he had read the article *On irregularities in the distribution of gas molecules and their influence on entropy and the equation of state* before publishing his own theories, leads one to wonder about the cause of such behaviour. If he believed that paper had no particular importance for his discovery of Brownian motion, because in reading it he was focused on issues of diffusion, his silence on the matter is incomprehensible. While writing the May 1905 paper, Einstein may not have known that Smoluchowski had for two years had a mathematical solution to the problem, because he never considered the issue, which was not the main subject of his interest. In December 1905, Einstein must have at least surmised the premise of Smoluchowski's work. Even if one assumes, as is unlikely, that reading the article did not inspire his findings, Smoluchowski's statements and comments must have reached him later. He never responded to them or expressed his opinion. The ambiguity of this situation makes it impossible to be sure of the true circumstances of the discovery.

Until May 1905, Einstein worked at a patent office and wrote scientific papers though his achievements were limited to just a few rather unimportant publications as he was not yet such a well-known physicist as he would be a few years later. He knew that his possible inspiration by Smoluchowski's paper would not be revealed as it was extremely unlikely at the time. The only documented trail

---

195 W. Sutherland, *A Dynamical Theory of Diffusion for Non-Electrolytes and the Molecular Mass of Albumin*, "Philosophical Magazine and Journal of Science", series. 6, vol. 9, London, Edinburgh, and Dublin 1905, pp. 781–785.

was the review of three articles for a book of limited circulation, to which only a narrow group of scientists would have access. In May 1905, Einstein could not have foreseen that a hundred years later his entire scientific legacy would be meticulously archived, many scientists would scrutinise the work he left with the aid of programmes and computer algorithms, and the fact that he was aware of Smoluchowski's paper would come to light.

Smoluchowski's exceptionally righteous nature meant that he never undermined Einstein's work. He never raised any objection or made any accusations, despite many thought-provoking comments. He was straightforward and not given to hasty judgements, especially when he could not be sure of their veracity. This led to unfavourable situations, such as when he held back in 1903 from publishing his work on Brownian motion due to the conviction that he should first ensure the validity of his conclusions experimentally. Having various doubts, he neither questioned or suggested anything publically. Einstein's respect for Smoluchowski is evidenced by the often-quoted words:

> "Everyone who knew Smoluchowski more closely valued him not only as a shrewd mind but as a noble, subtle and kind person"[196].

It is not insignificant that Einstein made this statement only after Smoluchowski had died. On December 14, 1917, a short article by Einstein, *Marian v. Smoluchowski*[197], appeared in '*Die Naturwissenschaften*' (The Science of Nature), referring to Smoluchowski's unexpected death and at the same time highlighting several of his research successes. The article is full of nostalgic recollections as well as respect and recognition for the deceased scientist. Einstein speaks highly of Smoluchowski – both as a person and as a physicist.

---

196   A. Einstein, *Wspomnienie o Smoluchowskim* (Recollections of Smoluchowski), "Problemy" (Problems) 1972, no 8 (317), p. 42.

197   Idem, *Marian v. Smoluchowski*, "Die Naturwissenschaften" (The Science of Nature) 1917, vol. 50, pp. 737–738

Einstein's article on Smoluchowski

However, on a more thorough reading, one gets the impression that something is lacking in the praise. The text was written eleven years after the publication of *An outline of a kinetic theory of Brownian motion and cloudy solutions*, the theory of Brownian motion had long since been proven and in the opinion of most scientists, Einstein was solely responsible for the discovery. As in the past, here too he did not respond to Smoluchowski's numerous comments on his 1905 articles, even though this was an apt, and perhaps the last, opportunity to do so. Remaining on a courteous level, he formulates his opinion, finally closing the issue of research on Brownian motion.

Why, when Smoluchowski discussed the contents of the 1905 articles many times after their publication, comparing the solutions proposed by Einstein with his own, assessing their pros and cons, did Einstein never express an interest in these comments or publish any substantive remarks? Einstein's rare comments were general in nature, not touching on important issues, which is strange as a few years later he took active part in discussions on the work proving the theory of Brownian motion, for example with Perrin and Svedberg.

The publication of equations describing Brownian motion never caused a public dispute between the two scientists over who was first, although it was without doubt mostly to Smoluchowski's credit. The Polish scholar's attitude inspires respect but also provokes a desire to stand up for the truth of the discovery. Einstein was extremely flattering of Smoluchowski's theory, calling it particularly elegant and illustrative. Although he wrote that Smoluchowski, by showing that a fluid's internal friction constantly slows velocity and that random collisions restore it, managed to explain the phenomenon of Brownian motion quantitatively, as has been stated many times, he never responded substantively to Smoluchowski's comments and he wrote that text after his death[198].

In 1914, in the essay *On thermodynamic fluctuations and Brownian motion*, Smoluchowski summarises the process of the theory's theoretical construction:

> When in 1905 and 1906, Einstein's theoretical work appeared as well as that of this paper's author, both of which, using entirely different methods of reasoning, arrived at compatible results as to the true essence of Brownian motion, experimentalists took up the subject, applying the only correct research method in this case, involving the compilation of statistics on the displacements achieved by particles over certain times[199].

In time, Smoluchowski reconciled himself to the imposed narrative. Circumstances had it that several years after making the discovery, Einstein's name appeared in bibliographies on the subject as the person responsible for it. He never corrected people when they cited him as the sole discoverer, nor did he ever write that there was a second or third person equally responsible for the discovery.

---

198 Idem, *Wspomnienie o Smoluchowskim* (Recollections of Smoluchowski), op. cit., p. 41.

199 M. Smoluchowski, *O fluktuacjach termodynamicznych i ruchach Browna*, (On thermodynamic fluctuations and Brownian motion) op. cit., pp. 298–299.

In reference to Smoluchowski's work, Duplantier is very impressed with the solutions the Pole proposes. He writes that the novelty and originality of Smoluchowski's approach lie in replacing an extremely difficult problem (of a Brownian particle colliding in a gas or liquid) with a relatively simple stochastic process. Every dynamic event, such as a collision, is considered as a chance event like the throw of a die, where elementary probabilities are to some degree determined by underlying mechanical laws[200]. This way of reasoning plays a fundamental role today in mechanics and statistical physics and – as Mark Kac notes – it is hard to imagine the degree of Smoluchowski's intellectual audacity in initiating the subject at the beginning of the last century.

In 1985, publishers Harper and Row released Mark Kac's autobiography *Enigmas of Chance*. Kac writes in it about the discovery of Brownian motion:

> One of the two historical papers was by Marian Smoluchowski. The other, which appeared somewhat earlier, was by Albert Einstein. It was Smoluchowski's bad luck that he had to share his first great discovery, as well as a number of later ones, including the explanation of the blueness of the sky, with so luminous a figure as Einstein.

There is probably no more extreme example of the "Matthew Effect," a wonderfully apt term invented by Robert Merton to describe the all too common phenomenon that the credit for a discovery made jointly or independently by two investigators of unequal fame is invariably given to the more famous one:

> For whosoever hath, to him shall be given, and he shall have more abundance: but whosoever hath not, from him shall be taken away even that he hath.
> [Matthew 13:12]

During his lifetime, Smoluchowski did not suffer from the Matthew Effect. He was universally recognized as one of the leading theoretical physicists of his day and he received many honors which he richly deserved. But with the passage of time the Matthew Effect took its toll. Few realize today what an important role Marian Smoluchowski played in bringing atoms to life and even fewer that it happened in Lwów [Lviv][201].

The hypothesis presents itself that the first person responsible for discovering the essence of Brownian motion, as well as proof of the kinetic-atomic composition of matter, was Marian Smoluchowski. This is only a hypothesis with quite strong premises, not undermining Einstein's merit in introducing the theory to physics, because those achievements cannot be denied. However, the

---

200 B. Duplantier, *Brownian Motion*, op. cit., p. 241.
201 M. Kac, *Enigmas of Chance, An autobiography*, Harper and Row, 1985, pp. 21–21

implication is that both names should appear before that theory, as they did in 1906. Duplantier emphasises that Einstein and Smoluchowski defined the activity of Brownian motion in the same way. Earlier, they had determined the "mean velocity of agitation" by following the path of a molecule as closely as possible. The values obtained this way always stood at several microns per second for particles of the order of a micron. But such assessments of the activity are absolutely wrong. The trajectories are confused and complicated so often and so rapidly that it is impossible to follow them; the trajectory actually measured is very much simpler and shorter than the real one. Similarly, the apparent mean speed of a grain during a given time varies in the wildest way in magnitude and direction, and does not tend to a limit as the time taken for an observation decreases.

Mark Kac (1914–1984)

Neglecting, therefore, the true velocity, which cannot be measured, and disregarding the extremely intricate path followed by a grain during a given time, Einstein and Smoluchowski chose, as the magnitude characteristic of the agitation, the rectilinear segment joining the starting and end points; in the mean, this line will clearly be longer the more active the agitation. The segment will be the displacement of the grain in the time considered[202].

The above arguments present in an entirely new light the problem of determining the contribution of individual scientists to the discovery of the essence of

---

202   B. Duplantier, *Brownian Motion*, op. cit., pp. 263–264.

Brownian motion. Firstly, it should once again be emphasised that Smoluchowski developed a theory of Brownian motion in 1900–1903, but did not publish it. The facts show that both the work of Sutherland and the duo of Einstein and Besso essentially started in 1903, at a time when Smoluchowski was just considering publishing the results of his work. A second issue demanding consideration is the application of Stokes' law to the research. In Smoluchowski's opinion, and in the accounts of many other scientists cited here, this fact is treated as an incomprehensible error while in the concepts of Sutherland and of Einstein and Besso, as presented by Duplantier, Stokes' law along with Nernst's molecular theory of diffusion and van 't Hoff's osmotic theory, is the key law used to discover the cause of Brownian motion. A comparative analysis by a physicist of various aspects of the research leading to Sutherland's Smoluchowski's and Einstein's discoveries would doubtless be interesting, but this requires a separate study. However, the question arises as to whether in science there should appear three names before this theory.

The analyses presented in these deliberations were conducted primarily from a position of the philosophy of science and philosophy of nature, hence the extracts directly related to physics may seem incomplete to the attentive reader, as they should be developed separately by a physicist. Today, more than a hundred years after those events, the considerations are significant only symbolically. They show how much a single decision on postponing publication weighed on Smoluchowski's life, Polish science, and even the history of physics.

Apart from scientific achievements worthy of a Nobel prize, broad interests, and a contribution to propagating physics and mathematics among the young, it is worth mentioning that there are still more important reasons why the name of the Polish physicist should appear in school textbooks and be known to all high school students, because the life and character of Smoluchowski set an example and role model to young people such as has appeared too seldom in Polish history.

He is not the type of hero dying on the barricades of some failed uprising or a patriot imprisoned in the Siberian gulags, nor an emigrant fighting for your freedom and ours. Should a role model today not be someone who built his success on work and science? A person functioning perfectly within the ranks of European scientists, collaborating on a daily basis with the elite of European physicists, at home in the universities of Vienna, Paris, Glasgow and Berlin and in addition possessing exceptional personality traits, being the embodiment of almost Platonic virtue; a pioneer of mountaineering, climbing and skiing, practising music and painting, speaking four languages fluently. And finally a good husband and father. Such a person should inspire new generations of researchers and scientists in Poland.

Chapter VIII Youth and family

Unfortunately, little documentary evidence of Smoluchowski's family relations has survived; some has been lost and what remains is mostly in the possession of his descendants. It can only be surmised, from letters and the descriptions of people close to the family, what sort of husband and father Marian Smoluchowski was. Some testament to the family atmosphere is to be found in the memoirs of Walery Goetel (1889–1972), first a student and later a friend of Smoluchowski's, rector of the Academy of Mining and Metallurgy in Kraków, a geologist and another lover of the Tatra mountains. Goetel, somewhat sentimentally, mentions that Smoluchowski's relations towards his family were exceptionally warm and evident of deep feeling.

Smoluchowski's life ran smoothly, peacefully, happily, it was a single wonderful harmony, developing and reaching ever more circles. There was no friction or falsehood in it, and the foundation of this for Smoluchowski was family life. (…) in the image of the great scholar, one of the most important elements would have been lacking could it not have been said that he was a person who achieved full personal contentment in family life. Whoever saw him at home, in the bosom of his beloved family, whoever could accompany him on a trip with his loved ones and admire how masterfully he brought them into contact with nature, whoever heard with what sincerity he talked of his family and how he yearned for its cheerful, sunny atmosphere, understood that family life was for him an eternal source from which an uninterrupted stream of happiness flowed and from which he drew the basis of the vast edifice of his aspirations and plans, of the whole boundless wealth of his life[203].

---

203   W. Goetel, *Zewspomnień osobistych o Maryanie Smoluchowskim*, (From personal recollections of Marian Smoluchowski) op. cit., pp. 226–227.

Wilhelm Ritter von Smolan- Smoluchowski (1831–1910), Marian Smoluchowski's father

Teofila Smoluchowska (Szczepanowska) (1847–1925), Marian Smoluchowski's mother

Marian's parents

Jan Smoluchowski – Marian Smoluchowski's grandfather, Wilhelm's father, Lord of Moszczenica

Wilhelm Smoluchowski

Teofila Smoluchowska

Marian Ritter von Smolan-Smoluchowski was born on May 28, 1872, in Vorderbrühl near Vienna. Marian's father, Wilhelm – a graduate of the Jagiellonian University, a doctor of law, employed initially as a lawyer and later as a high-ranking member of the Imperial Office of Emperor Franz Joseph's Privy

Council, dealt among other things with Polish affairs. For his services to His Imperial Majesty, he and his sons were granted the right to use the preposition *von* before their surname.

Antoni Popliński (1796–1868), Marian's great grandfather, father of Wanda Szczepanowska

Wanda Szczepanowska (Poplińska) (1829–1910), Marian's grandmother, mother of Teofila

Alleegasse in Vienna. The Smoluchowskis lived in the third house from the right, where they occupied the third floor. Near the Karlskirche, on the left, is a wall surrounding the Theresanium, which Marian attended.

Smoluchowski's mother Teofila, née Szczepanowska, much younger than her husband, came from a well-known family of long tradition. She was the granddaughter of the highly distinguished Poznań pedagogue and philologist Antoni Popliński – an editor, bookseller and owner of a Poznań printing house.

Teofila's brother was Stanisław Szepanowski, a Polish economist, engineer, intellectual, entrepreneur and member of the Austrian parliament and the Diet of Galicia and Lodomeria, an advocate of the industrialisation of Galicia and a pioneer of the nascent oil industry, an industrialist active in almost all of Europe[204]. He was a man of rare intellect and his deeds were so extraordinary that he aroused interest among the creators of Polish literature. Bolesław Prus and Henryk Sienkiewicz wrote about him and Stanisław Brzozowski dedicated his book *The Philosophy of Polish Romanticism* to him.

---

204 S.W. Szczepanowski, *Zarys życia i prac Stanisława Prus Szczepanowskiego* (Outline of the life and work of Stanisław Prus Szczepanowski), Wrocław 2020, http://www.rp-gospodarna.pl/Szczepanowski_2.pdf (access: 14.02.2022), pp. 14–15.

Stanisław Szczepanowski (1846–1900), Marian's uncle, Teofila's brother

Stanisław Szczepanowski

Marian's bother, Tadeusz, a doctor of chemistry at the University of Vienna, a well-known mountaineer and climber, four years older than our hero, was his most loyal friend. Together with his brother, he was the first to conquer sixteen peaks and summits in the Alps. Tadeusz, like Marian, was very highly regarded by his peers. Polish climbing and mountaineering pioneer Jerzy Maślanka wrote about the two brothers:

> The Smoluchowskis were wonderful characters truly as if cast from the same block of bronze, with an unusual straightforwardness that made them alpine companions that could be relied upon absolutely. They always related to people with extraordinary straightness and directness, with an endearing humility that only those above their surroundings can afford[205].

Jan Władysław Szczepanowski (1813– 1875), Marian's maternal grandfather

---

205 J. Maślanka, *Zaranie polskiego alpinizmu* (The dawn of Polish mountaineering) PAN and PAU Science Archive K III-36, typescript T. II 103.

Five-year-old Marian and his older brother, Tadeusz

In 1880, Marian Smoluchowski started attending Collegium Theresianum, Austria's best high school. The school was reserved for children born to parents of appropriate social status, including high-ranking imperial officials, of which Marian's father was one. Founded in the 18th century by Maria Theresa Hapsburg, the Theresianum was recognized as one of the best high schools in Central Europe. In light of the interests he brought from home, Marian was initially drawn towards the humanities and it was only at the end of high school, influenced by one of his professors, that he became interested in physics and mathematics. However, that early interest in the humanities was to stay with him for life.

The Favorita Palace, designated by Empress Maria Theresa as the seat of the Collegium Therasianum

Apart from his parents, Benigna Wolska – his mother's sister, who lived on the outskirts of Florence – had significant influence on Marian as he was growing up. The young Smoluchowski visited his family in Italy many times. It was a special house, always full of music, where concerts were regularly organised. The prevailing atmosphere in the Wolskis' Florence home is described by Armin Teske, who recalls that weeks-long stays in Fiesole were imbued with the atmosphere of a cult of beauty, which Benigna knew how to create and for which Florence provided the ideal backdrop[206]. She exerted considerable influence on the views of the schoolboy, and later student, including in shaping Smoluchowski's aesthetic taste.

---

206 A. Teske, *Marian Smoluchowski: życie i twórczość* (Marian Smoluchowski: life and works), op. cit., p. 13.

Marian Smoluchowski in his fourth year at school

The role of music during these visits is testified to by extracts of letters from "auntie Bena" encouraging her nephew to visit: "We played… many sweet notes, I have been missing them all year"[207].

A little later, during the course of studies at the Academy, influenced by his physics and philosophy teacher, Alois Höfler, with whom he stayed in close contact, the young Marian became fascinated with physics, mathematics and philosophy. Höfler was a philosopher and psychologist who studied with Franz Brentano and Ludwig Boltzmann. From 1903, he lectured at universities in Prague and later in Vienna.

Smoluchowski finished the Theresianum in 1890 and at the same time started his university studies at the University of Vienna. Three years later, still as a student of just 21 years of age, he published his first article in the *Vienna Academy Bulletin* on the internal attenuation of liquids.

---

207   Ibidem

University of Vienna Physical Institute (1912)

In it, he described his experimental work on internal friction in fluids. While studying the flow velocity of fluids in thin tubes, it was found that salts generally demonstrate greater internal friction during the flow than pure water, although some tested substances showed the opposite effect. Smoluchowski established that a relationship exists between the salt content of the solvent and the electrical conductivity of the solutions. He proposed changing the solvent flowing through the thin tubes, using alcohol instead of water. He predicted that salts that reduce the friction in water should cause an increase in friction in the flow of alcohol. It was possible to confirm this hypothesis with the aid of an apparatus proposed by Smoluchowski[208]. The student work foretold the future scientist's extraordinary career.

In the 1893/1894 academic year, Smoluchowski prepared a lecture for the University of Vienna Philosophical Seminarium in which he described a speech by the well-known philosopher and aesthetician Moriz Carrière given at the Bavarian Academy of Sciences in Munich titled *An increase of energy in the spiritual and organic world*. Carrière's lecture includes his explanations of

---

208 Ibidem, p. 16.

materialism, physiological chemistry and spiritualism. In it, the scholar presents various socials forms and – Smoluchowski claims – in an attempt to explain the notion of energy, evokes almost every field of philosophy. However, the Polish researcher subjected Carrière's work to criticism because in analyzing the concept of energy, he mixes scientific order with phenomena of the spiritual world. This is probably the only document bearing direct witness to Smoluchowski's participation in philosophical discourse[209].

In 1895, Smoluchowski defended his doctoral work in physics, titled *Acoustic research on the elasticity of soft bodies*, with distinction. The subject was connected with research conducted earlier by Jožef Stefan, a Slovenian professor, physicist and mathematician working at the University of Vienna. The research concerned considerations of the velocity of sound propagation in wax. Knowing the propagation velocity of a wave in a given body, it was possible to determine the elasticity of that body. In his work, Smoluchowski referred to Stefan's research but modified it slightly, making us of so-called torsion vibrations. He built a torsion pendulum comprised of a vertical glass tube, rigidly fixed on one side with a hard substrate to which he attached a piece of wax or paraffin on the other side. By making the tube vibrate, he also caused vibrations in the body embedded at its end, as a result of which the vibrations of the glass tube made a sound. The tone of the sound so produced enabled measurement of the velocity of the sound's propagation in the paraffin. By changing the temperature of the wax or paraffin, different tones were achieved. This consequently enabled determination of the modulus of elasticity of the body depending on its temperature.

---

209 M. Smoluchowski, *Vortrag im Philosophischen Seminar 1893/1894* (Lecture at the Philosophical Seminarium 1893/1894) University of Vienna, Vienna (1893/1894), Ed. M. Dziekan, "Zagadnienia Filozoficzne w Nauce" (Philosophical Issues in Science" 2017, vol. LXII, pp. 171–189.

Youth and family

Jožef Stefan (1835–1893)

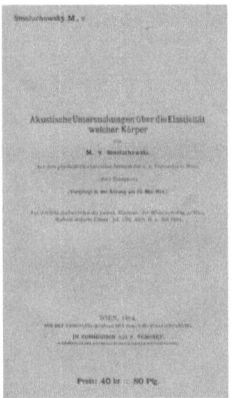

A kustische Untersuchungen über die Elastizität weicher Körper – Marian Smoluchowski's doctoral dissertation

Franz S. Exner (1849–1926)

A certain episode occurred in the same year that illustrates a characteristic personality trait of the young scientist. Smoluchowski received an offer from a university friend to travel around the world. The trip was to be supported by the friend's family, the Viennese Rothschilds, who offered to fund the journey in full. Marian's father, being a high-ranking official of the Imperial Chancellery, ensured his family a comfortable life, but such a long and exotic escapade would doubtless have been beyond his means. The offer was extremely attractive, especially for a young man beloved of the arts and full of curiosity about the world. However, Smoluchowski declined the offer. In November 1895, he wrote to his parents: "Without a doubt, such a trip would be unusually pleasant and interesting. I would see different countries, works of art which I would certainly otherwise never see, but in terms of my professional education I would gain nothing, just forget more, get out of practice and become unused to work"[210].

Almost every young person who has just successfully completed an important stage of their life would take advantage of such an opportunity. Who would not want to go on their annual holidays after the rigours of study and before

---

210   A. Teske, *Marian Smoluchowski: życie i twórczość* (Marian Smoluchowski: life and works), op. cit., p. 22.

embarking on their professional life, especially as it would cost them nothing? Knowing Smoluchowski's interests in nature, geography and art, that trip would have been an extremely interesting intellectual adventure. The young scientist knew, however, that such a long absence would hamper the path towards another, more important adventure – with science: "I want to start working properly in my profession, which I have already chosen for once and for all," he stated. "I really do not want to waste time, but to work seriously with all my energy"[211].

The young Marian got his interest in mountain tourism from the family home. His father, a doctor of law, headed the department of Polish affairs at the Imperial Office of Emperor Franz Joseph. As befitted the times, he was a very dignified gentleman. In the street he walked slowly, something that could not be expected of his much younger wife, Teofila, who, as Teske writes, often tore off ahead. "She would run ahead then, only to notice the distance from her casually strolling companion and hurry back again"[212]. This did not bother Wilhelm during trips to mountain peaks, even in his sixties.

In the period in which Smoluchowski spent his youth there, Vienna was one of the world's intellectual centres. It became a capital of modernism, where such figures lived and worked as: painters Gustav Klimt and Oskar Kokoschka, composers Gustav Mahler, Arnold Schönberg, Anton Webern, Franz Lehár and Emmerich Kálmán, architects Otto Wagner, Joseph Maria Olbrich and Alban Berg, the psychiatrist Sigmund Freud, economist Ludwig von Mises, graphic artist Koloman Moser, musician and poet Hugo von Hofmannsthal, writer Robert Musil, sociologist and philosopher Otto Neurath, philosopher Ludwig Wittgenstein and Dora Philippine Kallmus, known as Madame D'Ora or Madame d'Ora, an Austrian fashion photographer and portrait artist. This long list still does not include the names of all the people who created the intellectual atmosphere of Vienna at the time. As Ulam laconically had it: "The young Smoluchowski took advantage of the extremely conducive conditions in his childhood and youth. His mother, very well-educated, making use of the fact that the Austrian capital was at that time a world centre of art and science, created a home filled with an artistic and intellectual atmosphere"[213].

The city's artistic and spiritual climate and the atmosphere of an intelligentsia family engendered in Smoluchowkski a keen sensitivity for art and gave the

---

211 Ibidem, p. 23.
212 Ibidem, p. 9.
213 S. Ulam, *Marian Smoluchowski and the Theory of Probabilities in Physics*, op. cit.,p. 476.

young Marian a natural inclination towards practising it. He painted and drew and a number of his watercolours and sketches have been preserved, the main subject of which was nature. He produced pencil sketches and watercolours, in the mood of German Romanticism, chiefly for himself, for the sake of creating art.

Smoluchowski was extremely "sensitive to all manifestations of beauty in art and in science" – Klemensiewicz wrote – "he felt the beauty of alpine nature like no one else, (...) at a time when cameras had not yet come into use to easily 'immortalise' everything unselectively, he carried a sketchpad and a box of paints in his sack"[214].

Another watercolour artist was Smoluchowski's older brother Tadeusz, who started painting in 1891 and practised the art for the next forty years. His pictures are technically better than Marian's watercolours. It was probably he who passed on to his brother a passion for painting nature. Marian's watercolours differ from his brother's primarily through their use of colour. Marian used lively, cheerful, optimistic colours – lush greens, intense browns and reds, warm pastels, Tadeusz's watercolours are more subdued in hue, closer to natural colours, it could be said – closer to nature. Marian's watercolours are cheerful and warm, but Tadeusz's painting makes a more mature impression upon the viewer. Goetel recalled:

> he was very interested in paining. Here too he did not wish to content himself merely with the role of viewer. With joyful curiosity, entirely self-taught, he applied himself to painting landscapes and in a short time achieved excellent results. The watercolours of the region around the Carpathian mountains and Kraków, which he painted shortly before his death, shine with his passion for the sun, greenery and nature. He sometimes complained that his professional life did not permit him sufficient devotion to art. But he always knew how to find time to visit art exhibitions and on all his overseas trips he visited galleries and artistic monuments[215].

Marian Smoluchowski was a lover and connoisseur of music and played the piano, often playing in company as part of a duet preforming the works of Schubert, Schumann, and Mendelsohn. According to Armin Teske, the future scientist's favourite composers were Brahms and Franck but he had a particular interest in the music of Wagner and Bruckner. Of the Polish composers, he most enjoyed the music of Karłowicz, a representative of the late Romanticism movement who

---

214 Z. Klemensiewicz, *Marian Smoluchowski*, op. cit., p. 4.
215 W. Goetel, *Ze wspomnień osobistych o Maryanie Smoluchowskim* (From personal recollections of Marian Smoluchowski) op. cit., p. 225.

was also a passionate mountaineer and whose life ended at the age of 33 under an avalanche at the foot of the Mały Kościelec ridge. Smoluchowski's weakness for Romantic influences, both in life as well as in art, was not so obvious, as his loved ones describe – for example he did not go in for the music of Chopin, although it was the quintessence of Romanticism.

The family home, hobbies, practising sport – all this made Marian an exceptional young man in whom a love of science grew systematically. A passion arose in him of a scientist striving to learn and discover the truths of nature, to find pleasure in conducting research and participating in scientific life.

On June 1, 1901, Smoluchowski married Zofia Baraniecka (1881–1959), the daughter of Marian Baraniecki, a professor of mathematics at the Jagiellonian University. It was a marriage of love.

Zofia Baraniecka

Zofia Baraniecka

Wedding photo of Zofia and Marian Smoluchowski

Wedding photo of Zofia and Marian Smoluchowski

Smoluchowski's strong emotional involvement matched his Romantic nature. Love altered his definition of happiness, previously based on negation: "I firmly retract what I once wrote…, that happiness is just a lack of unhappiness"[216].

Two children were born of this loving marriage – a daughter, Aldona (1902–1984) and a son, Roman (1910– 1996), also a physicist and a university professor at Princeton and Austin.

Marian Smoluchowski with his sister-in- law, Jadwiga of the house of Braniecka

---

216 A. Teske, *Marian Smoluchowski: życie i twórczość* (Marian Smoluchowski: life and works), op. cit., p. 125. Quote from a letter to his parents in 1900.

Marian Smoluchowski with family in the Planty park in Kraków (around 1903)

The life of our great Pole was very warmly summed up by Kazimierz Grotowski, a retired Jagiellonian University professor, who for many years was the director of the university's Marian Smoluchowski Institute of Physics. It is worth noting that in his youth he was also a mountaineer and his brother was the late theatre director Jerzy Grotowski, one of the 20th century's greatest theatre reformers. In an essay recalling Marian Smoluchowski's achievements in mountaineering, Grotowski wrote that although Smoluchowski did great things in physics, in his short life of only 45 years he also had time to be a happy husband and father, play the piano, be interested in music and art, paint watercolours, ski and become a renowned mountaineer. He emphasised that for such a resumé, it is not enough to be lucky but that an appropriate scale of values is necessary, which is usually taken from the family home. The talents of a genius are also required.

Tadeusz Godlewski claimed something similar:

> As deep and comprehensive as he was in his research, so he was also a person of deep and comprehensive culture. Gifted also with rare linguistic talents, he classically mastered knowledge of four languages, in which he published his work. He had a great talent for painting, an outstanding ability for and a huge love of music, was a fanatical lover of nature, which he studied deeply in science and the beauty of which he learned to love and admire up close during his beloved mountain trips. In social life, he was extremely engaging and pleasant, drawing people to himself through a peculiar ease of manner and the depth and subtlety of his discernment. Talking to him, one felt the highest and most subtle Western culture against a background of the strange simplicity and naturalness of a truly great man. So much of the researcher's mental energies, so much of the creator's initiative, so many of man's ethical and cultural values were irretrievably lost with him![217]

---

217 T. Godlewski, *Marian Smoluchowski*, op. cit., pp. 28–29; see also: *Marian Smoluchowski (1872–1917). Fizyk, taternik, romantyk nauki*, (Marian Smoluchowski (1872–1917). Physicist, mountaineer, romantic of science), op. cit., pp. 18–19.

Marian Smoluchowski with his wife Zofia and her sister, Jadwiga

Those are beautiful words, describing extraordinary personality traits. But despite his many talents and artistic inclinations, the strongest passion of Smoluchowski's youth was mountaineering. A love of the mountains stayed with him for the whole of his life. Memoirs written by companions on mountain escapades pay great testament to the Pole's extraordinary personality.

Marian Smoluchowski with his wife at Błonia Park in Kraków

Chapter IX Mountaineer

Two passions essentially dominated Marian Smoluchowski's life. The first, which he spoke about many times, was science, to which he devoted his entire adult life. The second passion was formed in his teenage years – it became mountain climbing. He spent a significant amount of his youth in the mountains and returned there until his final days – they left an indelible mark on his personality. No study of Marian Smoluchowski's life would be complete without mentioning his fascination with mountains. Some episodes, successes and memories of that extraordinary passion would need to be recalled. Mountaineering as a sport was born in the 19th century. Initially, the English were the leaders at it, and the first mountaineering organisation, the British Alpine Club, was founded in London in 1857. In the middle of the century, Germans and Austrians made a series of daring ascents, unfortunately at the cost of numerous fatal accidents. As a reaction to this phenomenon, so-called legitimism emerged: a climber should be thoroughly prepared and does not have the right to attempt routes that are beyond his strength. The Smoluchowski brothers obeyed these rules, systematically improved their skills, and climbed their first routes in the company of experienced guides. From 1890, the Smoluchowski brothers started to conquer summits alone and made their first ascents. Marian was 18 at the time and had just passed his matriculation exams with distinction. For the next few years, they would devote every free moment to mountain climbing[218].

The safety rules of the time differed from those of today. It was recognised that in difficult terrain, climbers had to be roped together, but stone blocks were used to belay partners, and hooks had to be used on descents. Marian took the tradition of mountain tourism from the family home. His father, Wilhelm, took part in trips to summits even in his sixties.

> It is the year 1885. Wilhelm Smoluchowski is visiting Zakopane with his family. It is not easy to reach as there is no rail line for much of the way from Kraków, so they rent horse-drawn carts. But life was slower then and this means of transport added colour to mountain excursions. With his older brother Tadeusz, thirteen-year-old Marian walks through Zawrat and Polski Grzebień[219].

---

218 K. Grotowski, *Marian Smoluchowski – taternik i narciarz*, (Marian Smoluchowski – mountaineer and skier), in: *Marian Smoluchowski – od teorii atomistycznej do fizyki współczesnej*, (Marian Smoluchowski – from atomic theory to modern physics) op. cit., p. 51.

219 Idem, *Marian Smoluchowski – taternik i narciarz* (Marian Smoluchowski – mountaineer and skier), "PAUza Akademicka. Tygodnik PAU" (Academic PAUza, the

Until the age of 23, Smoluchowski lived in Vienna, a short distance from the Alps. To the west of Vienna lies the mountainous region of Kalkalpen – a national park of the Northern Austrian Limestone Alps, to the south is Carinthia, and even farther south the Dolomites. Smoluchowski grew up in a place that in itself encouraged mountain trips and created a great opportunity to commence alpine excursions.

As already mentioned, at the time of Smoluchowski's youth, mountaineering was just starting. The equipment used on climbs was heavy as it was often different equipment, merely adapted to the needs of mountaineering. Zygmunt Klemensiewicz, a pioneer of Polish mountaineering, writing about putting together the equipment, claimed that carrying a separate hammer to drive in hooks was redundant and that the first decent stone would suffice, dropped by a companion on a rope at the time of need. To today's mountaineers, this is quite bizarre advice, but Klemensiewicz wanted to eliminate the burden of "unnecessary" equipment, which due to its excessive weight often made climbing difficult. The safety rules of the time were also very different from today's.

Emil Zsigmondy, one of the most outstanding alpine climbers of the time, had a significant influence on the activities of young people. He said that only people physically prepared and with the necessary experience had the right to embark on independent mountaineering expeditions. This position, absolutely obvious from the current perspective, sounded revolutionary in the late 19th century when not overly-fit members of the urban intelligentsia were undertaking the challenge and the duty of handling the technical difficulties of climbing routes and ensuring an expedition's safety fell to alpine guides[220].

Unfortunately, many ascents were conducted in too risky a way and early climbing successes were paid for with numerous accidents, many of which proved fatal. The mountaineering community decided to adopt rules intended to prevent accidents and improve safety. It was recognised that every climber had to be thoroughly prepared, should not attempt a route beyond his abilities, and climbers should be roped together for safety. The adoption of these and several other basic rules improved the safety of alpine expeditions.

Weekly of the Polish Academy of Arts and Sciences) 2017, no 380–381, http://pauza.krakow.pl/380_381_8&9_2017.pdf (access 1.03.2022).

220 A. Palczewski, *Marian Smoluchowski – alpinista*, (Marian Smoluchowski – mountaineer) "Delta", December 1997, http://www.deltami.edu.pl/temat/roznosci/historia i filozofia/2017/08/17/Marian Smoluchowski-alpinista/ (access: 5.03.2022)

The Smoluchowskis attached great importance to the safety rules, and Zsigmondy's ideas were close to them. They ascended the first climbing routes with much more experienced guides, later systematically improving their skills throughout their whole time climbing. The subject was so important that Goetel raised it in his memoirs:

> Smoluchowski strove for a demanding alpine expedition to be as easy and safe as everyday life, so that the chances of an accident were no greater in the mountains than walking the city streets or undertaking everyday activities – and this aim, one of the hardest, he achieved. When the difficulties mounted excessively, when it became clear that his life was threatened disproportionately to the aim, he knew how to draw back in time[221].

Goetel emphasises that in the hardship of a hike or the effort of conquering a summit, Smoluchowski shaped his mountaineering temperament. He exhibited a great gift for orientation, a burning enthusiasm, and a talent for solving the toughest problems of an expedition. He took decisions quickly. His fearless attitude coupled with a calm prudence, boundless helpfulness and sacrifice towards companions on an alpine excursion, commanded understandable respect[222]. Other friends recalled him similarly – Jerzy Maślanka said there was no forgetting the captivating charm of Smoluchowski's personality, his incomparable calm, courage and gentle humour, phenomenal orientation in a terrain and the "almost superhuman superiority" that radiated from him[223].

Both brothers started mountain climbing in the Alps during their time at Collegium Theresianum and became seasoned mountaineers in their student years. Due to classes, they organised excursions during their studies mainly on Sundays and public holidays. They most often went to the mountains by train in the morning, regardless of the weather or time of year. Tadeusz organised the excursions from the logistic point of view, planning trips in detail.

---

221 W. Goetel, *Ze wspomnień osobistych o Maryanie Smoluchowskim* (From personal recollections of Marian Smoluchowski) op. cit., p. 221.
222 Idem, *Marian Smoluchowski – człowiek gór* (Marian Smoluchowski – man of the mountains), "Wierchy" (Peaks), 1953, vol. XXII, p.19.
223 J. Maślanka, *Początki narciarstwa i taternictwa polskiego*, (The beginnings of Polish mountaineering and skiing) Archiwum Nauki PAN i PAU, (PAN and PAU Scientific Archive) K III 36.

Marian Smoluchowski at lake Czarny Staw Pod Rysami

In 1891, they started mountaineering independently and were already superbly physically prepared for it. They also possessed plenty of experience, gained during numerous trips in the Alps.

Klemensiewicz mentions that Smoluchowski was a person:

> for whom physical strength, skill in various types of physical exercise, dexterity and courage, and on the other hand mental awareness, calm and circumspection, heralded a great mountaineering future. During this period he explores the whole of the Eastern

Alps while making a number of first ascents in the Dolomites. Soon, rock climbing trips are not enough for him. The classic developmental path leads Him in 1894 in search of wonderful adventures and fuller experiences on the peaks and glaciers of the Swiss Alps[224].

Marian Smoluchowski took up serious climbing when he started his university studies. Being in generally great shape due to intensively practising sport (swimming, horse riding, cycling), he started rock climbing. Around Vienna there are many limestone rock massifs, which provide perfect terrain for such sport. Tadeusz and Marian particularly frequently visited the Max and Schneeberg massifs, which abound with climbing routes of various types offering a range of difficulty. This rock-climbing training was preparation for more serious Alpine expeditions which the Smoluchowski brothers took part in as members of the Academic Section of the Alpine Association of Vienna (*Akademischer Sektion Wien des Alpen-Vereines*). In fact, Marian Smoluchowski's first serious Alpine expedition took place in 1889 (together with his brother Tadeusz and guide M. Nicolussi he went to the Brenner Pass in the southeastern Alps) and a year later, Marian took part in an expedition to Rieserferner in the Eastern Alps, however the real pinnacle of his mountaineering occurred in the years 1891–1894. This was also a period of intensive development of the Academic Section of the Alpine Association of Vienna.

In 1891–1894, the Smoluchowskis organized many alpine expeditions, mostly in the Rieserferner and Ortler groups as well as in the Dolomites. During this period, they discovered 24 new climbing routes and made 16 first ascents to summits. Among the most outstanding achievements of this period are summiting: Sas dal Léc (2,959 MASL (metres above sea level)) in the Dolomites (in 1892), Schluderzahn (3,255 MASL) in the Ortler group (in 1892), and Zehner (2,917 MASL) in the Sella group (in 1894). This last ascent, due to the length and difficulty of the route, has earned a place of merit in the official history of mountaineering.

---

224   Z. Klemensiewicz, *Marian Smoluchowski*, op. cit., pp. 3–4.

Marian Smoluchowski on skis in the Tatra mountains (probably Kalatówki, December 27, 1913)

On skis at Babia Góra

In the early 1990s, Marian Smoluchowski and his brother Tadeusz were recognised as leading European mountaineers and pioneers of climbing without guides. This was a period when expeditions were organised exclusively in the company of experienced highlanders. Meanwhile, the brothers conducted 16 new ascents to summits unassisted and discovered 24 new routes[225].

The Smoluchowski brothers' mountaineering activity was replete with spectacular climbing successes. When in 1890 they started independent expeditions, their first achievement was marking out a new route on the Ohrenspitzen – three peaks on the border of the Tyrol and South Tyrol, known as the "ear tips" due to their characteristic shape. The highest of the peaks (known as the "big ears") is 3,101 MASL. In 1891, the Smoluchowskis were the first to summit Rotstein (3,150 MASL), and in 1892 the southern (3,050 MASL) and northern (3,079 MASL) Gabelspitze. In 1892 in the Dolomites, the brothers achieved the first ascent of the Schluderzahn (3,255 MASL) and were the first to traverse it. In the same year, they recorded the first ascents of Sas dal Léc (2,950 MASL) and Sass Pordoi (2,950 MASL) – a plateau-like peak in the Sella group in the Dolomites.

In the most complete work to date on Marian Smoluchowski's climbing and skiing pursuits, Ewa Roszkowska writes:

> The year 1892 was the most fruitful in Marian Smoluchowski's mountaineering career. Together with Tadeusz he summited 13 virgin peaks and established three new routes. Among the most important of his achievements at this time must be considered: marking out a new route on Westlicher Fermedaturm (2,810 MASL) and the first summiting and traversing from Laaser Spitze (3,303 MASL) to Schluderzahn (3,255 MASL) in the Ortler group, achieved on July 27, 1892, with Hans Lorenz, Robert Lenk, the third team ascent of Wasserkofel (Sas dal'Ega, 2,942 MASL), conquering the virgin Östlicher Wasserkofel (at the crossing to Cresta di Campil, 2,880 MASL), an ascent with Robert Lenk and Hans Lorenz of the Zweite Cirspitze (2,545 MASL), and Höchste Cirspit (2,597 MASL) as well as the first ascent of the virgin Dritte Cirspitze. On August 7, these same men together with W. Merz conquered the virgin main peak of the Rotspitze and on August 10, Marian became the first along with Hans Lorenz and Victor Wessely to climb the smaller western peak of Rotspitze (2,379 MASL)[226].

Smoluchowski had a weakness for the Dolomites. This is understandable to anyone who has had the opportunity to get acquainted with these exceptionally picturesque mountains. Just before his death, on the 25th anniversary of his joining

---

225 A. Wróblewski, *Marian Smoluchowski: Polak, który stworzył nową gałąź fizyki*, (Marian Smoluchowski: The Pole who created a new branch of physics) op. cit., p.
226 E. Roszkowska, "Folia Turistica" 2012, no 26, http://www.folia-turistica.pl/attachments/article/402/FT_26_2012.pdf (access: 18.02.2022), pp. 220–221.

the Academic Section of the Alpine Association of Vienna, he received a silver edelweiss and the senior title – *Alter Herr*. To mark the occasion, he sent a letter to club members in which he recalled, among other things, unforgettable experiences of his youth while discovering the Dolomites:

> The terrain that appears before my eyes is above all the world of the Dolomites – and it seems to me that it used to be more beautiful than for today's tourist. After all, it was far more primitive and provided an opportunity for real expeditions of discovery, and if we tried new routes and new ascents, we felt not only the objective magnificence of the mountains but the subjective charm of discovery[227].

This "subjective charm of discovery" was characteristic of Marian Smoluchowski's two passions; both science and the mountains inspired him to discover secrets. On one expedition he experienced something falling within the scope of both his scientific and mountaineering interests. During a trip in April 1892, the participants witnessed quite a rare phenomenon. Smoluchowski recalled it thus:

> As we descended along the summit ridge of Little Ödstein in a southerly direction, we suddenly noticed the beautiful sight of the Brocken. The mists which had earlier surrounded us parted and the sun took on a strong blaze. Each of us saw only our head surrounded by a rainbow; in the middle was the shadow of the head, the arms could still be made out clearly. Towards the edge the shadow became less distinct, though the legs would have stretched more or less to the ring. But to partially see the shadow of a companion it was necessary to get as close to him as possible, e.g. touching heads. (...) We observed the phenomenon for about five minutes. It was five in the afternoon[228].

It was the so-called glory effect, or the dispersion of light by the mist droplets behind at an angle of close to 180° enhanced by interference – hence each person saw only his own head. In this description, the almost 20-year-old Marian exhibits the astuteness of a born researcher who draws attention to the important details of the peculiarity encountered. An identical phenomenon, but in the backscattering of de Broglie waves of material in atomic nuclei, was discovered in Kraków 70 years later. However, let's remember that in 1892 there was still no quantum mechanics, nothing was known about de Broglie waves, the existence of atomic nuclei was not suspected, and atoms still awaited their confirmation, by Smoluchowski himself – but in 1892 he was till however only a second-year student of physics.

The Smoluchowski brothers' appetite for conquering alpine summits was astonishing:

---

227   K.A. Grotowski, *Marian Smoluchowski – taternik i narciarz*, dz. cyt., s. 57.
228   Ibidem, p. 54.

August 13, 1892, – Marian and Tadeusz together with H. Lorenz and V. Wessely were active in the Mesules region. Here they were the first to climb the Gamsburg [correctly Gamsberg – ed] (2,990 MASL), the eastern peak of the Mesules (2,996 MASL), Furchia di Chamuzzi (2,919 MASL), Piz Rotice [correctly Rotic – ed] (2,966 MASL), Piz Beguz (2,972 MASL), Piz Mrara [correctly Miara – ed] (2,965 MASL), Piz Saliera (2,958 MASL), Piz Gralba (2,974 MASL), Piz Remiz (2,940 MASL), Piz Selva (2,941 MASL), from where they descended by a southeastern couloir to Valdelle Stries [correctly Val Delle Strie – ed] (…) on August 18, the two brothers, together with M. Binn, H. Bertram, H. Lorenz and O. Nafe, renewed their activity in the Sella group. This time they made the first ascent of Piz Ciavazes (2,836 MASL). The ascent to the summit was made by the southeastern face. It is worth emphasizing that despite its spread, the peak did not have a name as the inscription Piz Sella at the 2,814-m mark on the map referred to the peak situated about 200m to the northwest of the highest point of the massif summited by Smoluchowski and his companions. Hence the Smoluchowskis decided to give the highest point its own name, though unrelated to the term 'Sella' as it was not known or used by the local population. They also did not want to use the term 'Monte Pordoi', probably coming from the neighbouring Pordoi Spitze, situated to the north and separated from 'their peak' by the Stries valley, as suggested by an inscription on an old photo of the region. Therefore, after consulting with Prof. Gian [Giovanni – ed] Battista Alton, they gave the conquered peak the name Piz Ciavatzes (currently Piz Ciavazes), derived from town situated below[229].

Even on the way back, they did not pass up the chance to conquer another peak: "Four days later, probably while returning to Vienna, Tadeusz and Marian Smoluchowski stopped in the High Tauern and on August 29, 1892, were the first to summit the still-virgin Südliche and Nördliche Gabelspitze (3,050 MASL – difficult ascent, 3,079 MASL – easy ascent)"[230].

In 1893, the Smoluchowski brothers, in the company of Hans Lorenz and Victor Wessely, were the first to traverse the northern face and eastern descent of the Punta delle Cinque Dita ("five fingertips"), known to Austrians as the Fünffingerspitze.

Smoluchowski recalls the ascent of Sas dal Léc:

> Sas dal Léc, first ascent (…) we knew that the peak we had already used as an approach had been attacked unsuccessfully from the side of that gap. A high vertical wall (…) At 09:30 we achieved the main snow and scree- covered plateau, at 10:00 the northern face of Sas dal Léc. (…) A steep ice trough running through the face, falling from one side which is also unsuitable from this side due to a broad area of overhanging faults. After lengthy reconnaissance we noticed (…) a weakly formed shelf. To use it, climb

---

229 E. Roszkowska, *Alpejska działalność Mariana Smoluchowskiego* (The mountaineering activity of Marian Smoluchowski), op. cit. pp. 221–222.
230 Ibidem, pp. 222–223.

the steps until it was possible to make an easy crossing at a narrow spot. (…) We left our boots, ice picks (with one exception), crampons etc. under an overhanging rock. In socks onto the shelf. The shelf is breaking away, it's not possible to extend easily. It is narrow and exposed, the overhang repels a lot, strongly. After climbing onto a stone block (very difficult), a terrain is achieved ascending in steps, where altitude can be gained quickly. (…) Despite the existence of three different trails to the summit (…) only after 25 different ascents (…) did we manage to combine two different routes in the ascent and descent as a so-called traverse of the Fünffingerspitze. (…) we went out by lantern-light at 3 a.m. equipped with only sticks (…) easy to carry an ice pick and a 60-metre rope. (…) We chose the northern route for the ascent, first discovered by Norman-Neruda, and headed for the Grohmann glacier.
After overcoming the fissure at the edge and carving the necessary number of steps in the ice trough that rose on the other side of the fissure, we climbed up steeper and steeper rock steps. (…) as far as an overhanging step (…) where we put our boots on and tied ourselves together with rope. After traversing a short terrace we stood before an 80-metre chimney which cut through the yellow wall like a black dash closing the way ahead. (…) The exposure is so great that the climber sees the blue glacier at a dizzying depth between his legs. After passing a whole system of chimneys, we stand by a window in the ridge and a few minutes later at the cairns that adorned the broad and slightly sloping peak[231].

Seeking a traverse through 25 separate ascents – is it not reminiscent of the process of looking for solutions to a problem in physics?

---

231 K. Grotowski, *Marian Smoluchowski – taternik i narciarz*, (Marian Smoluchowski – mountaineer and skier), in: *Marian Smoluchowski – od teorii atomistycznej do fizyki współczesnej*, (Marian Smoluchowski – from atomic theory to modern physics) op. cit., pp. 52–53.

Fünffingerspitze in the Dolomites (2,918 MASL)

Smoluchowski's level of mountaineering rose incessantly.

In 1894, on July 27, Marian Smoluchowski, in the company of W. Merz, Oscar Schuster and the guide Kaspar Moser, accomplished the first crossing of Moserscharte and the first ascent of Zehner (2,917 MASL). This achievement was recognised as one of M. Smoluchowski's most important. The route marked out for the northeastern face demanded high rock climbing qualifications and mental endurance as it ran through exposed terrain with no opportunity for belaying and, due to the icy gully making the face difficult, also skills in using ice equipment. An even more serious undertaking was an ascent on August 1, 1894, by M. Smoluchowski with two companions: Victor Wessely and Hans Lorenz, from the south of the Grohmannscharte, considered at the time impossible to summit. The team's intention was to conquer Grohmannspitze (3,126 MASL) from that pass. The couloir that looked impressive from the south turned out to be an ice gully covered in hard ice. It branched off in the upper part, and the right-hand branch chosen by the team for their crossing was not only ever more vertical but also covered in hard ice stalagmites with the added risk of rock avalanches. In such conditions, the climbers advanced upwards very slowly and after reaching the pass, for safety reasons exacerbated by the lateness of the hour, withdrew, giving up on the ascent of the Grohmannspitze[232].

In 1894, Marian Smoluchowski set out for the Western Alps, in the region of the four-thousanders. During this one expedition he climbed many of the most famous peaks in the region. Together with Hans Lorenz, he summited the Zinalrothorn (4,223 MASL) and Monte Rosa (4,637 MASL), was also the first Pole to climb the Mattehorn (4,482 MASL) – a mountain that had for many years been considered inaccessible and in the 19th century represented a challenge both for mountaineers and professional guides. The dramatic story is well known of its first summiting by Edward Whymper in 1865, which cost the lives of four of his companions. Standing on the Theodul glacier or the Breithorn Plateau, we see to the east a snow-capped peak – the Breithorn (4,165 MASL) – and a little farther on the highest in this area, Monte Rosa (4,637 MASL). But the greatest respect was aroused by the steep rocky pyramid of the Matterhorn, situated in the west with an altitude of 4,478 MASL. Today we can get within the neighbourhood of the Matterhorn by cable cars from Zermatt on the Swiss side or Breuil in Italy. We can also stay at several campsites. In Marian Smoluchowski's day, climbers faced a march of up to dozens of hours, a bivouac at the base and usually another on the way to the summit[233]. While the Pole enjoyed mainly long climbs in the Eastern Alps, he was drawn to the Western Alps by great expeditions to ice-covered massifs.

---

232 E. Roszkowska, *Alpejska działalność Mariana Smoluchowskiego* (The mountaineering activity of Marian Smoluchowski), op. cit., pp. 224–225.

233 K. Grotowski, *Marian Smoluchowski – taternik i narciarz*, (Marian Smoluchowski – mountaineer and skier), in: *Marian Smoluchowski – od teorii atomistycznej do fizyki współczesnej*, (Marian Smoluchowski – from atomic theory to modern physics) op. cit., pp. 55–56.

In 1894, Smoluchowski became the first Pole to climb the Matterhorn

This ended Marian Smoluchowski's intensive period of mountaineering. In 1895, he gains a doctoral degree from the University of Vienna and goes on to further studies in Paris and England. He returns later to the Alps in the coming years (1897 and 1898), however these are mostly recreational trips, not connected with serious mountaineering expeditions. Of course this does not mean that Smoluchowski withdrew entirely from serious summit ascents.

It is known that the first skiers were mountain tourists.

Tadeusz Smoluchowski, who in 1894 moved to Peczeniżyn in the Podkarpackie province, and later to Wolanka near Borysław, like Marian not abandoning his outdoor pursuits, became a skiing pioneer.

> Only in 1904 and 1909, did Marian and Tadeusz go back to the mountains of their youth. For the second time, with Zygmunt Klemensiewicz and Tadeusz Kossowicz, they traverse the Finsteraarhorn (4274 m), Jungfrau (4159 m), and Lauterbrunner Breithorn (3782 m) peaks. These climbing aims testify to their perfect physical condition. But Lviv, and later, from 1903, Kraków, are far from the Alps. The Tatras and Carpathians are much closer. Marian takes up highland skiing. From Chornohora (…) to the Siedmiogród Carpathians[234]

---

234   Ibidem, pp. 57–58.

In 1909, Marian Smoluchowski went climbing in the Bernese Highlands, where together with Zygmunt Klemensiewicz and Tadeusz Kossowicz he summited the Jungfrau (4,185 MASL), Finsteraarhorn (4,274 MASL), Hugisattel (4,089 MASL), and Lauterbrunnen Breithorn (3,780 MASL). It was his last climbing season in the Alps[235]. Teske writes:

> In May, 1899, Smoluchowski made his first trip to the Eastern Carpathians, and from then on will hike through their peaks for many years, especially in winter. These journeys lead to many first winter ascents, among them Syvulya, Piekun, Poleński, Mihailcul, and Fărcăul. These were pioneering trips not only due to the first winter ascents but also for the use of skis, at the time still very rare in Poland. (…) Before the current style developed, skiing went through a phase which today seems rather humorous. People skied with one stick regulating their speed. The skier sat on the stick, like a witch on a broom. (…) It was also originally thought that skis could be used only on plains or undulating terrain; it was a surprise when the new equipment was found to be useful in mountaineering expeditions. (…) The skis of the time were very different from those of today, a bow-shaped cane being fixed to a nut in the middle of the ski, which gave support to the heel[236].

Writing these words in 1956, Teske could also not have imagined the changes that have occurred in skiing in recent times. The modern carving technique, involving riding on the edges of the skis, is diametrically different to the classical technique used in his day, in the 1950s. Teske himself surely skied differently to Smoluchowski, probably using telemarking, a fairly difficult technique with a twist of a 'free heel' (the telemark binding is similar to that on cross-country skis), which became popularized in the Austrian and Italian Alps at the end of the 20th century.

---

[235] E. Roszkowska, *Alpejska działalność Mariana Smoluchowskiego* (Marian Smoluchowski's alpine activity), op. cit., pp. 224–225.

[236] A. Teske, *Marian Smoluchowski: życie i twórczość* (Marian Smoluchowski: life and works), op. cit., p. 114.

Marian Smoluchowski and family at Kałatówki

A sleigh from Zakopane to Morskie Oko (December 21, 1913)

Marian Smoluchowski's activeness as a skier is connected with the emergence in Lviv of the Carpathian Skiing Society. It started with an ascent of Paraszka in the Skole Beskids in 1906. Among Smoluchowski's numerous skiing excursions in the years 1906-1912, a few first winter ascents can be listed, among others to Sewula (in February 1909), to Fărcăul and Mihailecul (in April 1911), as well as to Piekun and Poleński in the Maramurian Carpathians (currently situated in Romania and Ukraine).

Marian Smoluchowski, sister-in-law Maria Smoluchowska née Guttenberg, and nephew Bob (Wilhelm) Smoluchowski, photo taken during a family excursion from Hala Gasienicowa via Zawrat to Morskie Oko and Mnich (from the Smoluchowski family collection)

In the Carpathian Skiing Society, Smoluchowski met up with a group of young Polish climbers from the Himalaya Club, which included Zygmunt Klemensiewicz, Roman Kordys and Jerzy Maślanka, among others. This enticed him to return to the Polish mountains, which he had visited as a thirteen-year-old boy. In the years 1910–1916, Smoluchowski often went to the Tatras, both in the summer and the winter – taking over the physics chair at the Jagiellonian University in 1913 was conducive to this.

Marian and Tadeusz Smoluchowski on Królowa Rówień in the Tatras (the route from Kuźnice to Hala Gąsienicowa)

Marian Smoluchowski's mountaineering career is quite extensive, although it does not include very many first ascents. The chronicles record only crossing the trail on the northwest ridge of Smoczy Szczyt in the choice company of eminent mountaineers (Janusz Chmielowski, Władysław Kulczyński jun., Mieczysław Świerz, Wincenty Zakrzewski, Janusz Żuławski) in 1911 and two variants of the crossing of the western face of Mała Końjczysta in 1914.

By the Frozen Pond in the Tatra Mountains (seated: Marian Smoluchowski, behind him his nephew Bob Smoluchowski, standing Tadeusz Czezowski) (from the collection of the Smoluchowski family)

The fact that he was entrusted in 1911 with the role of president of the Tourist Section of the Tatra Society (established in 1902, the Section was a club grouping people practising sports mountaineering) testifies to the important role Marian Smoluchowski played in the climbing community more than any list of his first ascents. It is also testified to by the admiration expressed by Mieczysław Swierz, undoubtedly the best Polish mountaineer of the first quarter of the 20th century, who called him "a mountaineer above mountaineers". However, according to Klemensiewicz, Smoluchowski was little known to the wider public and did not exert such a great influence during the era of Polish mountaineering's emergence as he could have with his great experience and knowledge of the western climbing technique. On the contrary, Polish mountaineering, as soon as it reached the heights of the west, sought out Smoluchowski in order to give him his rightful place in the group. Because it is the privilege of that type of person in cultured society that they do not seek, but are sought[237].

---

237   Z. Klemensiewicz, *Marian Smoluchowski*, op. cit., p. 3.

Looking back at Smoluchowski's achievements as a climber, his significant contribution to the exploration of the Alps in the years 1891–1894 should be emphasised. Although only a few of his conquered peaks and first ascents made it permanently into the history of mountaineering, they still represent an important part of the Polish contribution to exploring these mountains. Among Polish climbers, Tadeusz and Marian Smoluchowski have the most first ascents in the Alps to their name. It should also be stressed that the Smoluchowskis practised sports mountaineering before anyone in Poland had thought of it. As Klemensiewicz recalls, Marian liked

> discovering difficult passages – new ones, great expeditions. That would maybe be enough for many appraisers to classify him as a gymnastic-competitive type. (…) As a companion, Smoluchowski was ideal. Confident, calm, understanding, he did not impose his opinion on anyone, seeming to say: I do it this way, and you do it the way you want, as long as you do it well. Despite an innate reluctance to act publicly, he did not shirk his obligations and devoted his invaluable time when it was expected of him[238].

A table compiled by Ewa Roszkowska presents Marian Smoluchowski's mountaineering career.

Year	Peak	Mountain group	Notes	Partners
1884	Hochobir (2,139 MASL)	Karavanks		T. Smoluchowski and probably parents
1887	Schwarzenstein (3,368 MASL), properly: Sasso Nero	Zillertal Alps		T. Smoluchowski and companions
1888	Nuvolau (2,575 MASL)	Dolomites – Nuvolau Group		T. Smoluchowski and companions
	Kesselkogel (3,002 MASL)	Dolomites – CatinaccioGroup (Rosengarten)		T. Smoluchowski and companions
1889	Cima Tosa (3,173 MASL)	Dolomites – Brenta Group		T. Smoluchowski, M. Nicolussi, G. Bernard, M. Bettega

---

238   Ibidem, p. 4.

Year	Peak	Mountain group	Notes	Partners
1889	Croz dell'Altissimo (2,338 MASL)	Dolomites – Brenta Group		T. Smoluchowski, M. Nicolussi, G. Bernard, M. Bettega
1889	Cima dei Lasteri (2,459 MASL)	Dolomites – Brenta Group		T. Smoluchowski, M. Nicolussi, G. Bernard, M. Bettega
1889	Brenta Alta (2,960 MASL)	Dolomites – Brenta Group		T. Smoluchowski, M. Nicolussi, G. Bernard, M. Bettega
1889	Wildspitze (3,774 MASL)	Ötztal Alps		T. Smoluchowski and companions
1889	Fluchtkogel (3,514 MASL)	Ötztal Alps		T. Smoluchowski and companions
1889	Weisskugel (3,746 MASL)	Ötztal Alps		T. Smoluchowski and companions
1889	Klobenwand-Jagdsteig	Rax		T. Smoluchowski and companions
1890	Hoher Dachstein (2,995 MASL)	Dachstein		T. Smoluchowski and companions
1890 – 10.09	Große Ohrenspitze (3,101 MASL) – south face	Rieserferner Group	new route	T. Smoluchowski
1891	Rax–Zerbenriegelsteig route	Rax	new route	T. Smoluchowski
1891 – 22.07	Große Rotstein (3,150 MASL)	Rieserferner Group	first ascent	T. Smoluchowski
1891	Monte Cristallo (3,126 MASL)	Dolomites		T. Smoluchowski and companions
1891	Cinque Torri (2,361 MASL)	Dolomites	Various routes	T. Smoluchowski and companions

Year	Peak	Mountain group	Notes	Partners
1891	Croda da Lago (2,709 MASL)	Dolomites		T. Smoluchowski
1891	Furchetta (3,025 MASL)	Dolomites		T. Smoluchowski
1891	Sass Rigais (3,025 MASL)	Dolomites		T. Smoluchowski i towarzysze
1891	Marmolada (3,343 MASL)	Dolomites		T. Smoluchowski
1891	Cima di Ball (2,802 MASL)	Dolomites – Pale di San Martino	not all partners established	T. Smoluchowski
	Cima di Val di Roda (2,791 MASL)	Dolomites	not all partners established	T. Smoluchowski
	Cima Cuseglio	Dolomites	not all partners established	T. Smoluchowski
	Cima della Madonna (2,752 MASL)	Dolomites	not all partners established	T. Smoluchowski
	Sass Maor (2,812 MASL)	Dolomites	not all partners established	T. Smoluchowski
	Zufrittspitze (3,439 MASL)	Ortler Group	not all partners established	T. Smoluchowski
	Madritschspitze (3,265 MASL)	Ortler Group	not all partners established	T. Smoluchowski
	Geisterspitze (3,465 MASL)	Ortler Group	not all partners established	T. Smoluchowski
	Naglerspitze (3,248 MASL)	Ortler Group	not all partners established	T. Smoluchowski
	Ortler (3,905 MASL)	Ortler Group	not all partners established	T. Smoluchowski
1891 – 30.08	Bärenkopf (2,937 MASL)	Ötztal Alps	first crossing of whole ridge	T. Smoluchowski

Year	Peak	Mountain group	Notes	Partners
1891	Rauriser Sonnblick (3,103 MASL)	Ötztal Alps	first crossing of whole ridge	T. Smoluchowski
1892	Großglockner (3,798 MASL)	High Tauern		
1892 – 24.07	Westlicher Fermedaturm (2,810 MASL)	Dolomites Geisler Group	new route, third ascent to summit	H. Lorenz, W. Merz, O. Nafe, T. Smoluchowski
1892 – 25.07	Cirspitze V (Punta Cir V – 2,520 MASL)	Dolomites		H. Lorenz, W. Merz, O. Nafe, T. Smoluchowski
1892 – 27.07	Laaserspitze (3,303 MASL) – Schluderspitze (3,231 MASL) – ridge	Ortler Group	first ascent	R. Lenk, H. Lorenz, T. Smoluchowski
1892 – 27.07	Schluderzahn (3,255 MASL) from south-east	Ortler Group	first ascent and traverse	R. Lenk, H. Lorenz, T. Smoluchowski
1892 – July	Wasserkofel (2,915 MASL)	Dolomites	Third ascent	R. Lenk, H. Lorenz, T. Smoluchowski
1892 – July	Östlicherwasserkofelok (2,900 MASL)	Dolomites	first ascent	R. Lenk, H. Lorenz, T. Smoluchowski
1892 – lipiec	Cresta di Campil (2,880 MASL)	Dolomites		R. Lenk, H. Lorenz, T. Smoluchowski
1892 – 31.07	Zweite Cirspitze (approx.2,545 MASL)	Dolomites	second ascent	R. Lenk, H. Lorenz, T. Smoluchowski
1892 – 31.07	Grosse Cirspitze (2,592 MASL)	Dolomites		R. Lenk, H. Lorenz, T. Smoluchowski
1892 – 31.07	Cirspitze I (approx. 2,515	Dolomites	first ascent	R. Lenk, H. Lorenz, T. Smoluchowski

Year	Peak	Mountain group	Notes	Partners
1892 – 7.08	Rotspitze– Hauptgipfel (2,837 MASL)	Dolomites	first ascent	W. Merz, R. Lenk, H. Lorenz, T. Smoluchowski
1892 – 8.08	Sas da Léc (2,936 MASL) – from north side	Dolomites – Sella	first ascent	H. Lorenz, V. Wessely, W. Merz, T. Smoluchowski
1892 – 10.08	Rotspitze – western peak (2,379 MASL)	Dolomites	first ascent	H. Lorenz, V. Wessely
1892 – 11.08	Westliche Cirspitze (2,518 MASL) – east face	Dolomites	second ascent	H. Lorenz, V. Wessely, T. Smoluchowski
1892 – 11.08	Dritte Cirspitze (2,565 MASL)	Dolomites	first ascent to summit	H. Lorenz, V. Wessely, T. Smoluchowski
	Fermedaturm	Dolomites		
1892 – 13.08	Gamsberg (2,990 MASL)	Dolomites	first ascent	H. Lorenz, V. Wessely, T. Smoluchowski
1892 – 13.08	Mesules (2,996 MASL) – eastern peak	Dolomites	new route	H. Lorenz, V. Wessely, T. Smoluchowski
1892 – 13.08	Furchiadi Chamuzzi (Gamssattel 2,919 MASL)	Dolomites		H. Lorenz, V. Wessely, T. Smoluchowski

Year	Peak	Mountain group	Notes	Partners
1892 – 13.08	Piz Rotic (2,966 MASL) – Piz Selva (2,941 MASL)	Dolomites	During the crossing of the ridge, they climbed:PizBegúz (2,972 MASL), Mrara (2,965 MASL), Piz Saliera (2,958 MASL), Piz Gralba (2,974 MASL), Piz Remiz (2,940 MASL), Piz Begúz (2,972 MASL), Piz Selva (2,941 MASL)	H. Lorenz, V. Wessely, T. Smoluchowski
1892 – 13.08	Piz Selva (2,941 MASL)	Dolomites	New route of descent to Val Delle Strie	H. Lorenz, V. Wessely, T. Smoluchowski
1893	Gran Sas de Mesdì (2,980 MASL)	Dolomites		H. Lorenz, V. Wessely, T. Smoluchowski
1893	Vilnösserturm (2,830 MASL)	Dolomites		H. Lorenz, V. Wessely, T. Smoluchowski
1892 – 18.08	Piz Ciavazes (2,828 MASL) – from south-eastern side	Dolomites	first ascent	H. Bertram, M. Binn, H. Lorenz, O. Nafe, T. Smoluchowski
1892 – 25.08	Östliche Cirspitze (2,538 MASL)	Dolomites	first ascent	H. Lorenz, W. Merz, V. Wesely, T. Smoluchowski
1892	Wildgall (3,272 MASL)	Rieserferner Group		T. Smoluchowski

Year	Peak	Mountain group	Notes	Partners
1892 – 29.08	Nördliche Gabelspitze (3,079 MASL)	High Tauern	first ascent	T. Smoluchowski
1892 – 29.08	Südliche Gabelspitze (3,050 MASL)	High Tauern	first ascent	T. Smoluchowski
1893 – 14.08	Rosengartenspitze (2,998 MASL)	Dolomites		F. Benesch
1893 – 14.08	Lauriswand (2,811 MASL)	Dolomites	first ascent	F. Benesch
1893 – 16.08	Sett Sass (2,578 MASL)	Dolomites	first ascent	F. Benesch
1893	Pordoi Spitze (2,952 MASL)			T. Smoluchowski
1893	Langkofel (3181 MASL)			T. Smoluchowski
1893 – 5.09	Fünffingerspitze (2,997 MASL)	Dolomites	first traverse and ascent without guide	H. Lorenz, V. Wessely, T. Smoluchowski
	Rax, Schneeberg	numerous summer and winter ascents in the years 1889–1893		T. Smoluchowski
1894 – 27.07	Moserscharte (1,913 MASL)	Dolomites	first ascent	W. Merz, O. Schuster i przewodnik H. Moser
1894 – 27.07	Zehner (2,917 MASL)	Dolomites	first ascent	W. Merz, O. Schuster i przewodnik H. Moser

Year	Peak	Mountain group	Notes	Partners
1894 – 1.08	Grohmannscharte from the south – while trying to summit Grohmannspitze (3,126 MASL)	Dolomites	first ascent	T. Wessely, H. Lorenz
1894	Matterhorn (4,478 MASL)	Valais Alps	was the first Pole at the summit	names of partners not known
1894	Dent Blanche (4,364 MASL)	Valais Alps		names of partners not known
1894	Zinalrothorn (4,223 MASL)	Valais Alps kie		names of partners not known
1894	Monte Rosa Massif	Valais Alps	names of peaks not known	names of partners not known
1909	Jungfrau (4,158 MASL)	Bernese Oberland		Z. Klemensiewicz, T. Kossowicz
	Finsteraarhorn (4,274 MASL)	Bernese Oberland		Z. Klemensiewicz, T. Kossowicz
1909	Hugisattel (4,089 MASL)	Bernese Oberland		Z. Klemensiewicz, T. Kossowicz
1909	Lauterbrunen Breithorn (3,780 MASL)	Bernese Oberland		Z. Klemensiewicz, T. Kossowicz

Source: E. Roszkowska, *Alpejska działalność Mariana Smoluchowskiego* (The mountaineering activity of Marian Smoluchowski), op. cit., pp. 229–234.

Unfortunately, distance from mountaineering centres in Poland meant that Smoluchowski's practice did not influence climbers domestically. When he appeared in the Tatra mountains, many mountaineers of the younger generation followed similar rules to him. However, he could still impress them with his excellent physical fitness, needed for a difficult climb.

Having a mountaineer's blood in his veins, Smoluchowski also added variety to his stay at the University of Glasgow in Scotland with mountain excursions, taking advantage of the opportunity to get to know the Highlands region. He recalled this expedition in a speech given at a mountaineers' club in Vienna:

The weather was rather typical for the British Isles. (..) Heavy rain poured at night, however, fortunately the sun shone again in the morning. So I set off (…) in the company of a young Englishman, greatly pleased that he had found a partner for the trip. I had a very strong desire to get to know this terrain, which was new to me, finally peering into the depths of Scotland's highland world[239].

Much of the route passed through bogs:

> mushy, wet terrain, covered with a dense undergrowth of heather and ferns mixed with brown reeds. (…) One of the unpleasant properties of these quagmires soon made an itself felt: the abundant water content of the terrain. (…) I had the impression I was walking on a water-soaked sponge, and the higher up the worse it was. Actually this was largely due to the strangely fickle sky, which would send torrential rain with incredible speed, then a bright sun would shine on the whole landscape, then again a downpour soaked the ground… and so on. My English companion taught me a new tourist tactic: advancing in leaps and bounds. As soon as the rain started lashing down, we huddled in our rubber raincoats, sheltering under a rocky block beyond the fold of the land that protected us from the gale. Usually the downpour lasted no longer then 5–10 minutes, after which we could peacefully walk on. (…) Right at the summit itself we suffered powerful hailstones; finally, shivering and teeth chattering from the cold, we stood on the summit. In these conditions it was not possible to enjoy the view as we had wished. However, the image will always stay in my memory of the blue-black Loch Lomond stretching out below on one side and on the other Loch Katrine, idyllically located and surrounded by forests, and the whole mass of jagged ridges and peaks of various shapes revealing themselves in the far north[240].

Smoluchowski's extremely picturesque descriptions of the mountains are reflected in his watercolours:

> (…) I went to Fort William (…) and immediately set off for Ben Nevis. (…) At the summit there is an equipped shelter, grandly known as the "Ben Nevis" hotel and connected to a small meteorological observatory. I resisted the temptation to enter the "hotel" and contented myself with platonic enjoyment of the wonderful view from the snow-covered terrace of the observatory. I was one of those chosen by fate to have a broad, clear view from the summit of Ben Nevis as that mountain is famous for the mists and clouds that usually shroud its peak. The view is indeed magnificent! To the south, east and north – a mountainscape far and wide, an entanglement of wide-ranging chains, steeply sloping cones and hillocks of the typical, delicate green-brown colouring in the foreground and farther afield displaced by a delicate mist; between them scattered a whole lot of dark blue, elongated highland lakes. Just to the side, towards the north, a deeply-set green

---

239 M. Smoluchowski, *Wycieczki górskie w Szkocji* (Mountain trips in Scotland), "Taternik. Organ Sekcji Tury- stycznej Polskiego Towarzystwa Tatrzańskiego" (Taternik, A unit of the Tourism Section of the Polish Tatra Society) 1915–1921, Kraków 1921, p. 5.

240

valley cut through lengthwise by the Caledonian canal. In the westerly direction, an entirely different view: an endless sea biting deep into the land in bays like fjords, breaking off countless islets from it… Trusting the compass and a map of the vicinity which I happened to have on me, (…) I started to climb in an adjacent direction up the mountain opposite. I stood speechless with amazement, suddenly seeing the cauldron of the lake at my feet. I can only compare the imposing scenery to Morskie Oko and Czarny Staw in the Tatras. But this view was even more wonderful![241]

Any trip to the mountains had more value to Smoluchowski when he had the chance to do a little climbing.

Sgurr-na-Gillean [correctly: Sgùrr nan Gillean – ed.] seemed to be the most interesting peak. They even call it the "Scottish Matterhorn" and photos of its toothed ridge were shown to me, of which some of the peaks in the Dolomites could be envious. I immediately decided to climb it, but the peak was covered in clouds all day (…) I wasn't easily put off (…) I got to the actual rocky core of the summit through bogs and traversed it in a southerly direction to the point where the smooth slabs meet the rugged ground so that I reached the southern ridge without great difficulty. (…) Now there was a pleasant scramble along the ridge, which filled me with real delight as it was the first proper climb of the year[242].

Pod Rysami

---

241 Ibidem, pp. 7–8.
242 Ibidem, pp. 8–9.

Along with Stanisław Barabasz, the Smoluchowski brothers were among the pioneers of skiing in Poland. They travelled to sparsely-inhabited alpine regions, between the peaks of Chornohora and the Siedmiogród Carpathians. In 1913, Smoluchowski recalled such expeditions in *Taternik* – the informational magazine of the Polish Mountaineering Association.

We were drawn by the mystery of these hard-to-reach sites that had never before been visited by skiers. We were enticed by the peaks belonging to the highest uplands of the Galician-Hungarian Carpathians, which we had never had the chance to see up close, even in summer. We looked forward to the adventurous touch of romance of our expedition, however modest[243].

Marian Smoluchowski with Adolf Guttenberg on Królowa Rówień on the descent to Hala Gąsienicowa

---

243 M. Smoluchowski, *Mihailecul (1926 m) i Farcaul (1961 m)*, (Mihailecul (1,926m) and Farcaul (1,961m)) "Taternik. Organ Sekcji Turystycznej Polskiego Towarzystwa Tatrzańskiego" (Taternik. A unit of the Tourism Section of the Polish Tatra Society) (1913, no 6, p. 103.

Smoluchowski recalls how he conquered Fărcăul, situated at 1,961 metres above sea level, on a skiing expedition to the Hungarian border with his brother.

What we see here surpasses our every expectation! There is an alpine scenery that is hard to find elsewhere in the Carpathians. A steep precipitous slope interspersed with rocky black cliffs stretches along the whole of the lengthy Mihailecul ridge and above them loom huge overhangs casting a long bluish shadow on the steep snows of that slope. Staying an appropriate distance from the edges of the overhangs, but still close enough to look out at the view, we glide lightly along the gently undulating gradient of the ridge. We stand at the summit with the feeling that it was well worth coming so far to meet such an original personality as Mihailecul. From here, we now see under us those valleys, gently sloping inclines and ridges stretching to the southwest. (…) However we are currently mainly occupied with our real aim for today, the steep, conical Fărcăul, rising before us in the north, as we have to scale it in order to descend later to the valley of the Kirva stream, where we intend to stay the night. Seen from here, Fărcăul strikes an imposing figure, but having gone down to the deep-cut pass between it and Mihailecul, we are convinced that climbing it presents no great difficulties. Proceeding along the ridge, we arrive at the highest point at three o'clock[244].

The description of the skiing expedition leads us into virgin territory for Polish skiing:

> now we will go down this steeply sloping western ridge of Fărcăul. The first run, at the necessary distance from the edge, goes perfectly up to the point of a steep bend. From there we could head north, below the ridge, but the hard top-frost covering that slope and the rocky ribs protruding from it do not look appealing; so we will go southwards, though the snow here has been softened by the sun all day and it shows a certain tendency towards creating avalanches. So exercising the necessary caution, we leave that way and travel really well until the 1,541 point, bearing the name Piciorul Dancu on the map, where we rest a while; later, keeping to the same direction, finally turning gradually to the left, we glide down the slope into the depths of a valley with dizzying speed. It was a wonderful run! Despite the air temperature being a little above zero, the face and hands froze stiff as a result of a current of air we cut through like an arrow. The polonyna reaches very deep here with a long, narrow tongue. Having passed through the forest by a path, leading along the ridge, we come to the bottom of the valley, still covered in winter snow, just at the desired spot: La feresteu. (…) Suddenly the whole of the valley was revealed to us along with the towering rocky Petros as the wind from this side of the mist was blowing to the south. But we had no time to enjoy the view. A wonderful run at the side of the ridge to the edge of the forest then an unbearable descent down an extremely steep path, winding and leading through the thick woods; the snow here

---

244 Ibidem, p. 105.

had been shamefully trampled by people bringing hay from winter haystacks and was quite hard but not hard enough to not sink into it having taken the skis off. Then at the bottom of the Kwaśny Potok valley we had a perfect route, but with no snow; so skis on the back and onward at a fast march towards Bogdan. I won't say that that valley caused me particular delight, especially as it stretches for a length of 9 km[245].

During this period, Marian Smoluchowski's frequent excursions were to the only high mountains in Galicia, the Tatras. He rode around them on a bicycle, climbed the peaks, but also journeyed through the empty valleys. He was interested in the geology, he painted watercolours. He was often accompanied on his trips by family or students, whom he treated then as a friend rather than as a professor. Mieczysław Swierz wrote years later in *Taternik* about one such climbing expedition, presenting the participants thus:

> this man with the gaze of a satyr, with a pleasant smile, breathing in the vapours of soup made with Maggi stock cubes, is Jerzy Żuławski, an unparalleled companion for rock expeditions or (…) examining the plum kompot with the solemnity of one discovering the rotation of heavenly bodies – a mountaineer above mountaineers – is Prof. Marian Smoluchowski, and finally that refined wanderer lying on the soft lawn enveloping himself in Havana cigar smoke with gentlemanly indifference – that's Janusz Chmielowski, of whom it was once said that he took over the sceptre of Zakopane tourism after Chałubiński[246].

The climbing goal was Smoczy Szczyt in the High Tatras.

Brief moments of preparation and the first one darts boldly over the seemingly smooth slabs, entrusting their existence only to the endurance of their fingers and hands, which cling without trembling to those barely perceptible rain-eroded crevices, nooks and slats and lift their body ever higher and higher above the cruel depths of the valley. (…) We stood at the summit. Our joy is brief. (…) The gaze, for long hours focused on the ropes, now thrown from the top into the vastness of free space, sees that it has faded, the blue above us has dimmed and the gloom of evening has fallen on the hallowed rocks[247].

During the descent, nightfall took them by surprise.

We're attempting the descent. To the left, to the right – it doesn't "release" us. For nothing the presence in the group of two authors of 'A Guide to the Tatras', for nothing Żuławski's familiarity with the Silver Globe which, regardless of the misery of its bard, has hidden itself somewhere in the heavenly abyss. For a moment

---

245   Ibidem, pp. 106–107.
246   Quoted from K. Grotkowski, *Marian Smoluchowski – taternik i narciarz* (Marian Smoluchowski – mountaineer and skier), op. cit., p. 59.
247   Ibidem, p. 60.

more, the will-o'-the-wisp of Smoluchowski's lantern flickers in the darkness of night as with the sense of smell of a Sioux from the distant western prairie, he seeks for a route among the ledges and steps – but this attempt too is soon met with failure. The way has been lost in the cliffs and the darkness. No one wants to go on lying in the jaws of misfortune. It's no use; we'll have to wait on the rocks until dawn. Thus Smoczy has avenged itself for the lateness of the visit. (…) Years have passed. (…) Unmourned comrades Zuławski and Smoluchowski have gone to the grave, news of Chmielowski is lost. So I will never stand with them, rock brothers, above the wild Cliff of the ridge and will not hurl a victory cry from the summit…[248]

Marian Smoluchowski in the Pięć Stawów valley

---

248   Ibidem, pp. 60–61.

Prof. Dr. Walery Goetel
założyciel krak. A. Z. S.

Walery Goetel (1889–1972)

The quoted recollections offer a sense of the atmosphere of those years. Of times when the mountains represented unexplored, virgin territory, decidedly more dangerous and demanding incomparably greater physical and mental exertion than today. The motives for practising mountain climbing were different then to those of most modern mountaineers – they were primarily attempts to challenge oneself. Achieving success, conquering peaks, became a source of satisfaction. Of course there was also rivalry then, which in itself was a positive phenomenon, but did not dominate in mountaineering to such an unnatural degree as can be seen today. A purely sporting spirit prevailed. It helped to form in young people extremely positive character traits that would stay with them for life. The integrity, reliability and diligence characterising Smoluchowski were grounded in him in part due to his mountain adventures.

The Smoluchowski family and companions…, descending the Thälihorn ridge in the Welsh Alps, Switzerland, after an unsuccessful attempt to climb the Thälihorn from the east (from the Smoluchowski family collection)

Marian Smoluchowski and wife on a trip to the mountains

Mieczysław Świerz (1891–1929)

Marian Smoluchowski on a trip with his wife and daughter

Janusz Chmielowski (1878–1968)

Years later, Smoluchowski summed up his experience in the following way: "Of what the mountains have given me, I consider three things the most valuable: getting used to undertaking hard tasks, joy in overcoming difficulties, the ability to beautify everyday life with the most sublime poetry: the poetry of the world of the mountains"[249]. As his son Roman writes:

> "these three comments can, *mutatis mutandis,* be applied to Marian Smoluchowski's relationship to science"[250].

One of the most outstanding mountaineers of the early 20th century, Mieczysław Świerz, in an article summarising the achievements of 25 years of the Tourist Section of the Polish Tatra Society, recalls Marian Smoluchowski thus:

> A professor of theoretical physics, a doctor 'honoris causa' of the University of Glasgow, rector of the Jagiellonian University, a scholar of European renown, Smoluchowski is at the same time the most outstanding – next to his brother Taduesz – mountaineer that Poland has produced. (…) His development curve does not even touch the development plane of Polish mountaineering. It is years ahead. Smoluchowski goes to the

---

249   Ibidem, p. 61.
250   Ibidem

Tatras, becomes one of the pioneers of new mountaineering trends, akin to Zsigmondy and Purtscheller, is a warrior in the fight to free hikers from the care of guides, discovers a range of new routes in the Dolomites, has under his belt expeditions to all the most prominent – with the exception of Mont Blanc – giants of the Eastern and Swiss Alps (…). When no one among us had yet thought about systematically preparing themselves for the increased efforts of trips to the Tatras, Smoluchowski turns his muscles and nerves to steel by practising various sports, exercising at climbing schools near Vienna. He becomes a master at perfecting highland technique. Memorable is the time when the section brotherhood were gathered in a group on the veranda of the Karpowiczówka, trying to do one-handed pull-ups on the doorframe. Many tried but none managed the feat. Meanwhile, sitting silently till then and not yet well known to us, the professor got up and taking no heed of the ironic smirks of the young, lightly hoisted himself three times by the hand. The ironic smirks had to give way to deep embarrassment and real wonder. This exceptional technical prowess did not make Smoluchowski merely a follower of pure climbing however. It became a tool for him serving to more effectively carry out his mountain plans, of which the essential feature, the most programmatic, was (…) versatility. Multidirectional versatility! This came to the fore in the quest to know all the main routes of the Alps, the Tatras, the Carpathians and other European ranges, and to climb in these the most prominent and characteristics peaks. It was externalised in the undertaking of expeditions demanding familiarity with various techniques, therefore rock techniques in limestone, dolomite, gneiss, and granite formations, snow, glacier and ice techniques. It sought an outlet in delving into the mountain landscape at any time of year, for Smoluchowski was both one of the earliest winter mountaineers and a pioneer of skiing in Poland, undertaking breakthrough expeditions for that sport to the Eastern Carpathians. That versatility contributed to the variety and richness of feelings that the mountains gave him. Among them were the pure joy of the battle and victory, there was the delight of clinging to terrifying cliff handholds and the terror of crushing snowfalls, there was enchantment at the splendour of landscapes, the adoration of nature's elements in their most powerful manifestations, there was forgetting about the fogs of everyday life and immersion in the sunny expanse, there was the thirst for extraordinary adventures, to experience terrifying depths…

As a mountain companion, Smoluchowski was unparalleled. Habitually laconic, seemingly closed within himself, he did not impose his opinion on anyone, he followed the others in a group of Tatra friends, but when he noticed dithering in the execution of a plan once conceived or when the spectre of danger stared him in the eye, then he seized the bull by the horns and always knew how to emerge from oppression victorious. In seeking out lost trails he had the sense of smell of a son of the far western prairie, on rocks and glaciers he maintained the most vigilant attention, belaying and letting himself be secured most meticulously, in bearing hardship and in adventures – e.g. surprised by darkness in the crags – he exhibited a joyful serenity, enjoying a powerful new experience. Modest to a fault, he almost never spoke about his expeditions, not giving any strictly highland descriptions of his trips; to the boastful, talking brazenly of their mountain deeds – there was never any shortage of them – he had only a smile, though hiding a paternal indulgence and understanding...[251]

Stanisław Ulam recalled that when he talked at the Scientific Laboratory in Los Alamos to physicist Hermann Hoerlin (1903–1983), the leading European mountaineer of the 1920s and 1930s, he heard from him that Smoluchowski was known in the community for being the first to conquer several peaks in the Tatras and Dolomites[252]. He evidently made a strong contribution to the history of mountaineering and the memory of his achievement lives on.

However, the presence of the great physicist in contemporary science is connected primarily with his achievements, as evidenced by the number of citations of his work, which not only do not diminish but grow significantly year after year.

---

251 M. Świerz, *W dwudziestopięciolecie Sekcji Turystycznej Polskiego Towarzystwa Tatrzańskiego 1903–1928*, (In 25 years of the Tourist Section of the Polish Tatra Society 1903–1928), "Taternik. Organ Sekcji Turystycznej Polskiego Towa- rzystwa Tatrzańskiego" (Mountaineer. A unit of the Tourist Section of the Polish Tatra Society) 1928, no 4–6, pp. 95–96.

252 S. Ulam, *Marian Smoluchowski and the Theory of Probabilities in Physics*, op. cit., p. 477.

Chapter X Citations

Many people outside Poland speak and write about Smoluchowski's achievements. How well are his accomplishments known globally? This can most easily be checked through the Google Scholar search engine.

Over the more than a hundred years since Smoluchowski's death, he has largely been forgotten in Polish culture and science and few people, including physicists, are in a position to cite *ad hoc* any of Smoluchowski's scientific achievements while outside of scientific circles he is not associated with physics at all. Scientific, biographical and commemorative publications concerning the Polish researcher have appeared over the last century so rarely that memory of him has faded. The centenary of his death passed almost unnoticed. Meanwhile, if not for an unhappy coincidence, Smoluchowski would probably have become a Nobel Prize winner. Although he never received a Nobel Prize, his presence in contemporary physics is impressive and the equations he calculated are among the fundamental equations in the theory of stochastic processes. In his book *Stochastic Processes in Physics and Chemistry*, Nico van Kampen writes that the Chapman-Kolmogorov equation related to homogeneous Markov processes is often called the Smoluchowski equation but he will not use the name as it is used interchangeably for various other related, but not identical, equations[253]. Elsewhere, he states that the Fokker-Planck equation describing the temporal evolution of the probability density function, is also sometimes called the Smoluchowski equation[254]. Smoluchowski's continuity equation is among the most frequently used in describing sedimentation and coagulation processes. This is an equation that includes Fick's first law (the law of diffusion) and takes account of the action of the external field. Smoluchowski noted that phenomenological laws of macroscopic diffusion can be applied to probability. The diffusion process is a result of the superposition of Brownian motion of individual particles of a substance[255]. Smoluchowski's continuity equation is invoked in the

---

253 See N.G. van Kampen, *Procesy stochastyczne w fizyce i chemii* (Stochastic Processes in Physics and Chemistry), Ed Ł.A. Turski, trans. M. Dudyński, M. Ekiel- Jeżewska, D. Śledziewska-Błocka, Warsaw 1990, p. 79.
254 Ibidem, p. 186.
255 See A. Fuliński, *Współczesne zastosowania równań Smoluchowskiego* (Modern applications of the Smoluchowski equations) in: *Marian Smoluchowski - od teorii atomistycznej do fizyki współczesnej* (Marian Smoluchowski - from atomic theory to modern physics), op. cit., p. 23.

most diverse contexts from theoretical work to industrial applications – it is used for example in industrial water purification, calculations defining the formation of soot deposits in aircraft engines, the coagulation of milk, the formation of gel barriers, the growth of nanotubes, the aggregation of granulocytes and the adhesion of leukocytes. This equation is often known as the *Smoluchowski equation* (for instance in the measurement of zeta potential or in real-time deviations in the relation of electro-osmotic flow). It also appears as the *Smoluchowski limit* in applications of the kinetics of diffusion- limited chemical reactions and in the study of a class of systems of stochastic differential equations describing diffusion phenomena[256].

Smoluchowski's work is cited in publications concerning such issues as the simulation of potassium channels in cell membranes, fluid mechanics (droplet collisions), protein electrophoresis, rotational diffusion, the influence of viscosity on electron transfer kinetics, cholera toxin impact analysis, the thermodynamics of information (numerical calculations), and Brownian dynamics.

Marian Smoluchowski is one of the most frequently cited Polish scientists. In world literature, between 1996 and 2001, he was cited 553 times. In the article *Współczesne zastosowania równań Smoluchowskiego* (Modern applications of the Smoluchowski equations), Andrzej Fuliński points out that between 1996 and 2002, Smoluchowski was cited 836 times[257], although these statistics do not take account of publications in which only the phrase 'Smoluchowski equation' was used. This shows that the number of Smoluchowski citations is not only not decreasing but is growing.

Publications in which Smoluchowski's work has been cited or his equations referred to have recently included the most diverse titles, from the most serious general physics periodicals (such as *Physical Review Letters* and *Physical Review*) and specialist ones (*Applied Physics Letters, Astronomy and Astrophysics, Colloid Journal, Journal of Aerosol Science, Journal of Chemical Physics, Journal of Crystal Growth* or *Journal of Fluid Mechanics*), through influential chemistry magazines (for example *Dyes and Pigments, Industrial & Engineering Chemistry Research, The Journal of Physical Chemistry, The Chemical Society of Japan* or *Journal of Colloid and Interface Science*), biology publications (including *Biochemistry, Biochemical Engineering Journal, The Journal of Biochemistry Molecular Biology and Biophysics* as well as *Journal of Biomedical Engineering*) to such publications as *Archive for Rational Mechanics and Analysis, Journal of Engineering for*

---

256 Ibidem, p. 25.
257 Ibidem, p. 24.

Gas Turbines and Power, Communications in Statistics – Theory and Methods, Computational Mechanics and *Journal of Computational Physics*[258]. This list of scientific titles is impressive, especially as the mentions concern a scientist who lived a hundred years ago and over the last century physics has been one of the most dynamically developing fields of science.

In the article *Potencjał ζ i równanie Smoluchowskiego* (Zeta potential and the Smoluchowski equation), Marek Kosmulski writes that the Polish scientist's name is associated in modern science mainly with electrokinetic potential ζ (zeta) and the Smoluchowski equation, which enables calculation of this potential on the basis of measurable quantities, e.g. electrophoretic mobility. This is not the physicist's only, or even most important, achievement but due to a renaissance of interest in colloidal particles – currently called nanoparticles – it is a context in which he is mentioned extremely frequently. Over the last decade, the correctly spelt name 'Smoluchowski' has appeared on average in 1,500 titles a year, published by Elsevier, Springer, the American Chemical Society or Wiley, to name but a few. A search of scientific literature for references to Smoluchowski's accomplishments is hampered by his name often being spelt incorrectly. According to Kosmulski, in articles indexed by Web of Science, the following spellings can be found, among others: Schmoluchowski, Schmolukowski, Smolucbowski, Smoluchoski, Smoluchovski, Smoluchowiski, Smoluchowsi, Smoluchowsky, Smoluhovski, Smoluhowski, Smolukhovski, Smolukhovskii, Smolukhovskiy, Smolukhovsky, Smolukhowski, Smolukowski, Smolushovski – and this list, as the above author states, is surely just the tip of the iceberg[259].

According to Web of Science, between 1894 and 2014, the number of citations of Smoluchowski's work stood at 7,235. For comparison, over a slightly longer period, between 1880 and 2015, the number of citations of the work of Maria Skłodowska-Curie (1867–1934) stood at 1,582[260]. The numbers given by Web of Science are growing constantly and will continue to grow. This is supported by the situation in industry, especially in chemistry and pharmaceuticals, where a large proportion of current research concerns coagulation processes. The foundation of these studies are the formulae and rules developed by Smoluchowski.

---

258   Ibidem, p. 25.
259   See M. Kosmulski, Potencjał ζ i równanie Smoluchowskiego, (Zeta potential and the Smoluchowski equation) op. cit., p. 7.
260   Wikipedia, *Web of Science*, https://en.wikipedia.org/wiki/Web_of_Science (access: 10.03.2020).

The most commonly cited is an article from 1917 in which he presented a new theory of colloid solidification, published shortly after the author's death in 1918 in *Zeitschrift für Physikalische Chemie* (Journal of Physical Chemistry) under the title *Versuch einer mathematischen Theorie der Koagulationskinetik kolloider Lösungen* (Attempt at a mathematical theory of the coagulation kinetics of colloidal solutions).

Jorge Eduardo Hirsch (born 1953)

This article was cited more than 5,500 times. *Drei Vorträge über Diffusion, Brownsche Molekularbewegung und Koagulation von Kolloidteilchen* (Three lectures on diffusion, Brownian motion and coagulation of colloidal particles), given in 1916 at the Wolfskehl congress in Göttingen, are still referred to. To this day, these are considered the best introduction to the topic of coagulation.

In 2005, Jorge E. Hirsch, an Argentinian physicist working at the University of California in San Diego, proposed an index reflecting the citations of a particular scientist's papers and the number of their best publications. The h factor is a means of measuring scientific achievement in terms of the number of publications and the number of citations. The h factor (also known as *index h, the h-index, Hirsch index*, or *Hirsch number*) for a given author is the number of papers cited $\geq h$ times[261]. This index also shows that Marian Smoluchowski is one of

---

261 See M. Kapczyński, *Indeks Hirscha – zastosowanie oraz metody obliczania*, (The Hirsch Index – application and calculation method) Thomson Reuters Scientific,

the most frequently referenced Polish scientists. The Hirsch index for his papers stood at 29 at the time of research, while Maria Skłodowska-Curie's was 16.

In 1990, two academics from the Max Planck Institute in Stuttgart, philosopher Werner Marx (1910–1994) and physicist Manuel Cardona (1934–2014), launched a research project under the working title of 'Blasts from the past.' The research was intended to answer the question of how to measure the significance or usefulness of a scientific paper. The researchers decided to check how many times a given paper had been included in the reference list of other papers. According to the researchers, the number of citations cannot be used to determine the overall significance or usefulness of a paper, particularly for recent work, the long-term significance of which is not yet clear, but also for many older papers which are not cited because their results are so well known that they appear in textbooks without a source. It would be easy to theorise and speculate about these matters, but the project's initiators assert that there is a much more satisfactory way to proceed: as is always the case in physics, the best way to make progress is to collect and analyse data. They analysed a huge number of papers that had appeared in modern times until 1930. Almost 10,000 scientists' names were considered – chiefly from the exact sciences – and citations of their papers were studied in modern scientific publications for the years 1990–2003.

The research results reveal an extremely high position for Marian Smoluchowski, whose work is very frequently referenced these days. Out of 10,000 scientists, Smoluchowski took sixth place and was the first person on the list not to have won a Nobel Prize.

(1) Einstein Albert (Nobel in physics) – 3,025 citations (died in 1955; lived 76 years);
(2) Debye Peter (1936 Nobel in chemistry) – 1,592 citations (died in 1966; lived 82 years);
(3) Born Max (1954 Nobel in physics) – 1,575 citations (died in 1970; lived 88 years);
(4) Langmuir Irving (1932 Nobel in chemistry) – 1,564 citations (died in 1957; lived 76 years);

July 2, 2012, http://biblioteka.ans.pila.pl/download/dVKic4GDFuImwDIHxbdnNw NHQ2KBkbfWU-Lhs_L3Y1KQs5J-3JbMGIRYBReaR8qPWwwYhxuJRkzMD I6Gw8sJDAqEzhmFFYjexFqBmY1S-HZzZThqJWNAXDYxND4cIFw8GjIeInEn Cjl8HGsOZxkHPyVsNX8fLhg-SNjwqOhk6Yno1LBlzKWoJJGQXYAR2GQM-czljMGxjFh8xOHJhVSNvdjg/indeks_h_zastosowanie_metody_obliczania.pdf (access: 10.03.2020).

(5) Strutt John William (1904 Nobel in physics) – 1,503 citations (died in 1919; lived 77 years);
(6) Smoluchowski Marian – 1,356 citations (died in 1917; lived 45 years)[262].

2 Most-cited authors pre-1930

First authors of the pre-1930 papers that have received the most citations in physics journals since 1990 (until 1 July 2003). The nine Nobel-prize winners in the list are indicated by an asterisk.

Einstein A*	3025
Debye P*	1592
Born M*	1575
Langmuir I*	1564
Strutt J W (Lord Rayleigh)*	1503
von Smoluchowski M	1356
Ewald P P	979
Maxwell J C	812
Weyl H	812
Dirac P A M*	779
Poincaré H	730
Mie G	698
Schrödinger E*	648
Klein O	610
Taylor G I	601
Fowler R H	573
Faraday M	559
Gibbs J W	552
Heisenberg W*	540
Bohr N*	532

List of most cited pre-1930 author names

Although Smoluchowski never received a Nobel Prize and lived on average 30 to 40 years shorter than the other scientists on the list, the number of citations of the first five scientists (after Einstein) is similar. In this context, an obvious question arises – what would our physicist's achievements have been had he lived to a ripe old age?

---

262   See W. Marx, M. Cardona, *Blasts from the Past*, "Physics World" 2004, vol. 17, no. 2.

Chapter XI Scientist

According to what people close to him have said, Smoluchowski had a romantic nature, but it is also indisputable that he was characterized by rationalism and common sense. Both his method of scientific and philosophical reasoning and his persistence in striving for the goals he set himself in his scientific life and later consistently achieved, prove that he was not lost in romanticism.

Straight after finishing his studies, Smoluchowski renounced all types of social, political and national activity. He recognised that science would become the guiding star of his life to which he would be faithful and which he would value above all other aspects of life. He kept his word.

Throughout his short life, the most important thing to him was his scientific work; he became an outstanding physicist, known and respected by the whole scientific world, and he achieved this position thanks to his outstanding abilities supported by hard work and his extraordinary personal traits.

When he was 23, immediately after his doctorate, he received a proposal to be an assistant to Boltzmann. This is significant as Smoluchowski was not a student of the Austrian physicist though he considered himself one. He attended his lectures, read his work and maintained a certain confidentiality with him, but Boltzmann took up the chair in Vienna in 1895, after Smoluchowski had finished his university studies. Nonetheless, apart from scientific esteem, the young Marian was warmly inclined towards Boltzmann and full of gratitude to him. This is confirmed by a letter he wrote to him: "In theoretical physics I owe the initial foundations to my venerable teacher Mr Stefan, and everything else to you, Hofrat[263]. You have exerted the greatest influence on me both through your lectures and especially through your textbooks, which I study with ever new satisfaction and benefit, and I am proud to count myself as one of your students"[264].

Despite these very good relations, Smoluchowski did not take advantage of Boltzmann's offer. Bearing in mind the great physicist's position, working beside him would have guaranteed a path towards an easy and a fairly certain scientific career. But that was not Smoluchowski's goal – he wanted to work in science, treating it as a life adventure, to become acquainted with various laboratories

---

263 Hofrat (imperial court councillor) – a title granted in Austria and Germany for outstanding professional merit, among others to university professors.
264 After: A Teske, *Marian Smoluchowski: życie i twórczość* (Marian Smoluchowski: life and works), op. cit., p. 15.

and the working atmosphere of European universities, and to collaborate with other great physicists.

In November he went to the Sorbonne, to the laboratory of Gabriel Lippmann, a physicist and member of the French Academy of Sciences and later a winner of the Nobel Prize, which he received for developing a method for reproducing colours, based on the phenomenon of light interference. This was the first method of achieving colour photography. Smoluchowski worked at the Sorbonne until July 1896. He conducted research into heat radiation, experimentally verifying the Kirchhoff-Clausius theorem, according to which the intensity of radiation emitted depends on the medium in which the emitting body is located.

In Paris he attended lectures by Henri Poincaré, Charles Hermite, Edmond Marie Bouty and others. In Lippmann's laboratory at the Sorbonne, he worked on experimentally testing Clausius's law on the dependence of thermal radiation on the refractive index of the medium in which the substance is immersed. He went on to develop the problem in a more general form than Clausius, assuming that the radiation comes not only from the surface but also from inside the body[265].

---

265 T. Godlewski, *Marian Smoluchowski*, op. cit., p. 26; see also *Marian Smoluchowski (1872–1917). Fizyk, taternik, romantyk nauki*, (Marian Smoluchowski (1872–1917). Physicist mountaineer, romantic of science) op. cit., p. 2.

Scientist

1417 PARIS. — La Sorbonne, M. le Professeur Lippmann, Membre de l'Institut (Laboratoire des recherches physiques)

Gabriel Lippmann (1845–1921)

Hotel Orfila, where Smoluchowski lived in Paris

A plaque commemorating Smoluchowski's stay at Hotel Orfila

The University of Glasgow

In his letters, Smoluchowski wrote with appreciation about the intellectual atmosphere of Paris, but also expressed some disappointment at the lack of contact between professors and younger scientists[266].

He spent the summer of 1896 in England, in Eastbourne and London, to improve his English ahead of taking up work in Scotland. The University of Glasgow and the laboratory of Lord Kelvin were another of Smoluchowski's scientific adventures. He worked there from September 1896 until April 1897. During Smoluchowski's stay, William Thomson was already a scientist of world renown with considerable achievements, at the forefront of which were the discovery of absolute zero, formulating the second law of thermodynamics, and discovering a thermoelectric phenomenon named the Thomson effect and the Joule–Thomson effect. He was the author of many valuable papers and dissertations in the field of theoretical physics. In 1866, he received the title of baronet for his scientific achievements and later a lordship, which obliged him to change

---

266 S. Ulam, *Marian Smoluchowski and the Theory of Probabilities in Physics*, op. cit., p. 477

his surname. He chose the name of a river that runs close to the University of Glasgow and hence is predominantly known in physics as Lord Kelvin.

Smoluchowski worked in Glasgow under Kelvin, studying the electrical properties of bodies and the properties of gases exposed to the recently discovered X-ray, ultraviolet light and uranium radiation. In this research he focused on the phenomenon known today as ionisation.

The passage of X-ray radiation through gases results in the release of electrons and simultaneously causes the emergence of positively charged atoms. When a voltage is applied to the two ends of a gas container, negative electrons and positively charged atoms move in opposite directions to the poles of the applied voltage, which was taken as the flow of electric current through the gas, even though under normal conditions gas is a good insulator. The research conducted in Glasgow did not fully explain the phenomenon but represented a valuable contribution to its further analysis[267]. It also initiated other similar research culminating in the publication in 1896 of a joint paper by Marian Smoluchowski, Lord Kelvin and Alfred Chester Beatty. From that sojourn came the Polish scientist's great respect for Lord Kelvin and warm appreciation of British culture. Their influences were apparent in his later life, many of his future theories having their origins here. Smoluchowski was deeply impressed by the laboratory's spirit, which hosted guests from around the world. For years he recalled the high level of experimental and theoretical work in this seemingly unremarkable place. In January 1897, he wrote to his brother: "Lord Kelvin seems always greatly excited. Every day some different problem seems most interesting to him and he forgets everything else. He is impatient and cannot conduct experiments himself but could keep busy with his ideas a whole army of experimental physicists"[268].

---

[267] A Teske, *Marian Smoluchowski: życie i twórczość* (Marian Smoluchowski: life and works), op. cit., p. 46.

[268] Quote after: S. Ulam, *Marian Smoluchowski and the Theory of Probabilities in Physics*, op. cit., p. 477.

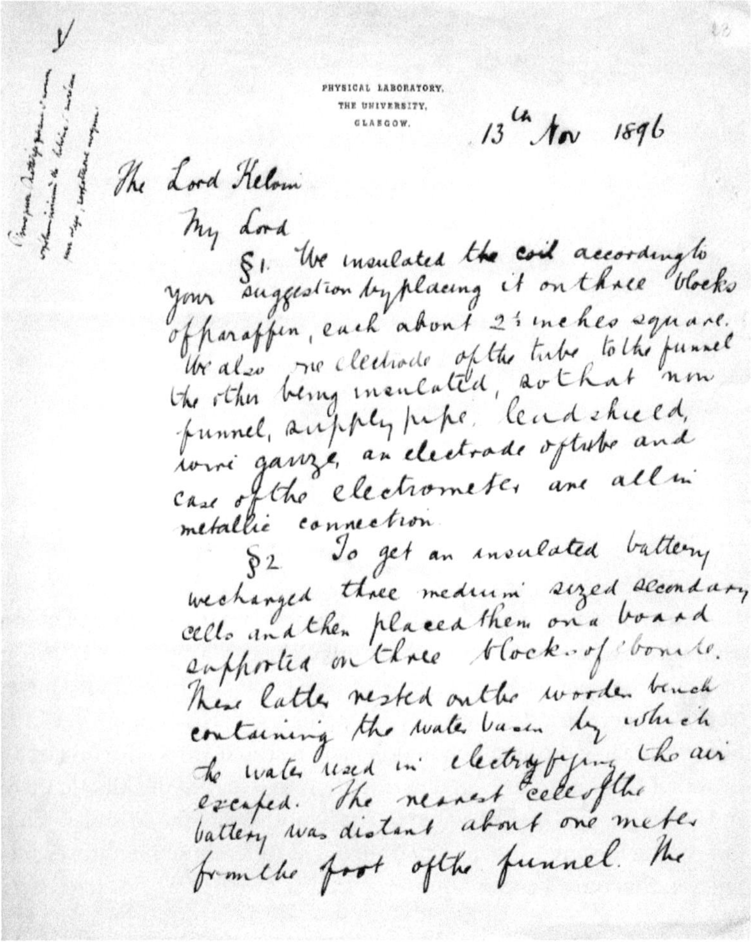

PHYSICAL LABORATORY,
THE UNIVERSITY,
GLASGOW.

13ᵗʰ Nov 1896

The Lord Kelvin

My Lord

§1. We insulated the coil according to your suggestion by placing it on three blocks of paraffin, each about 2½ inches square. We also ~ electrode of the tube to the funnel the other being insulated, so that now funnel, supply pipe, lead shield, wire gauze, an electrode of tube and case of the electrometer are all in metallic connection.

§2. To get an insulated battery we charged three medium sized secondary cells and then placed them on a board supported on three blocks of ebonite. These latter rested on the wooden bench containing the water basin by which the water used in electrifying the air escaped. The nearest cell of the battery was distant about one meter from the foot of the funnel. The

Letter to Lord Kelvin

William Thomson, Lord Kelvin (1824–1907)

Another research centre that inspired Smoluchowski was the Warburg Laboratory in Berlin, where he spent four months. Emil Gabriel Warburg, a German physicist of Jewish descent and a professor of physics at the Humboldt University, discovered temperature discontinuity in rarefied gas. His time in Berlin had a significant influence on Smoluchowski's later research work, the research conducted at that laboratory drawing his attention to issues of the kinetic theory of matter. He was particularly intrigued by the hypothesis of the so-called temperature jump, which occurs in when two media of different temperatures meet. It was believed that only a continuous temperature distribution occurred between the cold and warm parts. Smoluchowski noted: "To date we have not known one case in which (…) a temperature jump could be found or measured[269]. He assumed that in very large rarefactions, where the free path of molecules becomes very great, a temperature jump occurs between the gas and the wall, which apparently reduces conductivity.

> Smoluchowski tackles the problem in this way; he points out an analogy with the phenomenon of very rarefied gas slipping along a wall and subjects the phenomenon to detailed theoretical and experimental study (…). He therefore conducts a range

---

[269] Quote after: A. Teske, *Marian Smoluchowski: życie i twórczość* (Marian Smoluchowski: life and works), op. cit., pp. 62–63.

of precise measurements of the cooling rate of a thermometer placed in a heated gas vessel while the vessel is chilled externally with ice. From the known conductivity of the gas, he calculates the temperature jump at the gas boundary analogous to the slip coefficient. The same general conclusions and results are also reached (...) by another completely independent method, by measuring thermoelectrically the temperature of a copper plate suspended between two coaxial cylinders kept at constant and different temperatures so that a steady heat current flows between them. He approaches the whole phenomenon (...) in a detailed and theoretical way on the basis of the kinetic theory of gases, stemming from Clausius's assumption that kinetic energy (temperature) is a property of molecules that has no effect on the velocity distribution and that the molecules on average have that energy prevailing in the layer where the final collision takes place. On this basis, by calculating the heat flow between the walls enclosing the gas, he deduces that the temperature jump boundary must be proportional to the length of the free path. One of the reasons this happens is the incomplete alignment of the temperature with the wall when a single molecule strikes it and this factor must be particularly pronounced when molecules are very light, i.e. in the case of hydrogen. The theory he developed is fully supported by the results of his measurements as well as by data obtained first from experiments by Winkelmann and Schleiermacher and later by Brush. Smoluchowski would return several times in later years to theoretical and experimental research on the heat conductivity of gases. The discovered phenomenon of a temperature jump depending on the length of the free path can only really be explained on the basis of kinetic theory, from which stems the overall significance of the above research to the development and revival of this theory[270].

Smoluchowski solved the problem, demonstrating that a temperature jump exists and is inversely proportional to the pressure, therefore disappearing at higher pressures.

Stanisław Loria summarises the post-university period of Smoluchowski's scientific life in the following way:

the period of preparatory and specialist studies (...) not only broadened his horizons and scope of scientific interest, did not only enrich his theoretical knowledge and skills in using experimental and mathematical methods, but culminated in serious scientific achievement which earned the 26-year-old researcher widespread recognition among contemporary physicists.

---

270 T. Godlewski, *Marian Smoluchowski*, op. cit., p. 26; See also *Marian Smoluchowski (1872–1917). Fizyk, taternik, romantyk nauki* (Marian Smoluchowski (1872–1917), Physicist, mountaineer, romantic of science), op. cit., pp. 4–5.

Emil Gabriel Warburg (1846–1931)

The University of Vienna's Philosophy faculty also willingly awarded 'venia legendi' [right to lecture – Ed.] to its student of such outstanding qualifications[271].

Smoluchowski was a theoretical physicist so it may come as a surprise to some that he also possessed significant technical skills, which he used at the laboratory of the Jagiellonian University in Kraków, where he took up the chair of the experimental physics faculty. He was very conversant with the design and construction of equipment. It was the case that

> he happened to be exceedingly good at glass blowing, better than the professional mechanics at his institute. To illustrate anecdotally his liking for simple experimentation, the writer was told by Mrs Smoluchowski how one morning she was asked for a flat kitchen vessel, and gave her husband a salad bowl. Smoluchowski explained that he was trying to imitate a possible mechanism for the appearance of mountains on the surface of the earth. A few days later, one saw a layer of gelatine covered by a surface of mercury

---

271 S. Loria, *Marian Smoluchowski i jego dzieło* (Marian Smoluchowski and his work), op. cit., p. 9.

folding into mountain chains and ridges. This was the first step on experiments which gave rise to a series of papers, on a theory of mountain formation[272].

Smoluchowski's universal and multifaceted mind took in a range of quite specific sciences, an example of which is the above-mentioned theory of mountain formation. Godlewski writes further about it:

He works more closely on the problem raised by Kirchhoff of floating plates and demonstrates that such a plate remains flat when horizontal pressure is applied as long as the pressure does not exceed a certain value. If the pressure exceeds that value, a sine wave occurs corresponding to a permanent equilibrium for which the potential energy is at a minimum. The number of bends (semi-waves) depends on the dimensions and qualities of the plate but not on the pressure, the magnitude of which only determines the height of the wave. To test this theory, he performs experiments, captivatingly straightforward in their simplicity, on the formation of waves in plates of gelatine, gold, tinfoil, gutta-percha, shellac etc., floating in mercury or water. The number of waves, their length and amplitude transpired to be entirely consistent with the theory.

He takes on the problem of the erosion of glaciers, showing that not knowing the friction coefficient makes it impossible to develop and define a theory with certainty, both when the glacier is considered as a permanent plate and when its pressure is treated hydrostatically.

Starting with the principles of kinetic theory, he considers the atmospheric height of Earth and the planets. He demonstrates that certain gas particles must escape into the universe while others reach the Earth. Having taken account in his theoretical deliberations of the influence of radiation, and above all internal friction (regardless of pressure), he arrives at the conclusion that the whole of interplanetary space is filled with an extremely rarefied gas of a temperature close to absolute zero, which condenses in proximity to planets. As can be seen even from this broad presentation, Smoluchowski, who was primarily a kineticist, was extremely versatile and the scope of his scientific achievements covers the most diverse areas of the philosophy of nature, even ones quite distant from each other. (…) Smoluchowski presents the results of his creative-scientific research (in four languages) in the most varied national and foreign scientific journals[273].

---

272 S. Ulam, *Marian Smoluchowski and the Theory of Probabilities in Physics*, op. cit., pp. 477–478.
273 T. Godlewski, *Marian Smoluchowski*, op. cit., pp. 22–23; See also *Marian Smoluchowski (1872–1917). Fizyk, taternik, romantyk nauki*, (Marian Smoluchowski (1872–1917), Physicist, mountaineer, romantic of science) op. cit., pp. 14–15.

This was possible because, as Goetel adds: "He spoke English, French and German absolutely fluently and knew Italian, Russian and Swedish well".[274]

Smoluchowski presented broad arguments on geophysics, Earth morphology and seismology in a speech he gave on November 21, 1916, at a meeting of the Kraków branch of the Polish Copernicus Society of Naturalists, held to honour the memory of Maurycy Rudzki (1862–1916). He referred in it to the achievements of the eminent geophysicist, proving his perfect knowledge of the subject.

> *Fizyka Ziemi* [The Physics of the Earth – Ed.] is an absolutely original scientific work, perhaps the only one in the whole of world literature in which can be found the entire physics of the litho- and hydrosphere systematically, thoroughly and in accordance with the current state of science; for specialists, it is even more important that the author's work has been properly taken account of in it. The meteorology textbook, although accessibly written, without mathematical apparatus, in some ways is also at a much higher level than similar foreign works because meteorology does not appear in it as a collection of empirical, statistical rules, but takes the form, as far as possible in the current state of knowledge, of the science of the physics of the atmosphere. So this is clearly the direction in which that science is developing today and in which, by progressing, it will one day become a branch of the strict sciences, on an equal footing with other fields of geophysics. If we finally consider once again the whole of Rudzki's activity in the field of geophysics, we must recognise that the fruits of his labour were as great as were the problems he addressed. Closer study of his work heightens the impression of a deep sense of reverence for his mind, the real, deepest driver of which was an absolute, fanatical love of the truth. This was the cause of his distrust of risky hypotheses, his severe criticism of himself and others, the rigour of his reasoning, it was the source of his scientific ideas[275].

Stanisław Loria calls Smoluchowski's early years of Scientific work at the University of Lviv:

> a period of independently seeking his own paths, his own scientific problems. It is filled with extensive studies in various disciplines: he takes up issues in the field of aerodynamics, studies physical processes occurring in the atmosphere of Earth and the planets, systematically investigates progress in the field of the kinetic theory of matter, is interested in electroendosmosis and cataphoresis and publishes contributions to the theories of these phenomena; he researches experimentally and theoretically the creation process of the so-called 'veins' during the flow of fluids, and so on. An interesting and

---

274 W. Goetel, *Ze wspomnień osobistych o Maryanie Smoluchowskim* (From personal recollections of Marian Smoluchowski), op. cit., p. 215.

275 M. Smoluchowski, *Maurycy Rudzki jako geofizyk*, (Maurycy Rudzki as a geophysicist) speech given on November 21, 1916, at a meeting of the Kraków branch of the Polish Copernicus Society of Naturalists, held to honour the memory of M.P. Rudzki, "Kosmos" 1916, vol. XLI, pp. 122–123.

characteristic episode from this period includes, for example, his studies concerning the physical foundations of a theory that would explain the formation of mountains through folding of the Earth's crust. An avid climber and mountaineer, Smoluchowski clearly took up this problem while still a youth. Although his theoretical dissertations and interesting experiments, conducted in a very simple way with ingenious methods, were published only in the years 1909–1910, the familiarity with extensive geological literature apparent in them is evidence that the author had been dealing with tectonic issues and theories for a long time[276].

The outbreak of the First World War in 1914 seriously disrupted work at the university and working conditions at the Institute became hard to accept. The physics faculty building, built recently under Prof. Witkowski, was taken over by the Austrian army. Smoluchowski tried to find a way out of the awkward situation by moving his workplace to the former apartment of Professor Karol Olszewski, located in Collegium Chemicum, but unfortunately this was of little help because as a reserve officer he was called up to the army. He served for several months, after which he managed to return to Kraków. Coming back via Vienna, he visited his Alma Mater, where he was very well received. After returning to Kraków, he continued his scientific and teaching work though as the laboratory was in the institute building, still occupied by the army, there was no possibility of research.

Marian Smoluchowski

---

276 S. Loria, *Marian Smoluchowski i jego dzieło* (Marian Smoluchowski and his work), op. cit., p. 10.

In 1914 he gave a talk at a conference on limits to the validity of the second law of thermodynamics. That work was noticed and in 1916 Smoluchowski (together with Zsigmondy) was invited to the Wolfskehl congress in Göttingen where he presented his famous three lectures on diffusion, Brownian motion and the coagulation of colloidal particles. In 1917, he published a paper in *Zeitschrift für Physikalische Chemie* (*Journal of Physical Chemistry*) on the subject of a mathematical theory of kinetics and colloidal solution, which made him one of the creators of that theory (two years after the Polish physicist's death, Wilhelm Otswald, editor of *Kolloid-Zeitschrift* (*Colloid Journal*), published it separately in support of Smoluchwski's work in this field).[277]

Of the four years of Smoluchowski's sojourn in Kraków, three fell within the period of the First World War, when the Austrian army was present in the university, taking over among others the physics faculty building, and the scholar was torn away from his university work and called up to the army. These were conditions under which it was difficult to conduct scientific work, but Smoluchowski spent the time occupied with teaching work (he lectured in experimental physics) and writing the *Self-Study Handbook*. Of his work on this publication, he wrote to Stanisław Michalski, publisher and inspiration for the handbook: "This is my usual method of work: that I write a draft, then I put it aside and do something else, then only after some time do I read the draft anew. Then all the flaws present themselves much more clearly and objectively and are easier to fix"[278].

At this time, he also took up pioneering research in the field of statistical physics, laying the foundations of a new field of science. Fuliński noted that Smoluchowski created a new branch of physics, which is now among the leading fields – statistical physics[279]. At the beginning of the 20th century, he was in no position to foresee that what he started would develop so much or that his equations would still be relevant a hundred years later.

Research work in statistical physics conducted by Smoluchowski represents an important contribution to the development of the field thanks to which a new area of research emerged – the theory of stochastic processes. Mark Kac, a representative of the Lwów (Lviv) school of mathematics, believed that the calculation

---

277 S. Ulam, *Marian Smoluchowski and the Theory of Probabilities in Physics*, op. cit., p. 478.
278 M. Smoluchowski, *List do S. Michalskiego w sprawie „Poradnika dla samouków"* (Letter to S. Michalski on the matter of *"The Self-Study Handbook"*), Jagiellonian Library, BJ Rkp. 9412, k. 86.
279 A. Fuliński, *Współczesne zastosowania równań Smoluchowskiego* (Modern applications of the Smoluchowski equations), op. cit. p.

methods and theoretical foundations used by Smoluchowski became the basis of the development of statistical physics[280]. A similar view was aired by astrophysicist Subrahmanyan Chandrasekhar, who wrote that Smoluchowski's research work became the foundation of the modern theory of stochastic processes[281].

The statistical mechanics applied by Smoluchowski to research on Brownian motion inspired him to use probability theory in physics (of which there was wider discussion in Chapter I). The above conclusion was confirmed by Stanisław Ulam's mathematic reasoning in a 1956 article:

> One of Smoluchowski's, one might say, mathematical achievements was the clarification of the so-called ergodic hypothesis of Boltzmann. (...) This postulate of ergodicity is indispensable for the rigorous foundations of statistical mechanics. This property still remains unproved for most of the actual dynamical systems such as for example the n- body problem. General theorems, however, were proved by J. von Neu- manna and G.D. Birkhoff, asserting the existence of such averages in time of sojourn in sub-regions for "almost every" point of the phase space. This is the celebrated ergodic theorem. (...) One of the points of debate which was so crucial during Smoluchowski's activity was thus settled. That deterministic theories of phenomena, in general, lead to randomlike sequences of points in the representative space. Let us quote a sentence from the posthumous paper of Smoluchowski in *Naturwissenschaften* [*Natural Sciences* – Ed.] (1918). "It seems to us that it is very important even for the philosopher that one can prove, even though only in a limited part of mathematical physics, that the idea of probability in the ordinary sense of a regular frequency of random effects has also an objective meaning, namely, that the idea and genesis of randomness can be made rigorously precise also if one rigorously follows the determinism; the law of large numbers comes then not as a mystical principle and not as a purely empirical fact but as a simple mathematical result of the special form which the law of causality determines in such cases." This prophetic sentence certainly is justified by the ergodic theorems proved for the first time only in 1931.
>
> It is true that the work of mathematicians – even in the classical part of statistical mechanics (...) establishes a foundation merely for the beginning parts of thermodynamics. An analogous and rigorous treatment of the so-called H-theorem and the Boltzmann equations is not yet completed. It is interesting that Smoluchowski realized the difference in the logical structure of the theories of Maxwell, Boltzmann, on one hand, and the statistical mechanics of Gibbs. Again quoting from the same article, he says explicitly that "the difference between the two consists very likely of the fact that the former is based on certain ideas from the theory of probability, extremely plausible for these physical systems but not rigorously proved, whereas the latter avoiding these ideas, is based entirely on postulated physical properties." This feeling for the logical structure

---

280   M. Kac, *Marian Smoluchowski and the Evolution of Statistical Physics*, op. cit., s. 17.
281   S. Chandrasekhar, *Stochastic Problems in Physics, and Astronomy*, op. cit., s. 1–89.

of physical theories is extremely noteworthy when one remembers the time when these remarks were written – before the great vogue for axiomatization and the development of foundations of mathematics[282].

The cited quote raises a host of questions about references to statistical mechanics contained in Smoluchowsi's paper *O nieregularnościach w rozkładzie cząsteczek gazu i ich wpływie na entropię i na równanie stanu* (On irregularities in the distribution of gas molecules and their influence on entropy and the equation of state)[283], of which Einstein was probably aware prior to May 1905. At the same time, it again becomes apparent that Smoluchowski's highly mathematical mind was entirely subordinated to thoughts about physics.

Stanisław Ulam (1909–1984)

---

282 S. Ulam, *Marian Smoluchowski and the Theory of Probabilities in Physics*, op. cit., pp. 479–480.
283 M. Smoluchowski, *O nieregularnościach w rozkładzie cząsteczek gazu i ich wpływie na entropię i na równanie stanu*, (On irregularities in the distribution of gas molecules and their influence on entropy and the equation of state) – check – op. cit. p.

Smoluchowski was primarily a theorist. He mastered and embraced in a versatile way the enormous arsenal of mathematical knowledge he needed for this end and knew how to correctly apply it in a skillful way. However, in his research he solved extremely difficult and intricate problems mathematically with astonishing ease as in the essence of his mind he was never an abstract mathematician but always a physicist. To him, every differential equation, even the most complicated, had primarily a physical meaning, to him what lived directly in these equations was above all a change in physical quantities, based on true reality, which he calculated mathematically but whose reality he felt and guessed intuitively with the spirit of a naturalist. Gifted with extremely critical judgement, he could look at any problem most thoroughly, in depth, and with uncommon intuition feel the importance of its future consequences[284].

These extraordinary traits of Smoluchowski's mind distinguished him from other scientists not only in Poland, and meant he perceived every subject he worked on as separate, possessing unique qualities.

> A characteristic trait of his mind was simply its phenomenal clarity and an amazing capacity to approach a problem or phenomenon in the simplest way. This characterises all his theoretical work, both in the posing of questions and the means of solving them; this is plain to see in all his experimental work. I remember how he once told me in conversation: "experimental research in physics does not usually require expensive resources." And that was true, but true (...) particularly in regard to him. What others achieve very often with the extraordinary precision of complicated and expensive equipment, he achieved by conducting simply-conceived experiments. And indeed, the simplicity of the experiments he conducted is just astonishing. But then the whole essence of that powerful personality that was Smoluchowski was also simple, uniform and natural. Those lucky enough to have known him more closely in person could never forget him, not only as a great scientist, but as an exceptional man[285].

Maybe this is why, despite all the time that has passed since Smoluchowski's Kraków years, and despite science's dynamic development over the last few decades as well as now, in the third decade of the 21st century, Smoluchowski's research achievements still have significant influence in many fields. This is true mainly in specialisations that in the majority are a continuation of research that

---

284 T. Godlewski, *Marian Smoluchowski*, op. cit., s. 27–28; See also *Marian Smoluchowski (1872–1917). Fizyk, taternik, romantyk nauki*, (Marian Smoluchowski (1872–1917), Physicist, mountaineer, romantic of science) op. cit., p. 18.

285 Ibidem, p. 28; See also: *Marian Smoluchowski (1872–1917). Fizyk, taternik, romantyk nauki*, (Marian Smoluchowski (1872–1917), Physicist, mountaineer, romantic of science), op. cit., p. 18.

Smoluchowski was involved in such as, for example, molecular physics, fluid mechanics, or fluctuation theory. However, the impact of some of Smoluchowski's experiments exceeds the scope of the science of his time and touches on the most up-to-date issues, such as sub- atomic physics in the description of quark-gluon plasma or the use of research in the field of the physics of stochastic processes, which is being continued in mathematics in probabilistic methods of filtering out noise from corrupted signals.

Summing up the scientific significance of Smoluchowski's key works, Stanisław Loria wrote:

> Two dissertations in particular (...) deserve special attention: the first bears the title, "*Über Unregelmüßigkeiten in der Verteilung von Gasmolekülen und deren Einfluß auf Entropie und Zustandsgleichung*" [*On irregularities in the distribution of gas molecules and their influence on entropy and the equation of state*]; the second, "*Sur le chemin parcouru par les molécules d'un gaz et son rapport avec la théorie de diffusion*" [*On the path travelled by the molecules of a gas and its relation to the theory of diffusion*]. These papers mark the start of the third, peak period of Smoluchowski's scientific output. They concern topics around which the author's thought would revolved stubbornly for another 13 years. He later returned to them many times from different angles; he analysed them with ever more perfect mathematical methods, penetrating them ever more subtly, finding the deeply hidden relationships existing between them and illustrating their essential importance to the basic problem of the natural philosophy of the time. That problem was the relationship of thermodynamics, especially the so-called second law concerning entropy growth and the irreversibility of thermal phenomena, to the molecular-kinetic theory of matter. The subjects were: the phenomenon of fluctuation, and the stochastic process in the form of so- called Brownian motion[286].

Marian Smoluchowski was without doubt the greatest Polish physicist. He left behind achievements and the memory of an exceptional personality. He made a big impression on everyone who met him. Quiet and modest, possessed of many virtues, as Goetel writes, he became the central point of every circle he entered[287]. Selflessness, humility and diligence are the qualities mentioned in notes, letters and other texts written by the people with whom he had direct or indirect contact. He was an exceptional authority, as a scientist and sportsman and also an intellectual. At the same time, he himself receded into the background due to the modesty that was well known to all.

---

286 S. Loria, *Marian Smoluchowski i jego dzieło* (Marian Smoluchowski and his work), op. cit., p. 10.
287 W. Goetel, *Ze wspomnień osobistych o Maryanie Smoluchowskim*, (From personal recollections of Marian Smoluchowski), op. cit., p. 228.

His wife, Zofia, perceived him as a romantic, even claiming that he said that about himself, describing himself as a researcher who strove to solve many mysteries at once, never finding peace. His youthful working method was therefore very uneconomical as he directed his attention towards many different fields. Hence the more than 90 works collected in *Pisma Mariana Smoluchowskiego* (*The Writings of Marian Smoluchowski*) cover a very wide spectrum of research problems. Excessively simple and obvious things had no appeal to him, like the green hills compared to the rocky peaks[288].

Situations occurred in which "the sense of ownership of his idea or explanation was lost since the most important thing for him was to serve science and arouse sublime enthusiasm in others"[289].

Smoluchowski, as Ewa Wyka writes, was above all an outstanding physicist, had an unusually sharp mind, was a mountaineer, climber, skier, carrying out climbing plans with consistency and passion. All pleasures were subordinated to his scientific work and the professional duties he undertook. As a humanist, sensitive to nature, he recorded its beauty in quickly drawn pencil sketches and in the watercolours he painted purely for himself, from an internal need to capture the moment. In descriptions of alpine expeditions, he revealed himself as a scrupulous recorder of plain facts but also as a literary person, pouring onto the page moments of wonder and enchantment at alpine landscapes. He was a great lover of the arts and theatre and an aficionado of classical music, which he also loved to play[290].

Tadeusz Godlewski describes how Smoluchowski's progressive scientific achievements were noted and appreciated in European university circles.

Both the Polish and foreign scientific worlds were to assess the significance of Smoluchowski's work: in 1901, having travelled to Glasgow as a delegate of the

---

288 Quote from a statement by Zofia Smoluchowska for A. Teske, after: A. Teske, *Marian Smoluchowski: życie i twórczość*, (Marian Smoluchowski: life and works), op. cit., p. 140.
289 A. Gałecki, *Badania M. Smoluchowskiego w dziedzinie układów mikroskopijnych*, (M. Smoluchowski's research in the field of microscopic systems), speech given at a ceremonial sitting of the Kraków branch of the Polish Copernicus Society of Naturalists, devoted to honouring the memory of Professor Marian Smoluchowski, December 11, 1917, "Kosmos" 1917, yearbook XII, issues 5–12, p. 213.
290 See: *Marian Smoluchowski (1872–1917). Fizyk, taternik, romantyk nauki*, (Marian Smoluchowski (1872–1917), Physicist, mountaineer, romantic of science) op. cit.; E. Wyka, *Marian Smoluchowski we wspomnieniach bliskich i przyjaciół* (Marian Smoluchowski in recollections of relatives and friends) "Zwoje" 2003, no 2 (35), pp. 32–33.

University of Lviv, he received the honorary title of doctor of law of that University. In the year 1908, he was chosen as correspondent member of the Mathematical-Natural Sciences Section of the Academy of Arts and Sciences in Kraków, and in the same year received the Haitinger prize of the Austrian Academy of Sciences in Vienna for his theoretical work on Brownian motion. In 1917, he becomes an active member of the Kraków Academy and the Göttingen Academy of Sciences was considering electing him as a correspondent member when news came of his death[291].

It is also worth referring to a lecture given by Józef Moroziewicz three months after Smoluchowski's death, in which he describes the physicist's physiognomy, fitting the image of a noble man. Moroziewicz recalls the handsome and imposing figure of Marian Smoluchowski, his swarthy face as if "sculpted from bronze, always deep in thought with a tight mouth, a mouth that signified seriousness and the power of resolve, and which only occasionally bloomed into a cheerful smile"[292].

Goetel draws attention to an important personality trait of Smoluchowski.

> Scientific work often creates a certain one-sidedness. Especially when involvement in it is strong and hot, when it occupies every thought and a person's whole self, it happens that other areas of human life start to lie fallow, falling into the background until they disappear completely and a person of science becomes the type of scholar that sees nothing beyond their specialisation. This is an understandable phenomenon to a certain extent; it takes quite an exceptional person to master knowledge and not forget about other realms of the human spirit, to move freely within scientific work, but to also know how to live and to take from that life what is beautiful and great in it. Such a person was Maryan Smoluchowski, and he was so to an uncommon degree, as uncommon as everything he did, almost everything he touched[293].

Smoluchowski looked forward to his inaugural lecture as rector and prepared for it meticulously. He intended to talk about the *Uniformity of laws of nature*. The choice of subject was in line with the researcher's spectrum of interests in the final period of his life. They became more general and focused on the synthesis of phenomena[294]. Interestingly, Smoluchowski seemed to refer to Aristotelian discourse, from which he had departed so much earlier.

---

291 T. Godlewski, *Marian Smoluchowski*, op. cit., p. 27.
292 Speech by J. Moroziewicz delivered at a sitting of the Kraków branch of the Polish Copernicus Society of Naturalists, December 11, 1917, "Kosmos" 1917, yearbook XII, issues 5–12, p. 13.
293 W. Goetel, *Ze wspomnień osobistych o Maryanie Smoluchowskim* (From personal recollections of Maryan Smoluchowski), op. cit., p. 218.
294 S. Ulam, *Marian Smoluchowski and the Theory of Probabilities in Physics*, op. cit., p. 478.

Chapter XII Aristotle

According to Smoluchowski, both the ancient Greek era and later the medieval age were devoid of an important element characteristic of modern science, namely the experiment. Philosophical speculation dominated, in the best case supported by deduction. An experiment happened by chance, as in the case of Archimedes. Smoluchowski wrote: "We cannot conceive at all today of the disgust of the Greeks and learned scholars towards experiments or their attraction to groundless speculations and their application to the natural sciences. This trend seems to us a sort of madness hampering the development of these sciences, especially on account of Aristotle's authority throughout antiquity and the middle ages"[295].

Herschel expresses a similar view, claiming that the essential fault of Greek philosophy was the supposition that the same method that proved so effective in mathematical studies could be applied to natural research. The study of nature was therefore neglected and humble and careful examination of the facts was held in utter contempt as unworthy of the a priori approach a true philosopher should rely on[296].

As already mentioned, in antiquity and the Middle Ages, the deductive method dominated, and from general theorems about the essence of the world, with the aid of logical (deductive) reasoning, conclusions were drawn about various natural phenomena. People were not discouraged by the fact that each philosopher, depending on his starting point, arrived at different conclusions[297], Smoluchowski claimed. In *The Copernican Revolution*, Thomas Kuhn refers to the general views of the Stagirite, proving that they stemmed from speculative reasoning. "Aristotle was able to express in an abstract and consistent manner many spontaneous perceptions of the universe that had existed for centuries before he gave them a logical verbal rationale, but which education has suppressed in the adult world of the eighteenth, nineteenth and twentieth centuries"[298].

---

295 M. Smoluchowski, *Przedmiot, zadanie, metoda oraz podział fizyki* (Subject, task, method and the division of physics), op. cit., p. 177.
296 See J.F.W. Herschel, *Wstęp do badań przyrodniczych* (Preliminary Discourse on the Study of Natural Philosophy) trans. T. Pawłowski, Warsaw 1955, p. 107.
297 See M. Smoluchowski, *Poradnik dla samouków* (Self-Study Handbook), vol. I, op. cit., pp. 30 and 31.
298 Thomas S. Kuhn, *The Copernicus Revolution: Planetary Astronomy in the Development of Western Thought*, Cambridge, Mass: Harvard University Press, 1957, p.95.

Aristotle's system was burdened with many limitations, causing, as Kuhn writes, that "[t]he organic realm has a conceptual primacy, and the behavior of clouds, fire, and stones tends to be explained in terms of the internal drives and desires that move men and, presumably, animals"[299]. The remnants of Aristotle's thinking about the laws of nature can be seen in his mental speculations:

> the movements of the simple natural bodies (fire, earth and so on) show not only that there is such a thing as place, but also that it has a certain power. For unless prevented from doing so, each of them moves to its own place, which may be either above or below where it was. Above and below and the other four directions [i.e. right, left, forwards and backwards] are the parts or forms of place. (…) In the sense that they are relative to us, they are not always the same, but depend on our position (…) 'Above' is not just any random direction, but where fire and anything light move towards. Likewise, 'down' is not just any random direction, but where things with weight and earthy things move towards. So their powers as well as their positions make these places different[300].

The essential flaw in these analyses was the lack of experiments in support of the conclusions. However, Kuhn believes it is impossible not to perceive the breakthrough achieved under the Stagirite's influence in the "methodology" of thinking about nature and in creating the foundations of physics. He writes:

> Aristotle's world view was not the only one created in antiquity, nor was it the only one that gained adherents. But Aristotle's was far nearer to many primitive conceptions of the world than its ancient competitors and it corresponded more closely with the evidence of unaided sense perception. That is another reason why it was so immensely influential, particularly during the late Middle Ages"[301].

In the book *Medieval and Early Modern Science*, Alistair Cameron Crombie (1915–1966) notes: "Aristotle's *Physics*, translated in the twelfth century from Arabic to Latin, presenting scientific concepts in a common sense way, with faulty views on physics contained in it, turned out to be a work constituting a closed rational system explaining the universe on the basis of natural causes. Aristotle's system was more than natural sciences in the 20th century sense"[302]. Crombie asserts that, "[i]n the course of the twelfth century a wonderful concept appeared, which immediately enabled the development of science; this was

---

299 Ibidem, p. 96
300 *Aristotle, Physics*, trans. Robin Waterfield, Ed. David Bostock, Oxford University Press, 1999, pp. 78 –79.
301 T. Kuhn, *The Copernican Revolution*, op. cit. p. 99.
302 A.C. Crombie, *Nauka średniowieczna i początki nauki nowożytnej* (Medieval and Early Modern Science), vol. 1, *Nauka w średniowieczu w okresie V–XIII w.*, (Science in the Middle Ages, V–XIIcenturies), trans. S. Łypacewicz, Warsaw 1960, p. 77.

the idea of a rational explanation in the form of formal or geometric evidence, according to which a particular fact is explained when it can be deducted from a general principle. This occurred due to a gradual familiarity with Aristotle's logic and Greek and Arabic mathematics"[303].

Smoluchowski considered the deductive research method used by the Stagirite to be correct and effective but only in certain areas of study since – he argued – such a way of thinking hindered the development of science. According to the Polish scholar, the indisputable position of Aristotle's philosophy slowed the development of science.

Did the Aristotelian method of perceiving natural phenomena, called by Smoluchowski "a madness holding back the development of those sciences, especially as a result of Aristotle's authority throughout the whole course of antiquity and the Middle Ages"[304], really impede that development or was it rather an essential link making modern science possible, a stimulus fertilising the mind to think more astutely?

In the Middle Ages there existed schools of thought based on the Aristotelian conception of the universe, inspired not by metaphysics but by the mathematical, physical, astronomical and medical aspect of the Stagirite's science. The English Franciscan Roger Bacon (1214–1292) of Oxford was an opponent of the speculative resolution of philosophical problems. He considered the experimental research of phenomena with the use of mathematics to be important. He believed that the speed of light was finite, he foresaw the emergence of many inventions, such as ships without oars, carts driving without horses, and flying machines. In 14th-century Oxford and later in several other European university centres, there worked *Calculators* who used Aristotelian logic and physics in their calculations. They divided motion into uniform and uniformly accelerating and were able to determine the mean velocity of motion and the law of free-fall. They even attempted to calculate some physical properties like heat, colour and the density of light. Their research was not speculative in nature. The ideas of the Oxford *Calculators* did not spread to Europe however.

In the 19th century, the founder of the scientific method was considered to be Francis Bacon (1561–1626), who freed scientific inquiry from rational speculation. According to Herschel, there was no natural science in the scientific sense of the term prior to Bacon's *Novum Organum*. He is responsible for popularising

---

303 Ibidem, p.11.
304 M. Smoluchowski, *Poradnik dla samouków* (The Self-Study Handbook), vol. I, op. cit., p. 31.

the principle of induction as well as the dictum that the entirety of the natural sciences is comprised of a series of inductive generalisations which begin with the establishment of individual facts and lead to universal laws[305].

Smoluchowski notes that cases are known of the use of induction in scientific research in ancient times. Ptolemy made systematic measurements of the angle of incidence of the sun's rays and tried to establish the relationship of the results achieved to the angle of refraction of light rays[306]. Leonardo da Vinci used experiments, which he called "a question posed to nature"[307]. Some of Galileo's later studies show the application of induction. However, it is thanks to Bacon that induction became a scientific method consciously used to draw conclusions from experiments and that the experiment became recognised as the basic tool of science.

Smoluchowski argued that "as a consequence of applying the new method, a change occurred in the way scientific research was conducted, (…) it is also necessary to know how to pose a question in the appropriate way and performing experiments without a plan or the requisite preparation is merely a worthless game"[308].

Research conducted by some scientists in a way considered the only scientific one, i.e. when the researcher seeks simply to classify known facts without any preconceived theories or notions – rarely yields valuable results. Progress in science occurred when we started to approach the subject of inquiry with a definite plan, for example following some theory, even the simplest, that needs to be tested, or studying predicted relationships between figures[309].

In the context of the cited facts, the thesis that for almost nineteen hundred years the development of human thought was limited by the methodology imposed by Aristotle is indefensible. Could his philosophy have bound the minds of philosophers and thinkers to such an extent that they could not depart from the framework of perceiving nature imposed by the Stagirite? Smoluchowski accuses Aristotle, referring to his *Physics,* and claims that were it not for that school of thought, focused on philosophical speculation, most probably "some

---

305 See J.F.W. Herschel, *Wstęp do badań przyrodniczych* (Preliminary Discourse on the Study of Natural Philosophy), op. cit. pp. 103 and 104.
306 M. Smoluchowski, *Poradnik dla samouków* (The Self-Study Handbook), vol. I, op. cit., p. 32.
307 Ibidem, p.37.
308 Ibidem, p.38.
309 Ibidem, p.52.

Bacon" would have appeared much sooner and modern science would not have emerged only in the 17th century.

However, no one has accused Ptolemy (who lived in about 100–168 AD) of holding back the development of astronomy for close to 1,600 years by building a geocentric system according to which Earth was at the centre of the Universe. His theory put man at the centre of the world for a long time, which influenced the development of theology, philosophy, and especially the philosophy of nature, despite the fact that the Greek astronomer Aristarchus of Samos (approx. 310–230 BC) had proposed a heliocentric model of the Solar System much earlier. Nor is Euclid accused of holding back other geometric systems for many years by describing a system of geometry based on axioms in *Elements*. Smoluchowski could not have known whether "some Bacon" had appeared earlier and simply gone unnoticed.

An exemplification of this thesis are the *Calculators* or the theorems of a forerunner to Copernicus (1473– 1543) – Nicholas of Cusa (1401–1464). He maintained that Earth was not the centre of the Universe; that it is smaller than the Sun and larger than the Moon (which, he stated, could be observed during a solar eclipse), and rotates on its axis every 24 hours[310].

Until the 12th century, and in the Augustinian school of 13th-century philosophy and theology, mathematics was an introduction to metaphysics, or – in line with the terminology of the time – to theology. In the 14th century, under the influence of Ockham's views, mathematics became an indispensable tool of the science of nature. This heralded a new science[311].

Not until the late renaissance is a systemic change visible in the approach to natural philosophy. The Dominican Domingo de Soto (1494–1560), considered a forerunner of modern physics, is an example of that change. He established that a body in free-fall undergoes constant acceleration. This was 14 years before the birth of Galileo[312].

Aristotle's philosophy and physics were still present in the renaissance era. What then influenced the transformation of natural philosophy into science despite the clear continued presence of Aristotle? A revolution occurred in the system of thought and the way reality was perceived changed.

---

310 See. J.M.R. Morales, *Kościół i nauka. Konflikt czy współpraca?* (The church and science. Conflict or cooperation?) trans. S. Jędrusiak, Kraków 2003, p. 82.
311 See S. Bafia, *Fizyka* (Physics) in Quaestiones super octo libros "Physicorum" Aristotelis *John of Głogów*, Kraków 2013, p. 9.
312 J.M.R. Morales, *Kościół i nauka. Konflikt czy współpraca?* (The church and science. Conflict or cooperation?) op. cit., p. 83.

Marian Smoluchowski after his doctorate

The scientific and philosophical thought that had been forming since ancient Greece needed time to mature and take on the characteristics of a scientific project, based in the 20th century on induction and hypothesis, and also required a new methodology of studying nature. European thought, which had been developing until Galileo, matured under the influence of the Stagirite's philosophy, rather than in opposition to his eight books, to form a model of scientific reasoning. Without Greek philosophy and Aristotle, that process would probably have taken longer. It is significant that the scientific model we know emerged only in European culture, the foundation of which is Greek philosophy.

The Stagirite's contribution to building the scientific foundations is key. He is responsible for developing methods of the logical construction of thought, the creation of rules of inference, definition and classification. It was Aristotle who made us aware that concepts arise through abstraction and who gave importance to the ability to make observations and pose questions. His many- value logic left its mark on the whole of science (examples of which are modern computer programmes based on the binary system).

In early Christianity, philosophy was treated as pagan thought, in the Middle Ages studies were guided by different priorities – they were mostly theological inquiries and represent the foundation of deliberations. The situation in mathematics was a little different, seen in the Middle Ages as an exemplary science, as Crombie writes. Earlier in Arabia, and in the 12th and 13th centuries in medieval Europe, mathematics developed on the basis of Aristotle's logic among others, and though relegated to a role subordinate to theology, was a factor that significantly influenced the development of that discipline.

Aristotle's philosophy was a further step up in the study of nature. It opened the way for logical thought, inference, definition, classification, abstraction, observation and the asking of questions, bringing with it everything that we owe to Aristotle and which modern science strives for. However, for a long time, these important attributes of the methodology of science were not used because they were not useful. Thought was focused on theological aims for which the most important thing was the usefulness of theories relating to faith. The concept of Smoluchowski's utility criterion explains this situation persuasively. Knowledge in the field of natural philosophy did not make great progress since the utility of the theories applied created sufficient pressure to not strive for that change.

In his philosophical and scientific views, Smoluchowski, like Galileo, opposes a speculative analysis of nature, treating Aristotelianism as an intellectual basis for the study of the natural sciences. He does not see the philosophical and logical potential created by the Stagirite which enabled the emergence of Bacon's philosophy. It is worth emphasising that Smoluchowski's attitude to philosophy was strictly utilitarian. His writings do not contain unambiguous declarations or clearly defined philosophical assumptions. Being a physicist, he treated philosophy instrumentally, conducting philosophical deliberations when he needed to for his scientific research. However, by its nature, physics is a science in which philosophy is essential, hence there is considerable philosophy in the Polish physicist's work.

Chapter XIII Philosophy

Smoluchowski selectively used the whole spectrum of available philosophical concepts, usually not specifying which but merely using the tools that were derived from it. He also discussed certain philosophical theses – some he agreed with, others he rejected.

Portrait photo of Marian Smoluchowski – Lviv

In reading Smoluchowski it is worth first looking for ideas contained in his scientific work drawn from various philosophies as they are usually hidden in commentaries referring to the subjects of physics, mathematics or methodology. He is usually not interested in the philosophy itself, which was not his main area of interest, but at best in a useful theoretical tool helpful in understanding a reality that can be learned about chiefly through the theories and hypotheses of the exact sciences. Such a reading of Smoluchowski's work is an inspiration to review the main themes of nineteenth century natural philosophy both in terms of their presence in the Polish scholar's scientific writing and as an analysis of his references to the theses they put forward.

The positivism of Auguste Comte (1798– 1857) was a permanent presence in Smoluchowski's deliberations. Though his refections often went beyond that school of thought's remit, he agreed with the basic principles of positive philosophy, rejecting metaphysics, for example. In *The positive method in sixteen lectures*, Comte wrote:

> Currently I shall be able to define the nature of positive philosophy with ease. This philosophy regards all phenomena as being subject to immutable laws. It considers as futile the search for causes, whether primary or final. In positive explanations, the causes creating a phenomenon are not indicated, the circumstances in which they arise are analysed and these are connected, one to another, by a relationship of sequence and similarity. (…) The new philosophy differs from the old by striving to remove any search for primary or final causes, considering them barren[313].

Smoluchowski had a similar attitude towards causality and purposefulness, claiming that in antiquity and the Middle Ages natural phenomena were explained on the basis of purposefulness, which was questioned in the early 20th century by all naturalists. He wrote: "Explanation with the aid of the concept of purposefulness has been removed from the natural sciences today as a naïve anthropomorphism"[314].

Comte claimed that true knowledge was scientific knowledge, free from any metaphysical elements, which could be attained only via a theory's positive verification and which is achieved only through empirical research. A positive approach in science and philosophy comes down to the study of objects available to sensory cognition. Only the exact sciences have developed methods of study enabling the creation of scientific knowledge. Smoluchowski objected to pseudoscientific speculation and idealism in favour of what was certain, material and possible to study, i.e. empirically knowable through experiment. He drew attention to science's inductive nature and the leading role of the experiment.

In the process of forming positivistic philosophy, an important role was played by the work of Charles Darwin (1809– 1882). In the work titled *On the Origin of Species* Darwin wrote that he could see no evidence of design, still less of beneficent design[315]. Smoluchowski was a keen advocate of Darwin's theory, writing about it many times in various texts. He maintained that it was chiefly

---

313 A. Comte, *Metoda pozytywna w szesnastu wykładach*, (The positive method in sixteen lectures) trans. W. Wojciechowska, Warsaw 1961, pp. 15 and 301.
314 M. Smoluchowski, *Poradnik dla samouków* (The Self-Study Handbook) vol. I, op. cit., p. 18.
315 See F. Copleston, *A history of philosophy* vol. 8, *Modern Philosophy: Empiricism, Idealism, and Pragmatism in Britain and America* Doubleday, New York, 1994, p. 103.

due to him that science had departed from the concept of purposefulness in nature, which was a misleading *a priori* assumption, and that it was he who had shown in various examples how apparently purposeful adaptations arise that are *de facto* a result of natural causes[316].

Smoluchowski did not identify with every thesis of positivist philosophy. He conceived differently of the very essence of science – understanding science as a collection of hypotheses constituting a certain approximation to the laws of nature. His predisposition to pragmatism was evident here. He did not endow science with the indisputable position it had in positivism or empiriocriticism. He saw elements that differentiated scientific theories from the real laws of nature and the constant asymptotic path of science striving towards the knowledge of laws. His epistemological criticism, inspired by Kant, did not allow a recognition that scientific knowledge can reveal, in a duly justified way, the essence of reality[317].

Smoluchowski cannot be ascribed the position proposed by Jean Le Rond d'Alembert and Comte as he had a different understanding of the credibility of science. He wrote about scientific theories as more or less probable hypotheses which would never be fully verified. He believed that the leading role of the experiment in science did not translate into a claim that phenomena should be defined only on the basis of experience and that only empirical facts can constitute the real subject of knowledge. Similarly, he did not agree to limiting philosophy exclusively to its practical nature, as Comte postulated, so that it could be used mostly in the field of everyday life. He shared the view of the positivists, who questioned the metaphysical side of philosophy, but also saw a need for the existence of a philosophy of nature and a philosophy of science, which he himself practised.

Zenon Roskal notes: "Thanks to work by scholars like Smoluchowski, a breakthrough was possible in philosophy based on a departure beyond the positivist vision of science, but also beyond the barriers that positivist philosophy had erected on the path of knowing and mastering nature"[318]. The foundation of Comte's positivist philosophy was man and the society and science that

---

316   See M. Smoluchowski, *Poradnik dla samouków* (The Self-Study Handbook) vol. I, op. cit., p. 18.
317   I. Dąmbska, *O poglądach metanaukowych Władysława Natansona i Mariana Smoluchowskiego* (On the metascientific views of Władysław Natanson and Marian Smoluchowski), „Zagadnienia Naukoznawstwa" ("Problems of Science") 1979, vol. XV, book. 1 (57), p. 3.
318   Z.E. Roskal, *Mariana Smoluchowskiego ujęcie zasady przyczynowości w badaniach ruchów Browna* (Marian Smoluchowski's approach to the principle of causality in

surrounded him. With this approach, the French philosopher was written into the annals of history as a eulogist of utilitarianism. This was consistent with the later views of the pragmatists and close to Smoluchowski.

Scientism developed in the second half of the 19th century on the basis of empiricism and positivism. It assumed that science enables the prediction and direction in a deliberate way of natural and social processes occurring in the world. Smoluchowski was not an adherent of scientism but was a supporter of many scientistic assumptions. Nineteenth-century scientism, similarly to positivism, was somewhat naïve in its apotheosis of science and Smoluchowski saw this naivety. There are no direct references to scientism in the Polish researcher's scientific work because by conviction he did not enter into philosophical digressions, which to him were uncertain, remaining in the realm of methodology and the language of the exact sciences. In the philosophical conceptions he developed for himself, he distanced himself from the theories accepted in science and the legitimacy of scientific knowledge. He did not agree with the claim of scientism's adherents that the science they preached was "in itself" in line with reality[319]. This found expression in his work in the formulation of hypotheses that broke the accepted and binding standards of the science of the time.

Smoluchowski's sympathy with empiriocriticism – the so- called "second positivism", a philosophical movement of the turn of the 19th and 20th centuries whose most important representatives were Richard Avenarius (1843–1896) and Ernst Mach – was from the outset not uncritical and with time he distanced himself firmly from that school of thought. Empiriocriticists set themselves the goal of eradicating all fiction from science, which to Smoluchowski was obvious. The basic assumptions of empiriocriticism were rooted in Auguste Comte's philosophy. The empiriocriticists claimed it was necessary to reject non-scientific considerations which are not based on "pure" experience, and to take as much data as possible from experience in order to ensure its understanding. They postulated the elimination of the tendency to hypostasise reality onto things and their images, claiming that there is only one, monistic reality. According to Smoluchowski, such theses were too restrictive, limiting theoretical deliberations and hence the development of science. For example, both research on Brownian motion, culminating in scientific proof confirming atomic theory, and research in the field of kinematics, proved that scientific inquiries cannot

---

    research on Brownian motion) „Zagadnienia Filozoficzne w Nauce" ("Philosophical Problems in Science") 2017, vol. 62, p. 110.

319 See. B. Kotowa, *Scjentyzm jako światopogląd nauki* (Scientism as a worldview of science) „Nowa Krytyka" ("New Criticism"), 2004, no 16, p. 151.

be artificially limited only to experiments. Excluding the hypothesis from the process of creating scientific theories was unacceptable to Smoluchowski as he believed it was indispensable to the development of science and that its dynamic progress would be made impossible without theoretical deliberations formulated into hypotheses and subjected to experimental verification[320]. The aim of science cannot be merely the description of facts or phenomena devoid of attempts to explain them. The description should be formulated according to knowledge, but we should not limit ourselves to pure experience as knowledge cannot be based exclusively on empiricism[321].

Avenarius expounded a theory of the psychological nature of sensations, so-called introjection, or the projection of the perceived world into one's own interior, which led him to the conclusion that there are no separate physical and mental phenomena as they are merely two aspects of the same experience. Hence his epistemological monism. However, such an assumption led as a consequence to abandoning the category of truth in cognitive activities, which Smoluchowski, as a physicist, could not agree with.

Marian Smoluchowski – portrait photo taken at a photographic studio in Vienna

---

320 M. Smoluchowski, *Poradnik dla samouków* (The Self-Study Handbook) vol. I, op. cit., pp. 46–50.
321 Ibidem, p. 32.

He was an advocate of pragmatism, a philosophical system of the second half of the 19th century. The creators of this trend were two American philosophers: Charles Sanders Peirce (1839– 1914) and William James (1842–1910) and its basic thesis was the pragmatic theory of truth. The American philosophers assumed that utility constitutes the criterion of the veracity of judgements and concepts because truth is determined in relation to the goal achieved by cognition. A judgement is right to the extent that it contributes to effective action. Smoluchowski shared the assumption of the pragmatic theory of truth, according to which the veracity of theses is dependent upon their practical results[322]. James's theory of truth assumed that "True ideas are those we can assimilate, validate, corroborate and verify"[323].

The assumption that truth is not a permanent property of accepted theories but can at best only happen to them, while the most important property of knowledge is its ability to explain, making science not a collection of truths about the world but only a collection of explanations enabling us to understand it, was in line with Smoluchowski's thinking. He advocated such a conception of truth in science that would imbue it with usefulness to progress and make thought verifiable in terms of its practical consequences. Scientific progress cannot consist merely of discovering new laws but also of replacing existing explanations with better ones. He justified that opinion by developing the concept of the utility criterion[324].

Smoluchowski shared the Jamesian view, according to which science consists of generalities that give an abstract picture of the world rather than reality itself and that beliefs are more dispositions to action than a mental representation of reality. Although he did not refer directly to James's philosophy in his papers, he accepted the assumptions of pragmatism, namely that need and doubt direct us and make us want to know something new, which in turn leads to the creation of further beliefs and their formation with the aid of rules of conduct.

---

322 Ibidem, p. 51.
323 William James, *Pragmatism – A New Name for Some Old Ways of Thinking: Popular Lectures on Philosophy*, United Kingdom, Longmans, Green, 1907, p. 97.
324 Discussion of this theory is to found in the current book in Chapter XIV, *Utility*.

Marian Smoluchowski during the Kraków years

The instrumentalists argued that there is no subjective reference in the field of objective reality, and that problems are resolved through the formulation of hypotheses, their logical verification and then empirical falsification. They assumed that functioning theories and theorems play only instrumental roles and constitute a mere reaction to environmental factors, being tools enabling man to adapt to his environment, and that scientific theories are not assessed according to their trueness or falseness as these are merely rules of inference, constituting tools aiding adaptation to, and mastery of, the environment.

The essential theses propounded by the instrumentalists were applied by Smoluchowski in his scientific practice, but he did not accept the philosophical narrowing of the role of science. He assumed that the criterion of the truthfulness of a hypothesis is not an argument deciding whether it is accepted since truthfulness does not exhaust all the conditions for a theory's compliance with nature. Due to the prior assumption of the uncertainty of knowledge, he introduced a category which provides greater certainty of the realness of a scientific theory. This criterion is utility, as more substantial than the criterion of truthfulness; hence we do not differentiate between true and untrue theories, more or less probable, but only more or less useful. We can talk about their utility in three ways:

> The more useful a theory or hypothesis is: 1) the simpler and more illustrative it is in essence; 2) the greater the range of known phenomena it explains and makes accessible

to our mind; finally 3) the better guide it proves to be for further research. This last role of hypotheses, consisting of predicting yet-unknown things, is immensely important for a fair assessment of their value to science[325].

The constricting of science to an act aimed at both adapting and subordinating the environment to man would suggest a limitation of science to problems situated in the empirical realm, however Smoluchowski's utility covers a fuller spectrum with all manifestations of theoretical knowledge, which can fulfil various functions – not only empirical, but also, for example, explanatory, interpretative, and cognitive.

The trueness or falseness of theories, as well as all their instrumental functions, according to Smoluchowski, are verified by the degree of their use – the more a theory is used, the more in line it seems to be with the real laws of nature, which we will probably never understand in full.

> "Fanaticism for veracity, (…) fanatical strivings for integrity and truth – this is the ethical bedrock appropriate to the natural sciences and reinforcing them. They are doing battle against bluff and cliché, the diseases that afflict our society and distort our language"[326].

This statement concerned issues of ethical direction in the research conducted, however Smoluchowski's attitude to truth in science remained critical as he believed that physical theories striving to embrace material phenomena in a general mechanical worldview cannot claim to be absolute truths. They can only be an "image" of the phenomena revealing themselves to us[327]. This idea is close to another philosophical view, which assumes that beneath the surface of observable phenomena lies hidden the true, essential reality. To Smoluchowski, essentialism meant at its core the hiding of the essence of natural laws and in fact their non-perception, so assumed that functioning hypotheses and scientific theories may prove false and it will never be possible to definitively state the truth of some of them.

Essentialism, a school of thought created by Karl Popper and intended to answer the question of what laws of science are, corresponds with Smoluchowski's views. Following a suggestion by Einstein, Popper claimed that the laws of

---

325 M. Smoluchowski, *Przedmiot, zadanie, metoda oraz podział fizyki*, (Subject, task, method and the division of physics) op. cit., p. 195.
326 See idem, *Znaczenie nauk ścisłych w wykształceniu ogólnym* (The importance of the pure sciences in general education), op. cit. p. 130.
327 Idem, *O nowszych postępach na polu kinetycznych teorii materii* (On newer progress in the field of kinetic theories of matter), in: *Pisma Mariana Smoluchowskiego* (The writings of Marian Smoluchowski), vol. 1, op. cit., p. 279.

science describe structural properties of the world as essential properties which "cannot be seen with the naked eye"[328].

Karl Raimund Popper (1902–1994)

He argued that if we want to have knowledge, we must seek laws of nature and the regularities present in them. We cannot, however, assume that certain strict regularities exist; it is sufficient to realise that our knowledge consists of seeking universal regularities, as if they existed[329]. This is a thesis close to the convictions of Smoluchowski, who assumed the existence of such universal regularities, though with the important proviso that we will never know them. Like Smoluchowski, Popper did not believe in the possibility of reaching a final explanation which would not need any further modifications, but at the same time was convinced that we can reach ever deeper into the structure of our world, towards its ever more essential or deeper properties[330]. Popper's essentialist views, expressed several decades after Smoluchowski, remain in many ways convergent with the Polish physicist's theses.

Conventionalism is a school of French philosophy from the early 20th century whose main proponents were Henri Poincaré, Pierre Duhem (1861–1916)

---

328 See S. Wszołek, *Esencjalizm transcendentalny K.R. Poppera*, (The transcendental essentialism of K.R. Popper) „Zagadnienia Filozoficzne w Nauce" ("Philosophical Problems in Science"), 2002, no 31, p. 127.
329 Ibidem.
330 Ibidem, p. 126.

and Édouard Le Roy (1870–1954). The conception of science up until that time, expressed in the accumulation of judgements about facts and the formulation on their basis of general theorems, justified inductively, could not unambiguously and persuasively defend itself against the criticism proffered by David Hume (1711–1776). Research conducted by the Scottish philosopher was intended to negate the possibility of making judgements both about the world and about the subject learning about it. In turn, the emergence of non-Euclidean geometries undermined the conception of synthetic *a priori* judgements to which Immanuel Kant (1724–1804) had included the axioms of Euclid's geometry. The cited arguments contributed to the verification of the hitherto understanding of science. According to Poincaré, in the formulation of scientific laws and the description of facts, apart from individual statements about facts considered true, an essential role is played by the decision of the scientist or the agreement of the scientific community concerning so-called conventions, meaning statements introduced to science on the basis of an agreement. Due to the essence of human nature, scientific theories and theorems are contractual in nature and the attitudes adopted towards perceived reality are changeable and tend to develop. The reason for their acceptance is not so much their veracity as their cognitive value, relative simplicity, convenience, economy of operation, and even aesthetics[331].

Henri Poincaré (1854–1912)

Reflections indicative of an acceptance of this point of view appear many times in the work of Smoluchowski, who – like Poincaré – believed that the changes taking place in science, especially the radical ones, are made possible by the conventions present in it. In Smoluchowski's investigations related to issues of chance and

---

331   See I Szumilewicz, *Poincaré*, Warsaw 1978, p. 26.

probability calculus, the clear influence of the French philosopher can be seen. Adopting Poincaré's position, the Polish physicist used the concept of chance as a certain type of causal link and in this respect sought an explanation of the possibility of chance occurring and the application of statistical methods in a world governed by deterministic laws. Smoluchowski was not a declared conventionalist and did not agree with some of Poincaré's conjectures, in particular with the thesis of the independence of the principles of science from experience, although the French physicist and mathematician's works significantly influenced his philosophical beliefs. Frequent references to Poincaré's theses and arguments can be seen, especially in two articles: *Nauka i metoda* (Science and Method) and *Nauka i hypoteza* (Science and Hypothesis), on which Smoluchowski wrote a separate, exceptionally kind, essay titled *Dwie książki z dziedziny „filozofii przyrody"* (Two books from the field of "the philosophy of nature").

In Smoluchowski's philosophy we also find some elements of neo-Kantianism after the Marburg school. This school of thought emerged in the second half of the 19th century and represented a return to Kant's philosophy not only in the field of methodology, but as a reaction to the speculative systems of German idealism and the empiricist materialism of the developing natural sciences. Represented by Hermann Cohen (1842–1918), Paul Natorp (1854–1945), and Ernst Cassirer (1874–1945), neo- Kantianism featured some assumptions convergent with Smoluchowski's views. For example, he recognised the view on the active role of the intellect in the learning process and focused on logical and epistemological-methodological issues. According to proponents of this school, the task of philosophy was to establish a logical structure to scientific knowledge and to show the logical conditioning occurring between real material relationships. Certain analogies to theses assuming that practising philosophy should start with a critique of knowledge, that knowledge is of a subjective nature, concepts do not describe reality but are a creation of the intellect, knowledge is only recognised as that which the mind constructs, or that 'knowledge' should be treated as an analysis of how it is possible, are also found in Smoluchowski's writings[332].

---

332 See idem, *Poradnik dla samouków* (The Self-Study Handbook) vol. I, op. cit., pp. 13–14, 16–17 and 50.

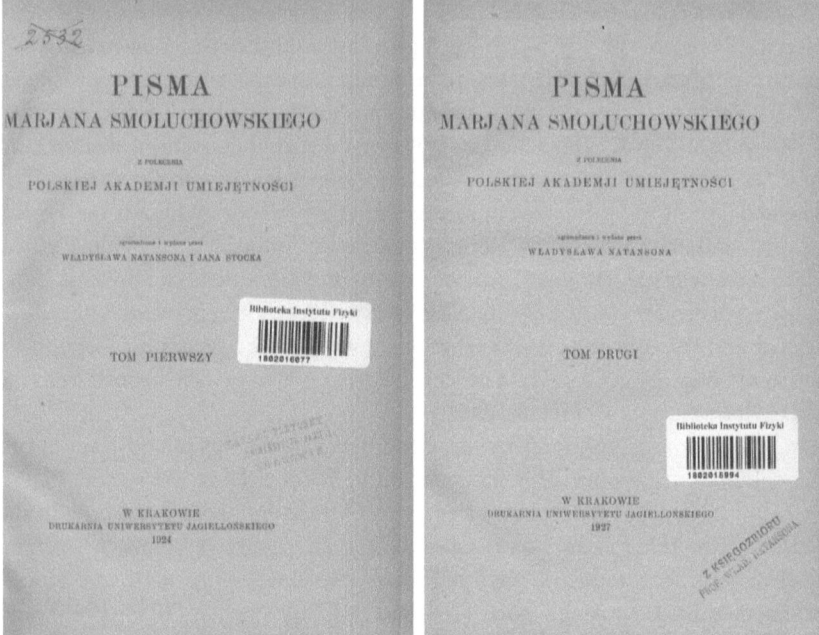

The Writings of Marian Smoluchowski, commissioned by the Polish Academy of Arts and Sciences

However, he did not agree with all the assertions of the neo-Kantianists, especially with those that as a practising physicist he could not accept, like for example the thesis that knowledge could not exceed the boundaries of experience, or that thought is the only true subject of scientific knowledge while data from the external world possess no cognitive value. Smoluchowski had differing views on this subject. In conclusion, it should be stated that his relationships with neo-Kantianism were sporadic and superficial, but perceptible.

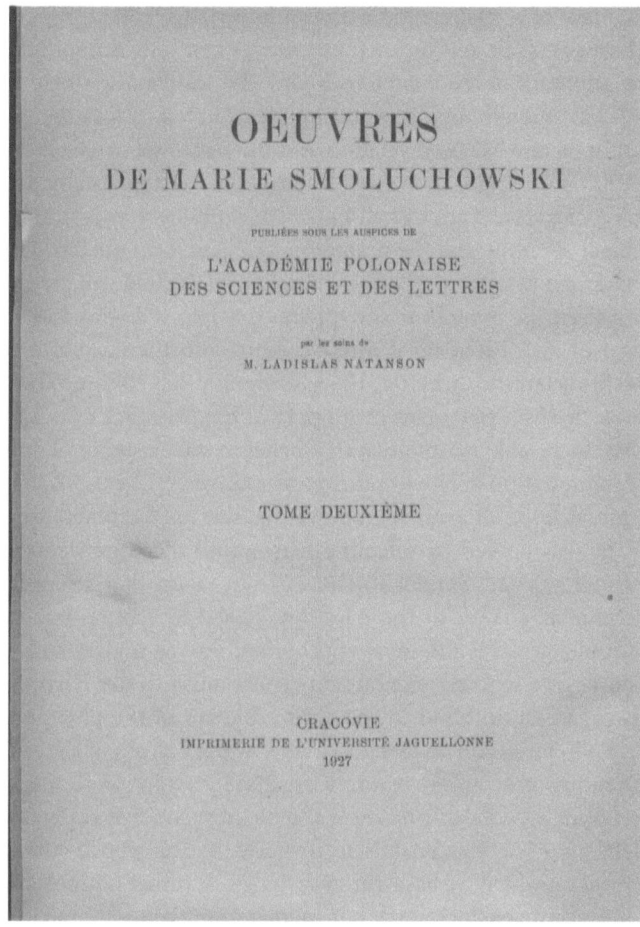

The Writings of Marian Smoluchowski, commissioned by the Polish Academy of Arts and Sciences – French version

In the second half of the 19th century, a renaissance occurred in the thought of St Thomas Aquinas (1225–1274) in the form of neo-Thomism. Its founders were influenced by the changes that had occurred in the perception of reality over the several hundred years since the emergence of Thomism. In several of Europe's intellectual centres, research began on neo- Scholastic thought. The encyclicals *Aeterni Patris* of Leo XIII (1879) and *Pascendi Dominici Gregis* of Pius X (1907) confirmed this state of affairs. A key tenet of neo-Thomism was the establishment of a relationship between knowledge and faith. According to the neo-Thomists, faith and rational knowledge not only do not contradict or exclude one another but are complementary. The source of rational knowledge is human reason and although it is an imperfect tool, it cannot be rejected. The source of faith, meanwhile, is revelation and the truths we arrive at on this path are absolute in nature. Neo-Thomism established a strict hierarchy, according to which theology forms the apex of all knowledge, philosophy is in the middle of the hierarchical pyramid, and the remaining sciences form the base. The boundary of scientific knowledge was to be the world of created things[333], and interference from scientific thought in the sphere of religious dogma was ruled out.

The theses of neo-Thomism did not concur with the general principles of Smoluchowski's scientific thinking. The assumption that all thing change in line with divine design is a statement in support of the existence of purposefulness in nature. Smoluchowski maintained that belief in the existence of final cause in nature is a manifestation of naïve anthropomorphism[334]. The thesis that theology forms the pinnacle of all available knowledge was unacceptable to the Polish physicist as he was guided by scientific reason and theology went beyond the methodology of science. Smoluchowski did not claim that scientific theories provide a complete answer to the questions posed by nature, but he believed that the path of science is the most appropriate way to understand nature. He wrote: "Who knows whether the human race, bound to the Earth, is not as a result of its organisation blind to the entire domain of the phenomena of the universe, like a holothurian attached to a rock at the bottom of the sea"[335].

The limitations on scientific research created by evolution in man raised his doubts. He asked: "Are we able to know the phenomena that really take place at all, or is there some way that enables us to study the real world? After all, all the factual material on which we base our knowledge, or rather notions, of the external world, is made up exclusively of our mental impressions"[336].

Smoluchowski's philosophical interest resulted from questions posed to the strict sciences, primarily to physics and mathematics, inspired by relationships occurring in nature, frequently exceeding the field of cognitive possibilities nature has armed us with, enabling us to know the Universe we live in only locally. As

---

333 See A. Maryniarczyk, *Tomizm* (Thomism), in: *Powszechna encyklopedia filozofii* (The Universal Encyclopedia of Philosophy), vol. IX, Lublin 2008, ps. 503.
334 See M. Smoluchowski, *Poradnik dla samouków* (The Self-Study Handbook), vol. I, op. cit., pp. 18–19.
335 Ibidem, p. 16.
336 Ibidem, p. 13.

Heller put it: "This field by some miracle emerges – like a communicative centre in non-commutative algebra – from the ocean of the Universe, which surpasses the power of our mind and imagination"[337]. Smoluchowski argued that we do not go beyond certain fields limited by the possibilities of empirical knowledge, "the explanation of phenomena affecting the functioning of nature is one of the tasks of physics – he explained – without invoking final cause"[338].

Since ancient times, the functioning of nature has been explained using final cause. Situations concerning people and animals have been carried over to natural phenomena. Smoluchowski noted that "the naivety of such reasoning leads to the ridiculous suggestion that final cause in the functioning of nature boils down to it working for the benefit of humanity"[339]. In Smoluchowski's texts there is no direct reference to neo-Thomism although we regularly find deliberations in them on the subject of causality and purposefulness. The contention of this school of thought's main theses enables a look into the sphere of the philosophical views of the Polish scientist, putting him in opposition to metaphysics.

In writing about Smoluchowski's philosophy, a school of thought should not be omitted that had a significant influence on the development of science, especially of physics and chemistry, which was characterised by a phenomenological view of matter.

The classical thermodynamics that dominated in the 19th century, also known as phenomenological thermodynamics, treated solids, liquids and gases as continuous media having no molecular structure. A group of scientists, known as the energeticists, dealing with macroscopic thermodynamic phenomena, postulated that scientific research be based on pure experience, on the description of facts, rather than on explaining phenomena. Smoluchowski was opposed to the phenomenological approach to the material world, as well as to the energeticists' philosophical conception, as he wrote many times.

Another important issue of the Poles' philosophy is his attitude to materialism, however not because materialistic philosophy was significant to his views. A problem arose in the 1950s, when for political reasons Smoluchowski started to be ascribed materialistic views in numerous publications, with no supporting basis in his scientific work. As a result of this manipulation, in the following years the label of materialist was pinned to the philosopher. This issue – as

---

337 M. Heller, *Czas i przyczynowość* (Time and causality), Lublin 2002, p. 42.
338 See M. Smoluchowski, *Poradnik dla samouków* (The Self-Study Handbook), vol. I, op. cit., pp. 18–21.
339 Ibidem, p. 18.

an important point demanding study – has been expanded on in greater depth below (see Chaper XVI *Materialism*).

Following publication of the paper *On Thermodynamics and Brownian Motion* in 1906, and after general acceptance by the European academic community of the kinetic-atomic theory of matter, a change took place in Smoluchowski's philosophical outlook, which is visible in both the scientific problems he took up and in his publications. An important issue is Smoluchowski's attitude to such philosophical problems as, for example, the principle of economy of thought or the sense of beauty in the theories and formulae created in the exact sciences. They exerted significant influence on the development of science and on the work of the Polish physicist.

Chapter XIV Beauty in physics

Smoluchowski considered the basic aim of physics to be the most accurate study of empirical phenomena, in particular discovering their regularities, relating them to a comprehensible whole and presenting them in the simplest way possible in the form of strict functions mapping the relations of real quantities. He argued that the advantage of the "educated view" over the "naïve" one was primarily not that the former is true and the latter false, but that the former is emphatically more accurate and understandable. The naïve view is usually the view of the majority and even the greatest scientists are guided by it in the activities of everyday life as it represents the result of direct daily experience and enables everyday existence. A straightforward perception of reality is uncertain and can lead the senses astray because they may be subject to illusion (visual, thermal, acoustic illusion or others).

> As early as infancy – Smoluchowski noted – a child gets accustomed to associating certain sensory impressions: muscular sensations, pains, thirst, hunger and so on, with its own self, with what in time it calls "me" and which later may be divided into "my body" and "my mental world", while other sensory impressions, forming entirely separate chains of association, it involuntarily relates to a separate whole, from which in time it forms the notion of the external world. The boundary between these two types of sensory impression is very clear and even a completely uneducated, utterly naïve person differentiates perfectly between what is related to their own body and what is the external world, apparently entirely separate and independent of the person. He ascribes independent existence to this external world and imagines it to be essentially the way it appears to him. He believes that there exists, for example, a hard body, so-called gold, which is yellow, shiny, relatively heavy: after all he can see it, touch it, pick it up; he believes there exist soft, green objects of certain shapes, called leaves; he believes that there exist luminous objects: stars, the Sun; it never occurs to him that these may be delusions of his own senses[340].

Awareness of the fact that subjective perception is limited by our senses, and very often the additional burden of their faulty functioning, is the reason why we strive to rid our world view of all anthropomorphic traits. The theorem of the intelligent person that light is a transverse undulatory phenomenon – Smoluchowski believes – is superior to the statement of the naïve person: light is that which illuminates, that it is a step away from anthropomorphizing our

---

340 Idem, *Przedmiot, zadanie, metoda oraz podział fizyki* (Subject, task, method and the divisions of physics), op. cit., p. 162.

perception[341]. The task of physics is to discover through research the reality that is hidden beneath the illusion of our senses.

However, if perception of reality is hampered by the imperfection of our senses, and anthropomorphisation projects onto the laws and hypotheses created, can we believe in it? According to Smoluchowsk, this is a personal matter having nothing to do with physics. As a rule, people do not like to have doubts, or make the systematic mental effort that critical thinking requires; they prefer to rely on a safe and undisputed belief in reality. Everyday existence inclines us to create certain frameworks in which our mind can function efficiently.

> It would be ridiculous to demand that an astronomer keep repeating in his mind: I cannot know anything about the true universe, meanwhile I stick to Copernicus's system since it is the simplest; for the chemist to constantly say to himself: I do not know whether atoms exist; all I know is that everything happens as if matter is really made up of atoms. Over time, a certain faith is involuntarily created in these theories, which so often prove to be infallible signposts. If we really need to believe something, it is obviously the system that is the latest fruit of science. Let's just remember one thing: let's beware of the stubborn conservatism usually associated with such faith; if a theory turns out to be flawed in the face of scientific progress, let us not hesitate to replace it with another and let us not becry the bankruptcy of science; science does not demand that we believe fully in the reality of its view of the world[342].

This would seem to be a declaration of quite a controversial message. Smoluchowski – a scientist, a physicist – argues that science does not represent certain knowledge, based on unwaveringly stable foundations, as many other scientists have argued, but rather the opposite – it demands trust, and even faith, in the currently accepted laws. However, as is the way of faith, scepticism is essential in relation to the subject of faith, enabling the creation of new hypotheses and laws. Smoluchowski pointed out that we should not see theories of physics as the permanent contents of science but rather as an instrument of research.

> Nobody is offended by the boldness of new theories. We accept even seemingly strange ideas with enthusiasm, like a revelation if it proves useful as a signpost to new research. This does not at all mean that uncritical fantasists are currently winning. Let he who is not schooled in strict mathematical thinking, who is not accustomed to precision in experimental work or in logical reasoning, who does not possess a thorough awareness

---

341 Idem, *Poradnik dla samouków* (The Self-Study Handbook) op. cit., vol. 1, p. 17.
342 Idem, *Przedmiot, zadanie, metoda oraz podział fizyki* (Subject, task, method and the divisions of physics), op. cit., p. 166.

of the whole field of physics, stay far away from creative scientific work; physics remains, as it was, the prototype of strict natural science[343].

Sanctified traditional dogma needs to be demolished along with hitherto unshakable principles, if they even seem to us to be inappropriate. The principle of economy of thought represented a key methodological feature in Smoluchowski's science[344]. The Polish physicist's work contains numerous statements related to the theory of Ernst Mach, according to whom the description of facts should be approached in the simplest possible way in order to make them understandable with the least mental effort. The history of science is a constant striving towards this ideal, it is a range of images depicting reality in an ever more economical way.

Some philosophers considered the claim of the unity or simplicity of nature to be an obvious truth while others treated the formulation of the simplest laws of nature as obvious dogma. Smoluchowski, not believing in the objective reality of the laws proclaimed by physics, considered them hollow and baseless. He wrote that in the history of science, various laws, adopted initially with faith in their simplicity, were replaced by more complicated formulae as observations improved (for example, the Boyle- Charles law or Newton's law of cooling)[345].

To Smoluchowski, the assumption of applying the principle of economy of thought to the creation of science was important, but did not represent a basic element shaping its image. He was inspired by Mach's philosophical concepts, claiming for instance that natural phenomena can be divided into their components in order for their laws to take the simplest form as only such a form meets the needs of science. Mach complemented this idea with the statement that this simplification enables a problem to be explained and understood, but at what level of simplification we stop – this is a matter of economy of thought on the one hand and of taste on the other[346].

Smoluchowski developed his own concept, the so- called utility criterion. A scientific system is useful to the extent that it helps our mind to grasp a whole

---

343   Idem, *Kierunki i zagadnienia fizyki dzisiejszej*, (Themes and issues in today's physics) in: *Pisma Mariana Smoluchowskiego* (The Writings of Marian Smoluchowski), vol. 3, op. cit., p. 222.
344   Idem, *Poradnik dla samouków* (The Self-Study Handbook), op. cit., vol. 1, pp. 45– 46.
345   Idem, *Uwagi o roli przypadku we fizyce* (Observations on the role of Chance in physics), op. cit., p. 45.
346   I. Szumilewicz, *Koncepcja przyczynowości u Macha w świecie współczesnego determinizmu* (Mach's concept of causality in the world of modern determinism), „Studia Filozoficzne" ("Philosophical Studies"), 1959, no 4, p. 129.

and satisfies the striving for economy of thought. However, the Polish physicist had one reservation to this position – namely, he believed that this is a perception more in line with nominalism, according to which laws of nature are a subjective creation of our mind rather than something objective, existing in nature[347].

John D. Norton, an Australian philosopher at the University of Pittsburgh, asserts that our decisions concerning what is simple or simpler depend essentially on facts or laws that in our opinion have precedence over others and it is these facts that dictate which theoretical structures we can use. Invoking simplicity is actually an attempt to avoid introducing theoretical structures ill- adapted to physical reality governed by these facts and laws. This can be quite clearly seen in the popular example of adjusting a curve, when data presented as a finite number of points on a piece of paper are related as a linear or square equation. We routinely adjust the marked points to an equation we know because in that way we acquire a simpler solution. Our choice is satisfactorily and silently accepted as it is necessary for us to gain the right result. That is so natural that the emerging theories are created based on the knowledge currently embedded in the existing culture of science. The tendency to create theoretical structures guided by the attainment of maximum simplicity and related to the use of rules known to us is a manifestation of an attempt to merely avoid equations ill-adapted to physical reality but also to our understanding of that reality. Laws must function in a way that is comprehensible to us but there exists a tendency to build them based on formulae according to which, in our general imagination, reality functions. Nature is "mathematical", hence we seek relations of a linear, square, exponential or sinusoidal type which hold true, but we do not know whether they hold true because that is the way nature is or because we have imposed a matrix upon it that meets our demands and expectations. It is therefore possible that the principle of economy of thought inclines our comprehension towards functioning in the scientific space created by us ourselves and for this reason is susceptible to the imposition of new hypotheses fitting the matrices functioning in science.

Slightly different, quite surprising conclusions emerge from Sabine Hossenfelder's book *Lost in Math: How Beauty Leads Physics Astray*. These concern the verification of theories created in modern physics. A hundred and twenty years after a crisis in physics, due to which many scientists assumed the end of this field of science, we are again faced with a fairly complicated situation. Hossenfelder asserts that in certain branches of physics, no important new data

---

347   M. Smoluchowski, *Poradnik dla samouków*, (The Self-Study Handbook), op. cit., vol. 1, p. 46.

has appeared for decades. The Higgs boson, independently proposed by several researchers in the early 1960s, was the last fundamental particle discovered (its detection occurred in 2012). Since 1973, no new accurate prediction has been made that would go beyond the standard model[348]. The Italian physicist Guido Altarelli put it more bluntly at a 2011 symposium in Vancouver: "It is not time to be desperate yet... but maybe it is time for depression already"[349].

Faced with a lack of indicators from experiments, physicists are becoming disoriented and are applying aesthetic criteria to the assessment of theories, such as the criterion of beauty[350].

When new data are insufficient, theoretical physicists rely on their sense of simplicity and beauty in the evaluation of theories. Beauty is not a scientific criterion but can have its basis in experience[351]. Without deprecating the role of the beauty criterion in accepting new theories, we must be aware of how its adoption affects the development of science. Many physicists have believed that beauty is closely related to truth. Werner Heisenberg argued: "If nature leads us to mathematical forms of greater simplicity and beauty, (…) we cannot help thinking that they are 'true,' that they reveal a genuine feature of nature"[352], while Paul Dirac wrote: "A physical law must possess mathematical beauty"[353]. Hence theoretical physicists also use simplicity, naturalness and elegance as criteria to evaluate theories[354]. Quantum mechanics works fine though many physicists complain of its ugliness and lack of intuitiveness[355]. Such a perception of this field has resulted in its lack of acceptance by many theoretical physicists who believe that being guided by beauty can be justified by acquired experience.

According to Smoluchowski, we have an overarching universal principle enabling the definition not of what is simpler or beautiful, but what is most useful; a useful thing or theory contains simplicity and beauty because while defining what is simpler assumes a degree of postulation, usefulness can be verified in everyday life.

---

348 S. Hossenfelder, *Lost in Math: How Beauty Leads Physics Astray*, 2019, pp. 85–86
349 X. Portell Bueso, *SUSY Searchesat the Tevatron and the LHC*, paper given at the 'Physics in Collision ' international symposium, Vancouver, Canada, August–September 2011, slide 41.
350 S. Hossenfelder, *Lost in Math*, op. cit., p. 32.
351 Ibidem, p. 100.
352 Ibidem, p. 96.
353 Ibidem, p. 39.
354 Ibidem, p. 172.
355 Ibidem, p. 200.

The way we perceive reality, determined by biology and culture, affects the process of creating scientific theories through giving precedence to ideas of simplicity or beauty, although we cannot define these concepts. Meanwhile, to paraphrase Poincaré, not only does nature know nothing of our ideas of simplicity and beauty but – very probably – may "understand" that simplicity entirely differently. This has been proven in pre- Copernican and Copernican astronomy in which it was assumed that planets' orbits must be in a perfect circle because a planet must move in a perfect, beautiful arc as described by a circle. However, planets are objects that formed in the ancient history of the solar system whose orbits have various shapes while there exists no particular reason why their orbits should be circular[356].

Emil du Bois-Reymond (1818–1896) used the maxim *Ignoramus et ignorabimus* ("We do not know and will not know"), which prompted Smoluchowski to pose the following questions:

> Does this mean that man's capacity to know nature has limited and unbreakable boundaries? Do phenomena exist that are always inaccessible to us and that do not influence our experience, which do not concern us because we will never notice their influence? Are we totally indifferent to rays that we will never be able to perceive and whose influence will never be manifest in the phenomena available to us? Do we really losing something more than a beautiful fantasy? Are events that affect our experience partly knowable to us, albeit indirectly?[357]

If we answer some of these questions in the affirmative, more emerge. On what basis does science function? By what principle do we verify hypotheses and theories functioning within the realm of science? If our knowledge is relativistic in relation to the truths concerning reality, what criterion enables acceptance of the current binding science?

---

356 *Newton's Dream*, ed. M.S. Stayer, Montreal 1988, pp. 37–38.
357 M. Smoluchowski, Uwagi o roli przypadku we fizyce (Notes on the role of Chance in physics), op. cit., p. 16.

Chapter XV Utility

In seeking answers to the above questions, Smoluchowski arrived at the conclusion that it is not possible to distinguish between true and untrue theories, more or less probable, but rather between more and less useful[358]. This bold thesis, enabling the verification of accepted theories, is supported by the principle of economy of thought, which forces the constant verification of rules of science through the everyday experience of their application. The category of utility, often applied entirely unconsciously, can be a basic instrument in the work of a scientist in recognising which among the competing theories is more congruent with reality. At the same time, it uses the principle of economy as a tool facilitating the verification of proposed hypotheses. The more categorically the principle of economy of thought s applied, the easier it is to establish the usefulness of a theory. The utility of theories and hypotheses enables thorough and clear knowledge of the world of phenomena available to us and their combination into a whole comprehensible to our mind[359].

In difficult conditions, cultural burdens make their presence felt more than we might expect. Smoluchowski's criterion of utility explains this mechanism. Accepted scientific theories are verified through their current usefulness, in which they prove true. The mathematical analogies used, giving priority to already accepted solutions, dictating the theoretical structures used to create a theory, are valid within the boundaries verifiable by the principle of utility. The foundation of knowledge we believe science to be is not as stable as we would expect. The Polish physicist ascribed to utility both methodological and epistemic significance – constituting the basic criteria for verifying theories.

Utility references the anthropomorphism of science, assuming that in the first step of creating elements of science, the most important argument is the effectiveness of their application. If the characteristic of anthropomorphism of knowledge has been a burden through the ages on effectively learning the truth of reality, then in utility theory anthropomorphism becomes an epistemic attribute enabling verification of the value of accepted knowledge. The principle of economy of thought has been an element contributing to scientific progress though it has also happened that it has limited that progress, but it has above all

---

358 M. Smoluchowski, *Uwagi o roli przypadku we fizyce* (Notes on the role of chance in physics), op. cit., p. 51.
359 Ibidem, p. 17.

made it more probable and, importantly, more useful. Both concepts – the principle of economy of thought together with utility theory – represent in tandem a driving force of scientific progress.

New Zealand philosopher Gregory W. Dawes believes that contemporary physicists generally accept the theory of quantum mechanics developed in the 1920s by Erwin Schrödinger, Werner Heisenberg and Max Born. This also includes the intuitively accepted but surprising uncertainty principle, which states that it is impossible to simultaneously determine both the position and velocity of a sub-atomic particle. A physicist could accept this theory because it is distinguished by a high degree of verifiability, but at the same time not really manage to believe in it. It was the best available explanation of the behaviour of sub-atomic particles but it is hard to believe that every scientist accepting its assumptions was fully convinced of its veracity. They could believe it to be the best available theory and therefore use it regardless of its truthfulness. If the acceptance of a theory has occurred with the aim of broadening knowledge, and that knowledge is consistent with reality, it may seem that a theory cannot be accepted that is considered false. When dealing with straightforward situations such thinking seems correct, but in the case of scientific theories the situation becomes more complex[360].

Introducing the category of utility to the context of Dawes' considerations changes the view of things because a theory's acceptance is connected with a perception of its usefulness, which most often leads to its acceptance as the best available theory, true, and one which should be worked with in striving to broaden knowledge. However, convictions concerning the veracity of theories do not have to be entirely correct as it is not they but the currently understood utility of a theory that has decisive importance. False theories are known which functioned as true for many years, such as the concept of ether[361], or even for centuries – like the theory of geocentricity. We create science by pragmatically adopting the best possible explanations of phenomena occurring in nature. Dawes notes that these theories, with all the baggage of characteristics defined by the methodology of science, become a useful tool in developing our knowledge

---

360 G.W. Dawes, *Belief Is Not the Issue: A Defence of Inference to the Best Explanation*, "Ratio. An International Journal of Analytic Philosophy" 2013, vol. 26, no. 1, p. 69.
361 The concept of ether returns in some hypotheses but has a different formula than that functioning in the 19th century, see L. Kostro, *Alberta Einsteina koncepcja nowego eteru: jej historia, sens fizyczny i uwarunkowania filozoficzne* (Albert Einstein's concept of the new ether: its history, physical sense and philosophical background), Gdańsk 1999.

of the world. The dependency is two-way: accepted theories must demonstrate their utility in order to be included in the realm of science, and at the same time by their usefulness they assist in the creation of new theories, which can eliminate previously accepted theories from science. This can be seen among other things as the cause of progress in science.

A question arises as to whether perception of the end of utility inspires the search for new theories or whether a new theory makes us aware of the decline in usefulness of the previous one. There is probably no unambiguous answer since both in the search for the truth about our reality and in the quest for more efficient theories, too many other factor are at play.

An assessment of the value of the hypotheses and theories created is important in a researcher's work. Smoluchowski treats hypotheses as assumptions that seem likely to us in advance, calling them thought experiments. They are taken on a trial basis, as a foundation intended to lead to experimentally verifiable conclusions. On the other hand, he understands theories as the entirety of well-founded hypotheses together with all conclusions related to the phenomenon[362]. In analysing the value of theories in science, he notes that a hypothesis or theory has been verified when the conclusions we draw from it match experience. When at least one conclusion is not confirmed we consider the theory untrue. If veracity were to mean conformity with truth, such a state of affairs would have to raise serious doubts. We never know if there will ever be evidence refuting a given hypothesis although it seemed mostly firmly grounded. Even if no evidence were found against it, the conclusion can still not be drawn that the hypothesis reflects reality – we do not know whether it may be possible to posit other hypotheses that lead to the same conclusions[363].

Smoluchowski does not write about the search for arguments confirming the correctness of a theory but about the search for a correctly derived conclusion negating a theory. An association of Smoluchowski's reflections with Popper's falsificationism is quite obvious here. A scientist does not have to claim that he believes in a theory but must consider whether there is reason to accept it. Such cases represent an argument for differentiating between faith and acceptance, while not undermining the idea that the acceptance of scientific theories is guided by a broadening of knowledge. The scientist may admit that a partial truth is the best that can be achieved at a given moment. By putting forward a

---

362 M. Smoluchowski, *Uwagi o roli przypadku we fizyce* (Observations on the role of Chance in physics), op. cit., pp. 47–48.
363 Ibidem, p. 49.

new theory and applying the principle of utility within the framework of prevailing knowledge, he is in a position to accept a new theory which is the best he can have at the moment in terms of its level of utility.

The criterion of truth, faith and acceptability of created theories, despite its departure from the radical falsification of a theory, in relation to what Smoluchowski wrote has essentially not changed, it is still the degree of these theories' usefulness. If the utility of a prevailing theory is satisfactory, then despite flaws and shortcomings it still functions in science. If we cannot objectively relate scientific theories to reality without reservation, what makes them believable? What is the criterion by which we could assess a theory in terms of its conformity with reality? We cannot objectively establish that conformity since we do not know the real laws of nature, we are not in a position to define them and – equally importantly – we do not know if we will ever be able to know them. However, in the longer term, theories are defended that are closer to a true description of reality.

The compatibility with reality of scientific assertions is without doubt one of the aims of science, though not the only one. In the case of many theories we can talk only about the degree of probability of that compatibility. According to Smoluchowski, it is not about learning the essence of things hidden beneath appearances – the task of the physicist is, as far as possible, to understand the world of phenomena presented to us. It is about the most thorough examination of these phenomena and combining them into a whole that is comprehensible to our mind[364]. The Polish scholar could be accused of underestimating science or even of a fairly specific relativisation of scientific knowledge itself, though that would not be a valid suspicion. Smoluchowski, as an ambitious scientist, was convinced of the unceasing development of science and its indisputable role in supporting almost every aspect of our life, although at the same time he also had a philosophical distance towards science treated as a reservoir of certain knowledge.

Charles Sanders Peirce and William James had earlier argued that utility represents a criterion of the veracity of judgments and concepts[365]. However that was not an identical idea to Smoluchowski's. According to James, an idea that comes true can be verified. The statement that it is useful, because it is true,

---

364 M. Smoluchowski, *Przedmiot, zadanie, metoda oraz podział fizyki* (Subject, task, method and the division of physics), op. cit., p. 165.
365 J. Herbut, *Pragmatyzm* (Pragmatism) in: *Powszechna encyklopedia filozofii* (The universal encyclopedia of philosophy) vol. VIII, Lublin 2007, p. 442.

means at the same time that it is true because it is useful – both expressions mean exactly the same thing. James argued that "[t]he practical value of true ideas is (…) primarily derived from the practical importance of their objects to us"[366]. This means that the truthfulness of an idea is determined by the practical relationship of truth and utility in relation to the objects to which the idea pertains[367]. However, Smoluchowski claimed that we do not differentiate between true and untrue theories but between more and less useful[368].

The American pragmatists argued that truthfulness is that property of a judgement that leads to effective actions. The truth of an idea is not a permanent property inherent in it – truth happens to the idea. It becomes true when events make it true[369]. Such reasoning had a deductive nature, in which the effectiveness of creating science was conditioned by the assumption of the veracity of previous judgements. We arrive at a conclusion, which is a new theory created on the basis of the assumed premises that are previously verified judgments. The reasoning of Smoluchowski's argument was inductive in nature; the utility criterion is open to every new hypothesis that can displace a previously functioning theory. A new theory is also based on the veracity of previous judgements but it may happen that it contradicts those judgements though this does not disqualify it as the most important thing is that its usefulness is greater than the usefulness of the previous theory.

The truthfulness of Newtonian theories, according to pragmatism, occurs in the context of events that make them true, which relates mostly to the macroworld on the scale of our planet. The truthfulness of Einstein's theories occurs in the context of events that take place on the scale of our planet but also on a scale far beyond it. The conclusion that results from these comparisons would incline one to abandon Newton's theories due to the much greater concordance of events making Einstein's theory more true. Why then do institutes, workshops or production companies in many areas of technology base their work on Newton's theories? This happens because the utility of Newton's theories is incomparably greater in these fields. This fact does not undermine the pragmatic definition of the truthfulness of a theory since Newton's theory still holds true in terms of the events that make it true. According to Hawking, Newton's theory is

---

366 See W. James, *Pragmatism*, op. cit. p. 163.
367 Ibidem.
368 M. Smoluchowski, *Uwagi o roli przypadku we fizyce* (Observations on the role of chance in physics), op. cit., p. 51.
369 W. James, *Pragmatism*, op. cit. p. 161.

a fair approximation of the theory of general relativity at the boundary of weak gravitational fields[370].

In his book *Stephen Hawking: A Memoir of Friendship and Physics*, Leonard Mlodinow writes that Hawking demonstrated that science is based chiefly on approximations. Galileo eloquently argued that the book of nature is written in equations, but he neglected to mention that they are equations we can't solve. Newton's theory of gravity famously explains the orbits of planets, but we can solve his equations only for an unrealistically simple solar system with just one planet. In the much-celebrated quantum theory of atoms, all of chemistry emerges from a single equation, but the only element whose behavior we can calculate exactly via that equation is hydrogen – the simplest of them all. If we want to describe the planetary orbits in our actual solar system or derive the chemistry of elements other than hydrogen we have to settle for an approximate picture, (…) Those approximations and guesses do not come with any mathematical guarantee of correctness, but working physicists develop a good feeling for what is legitimate and what isn't. In physics we accept mathematical manipulations that we think 'ought to work'. (…) In all but the simplest investigations in theoretical physics, we alter, assume, and approximate, and then we argue why our simplified model and the conclusions we've drawn from it are nevertheless valid. (…) The argument over that is part of the professional physicists' (sometimes heated) dialog, which is a lot messier than the stereotypical portrait of science. Yet the fact that our airplanes fly, our lasers shine, and our computers compute suggests that a lot of our muddling works out in the end[371].

Smoluchowski's criterion explains the application of theories proving themselves useful. The definition of truth as conformity with the criteria of truth only partially fits in with Smoluchowski's concept as the utility criterion does not require a definition of truth in order to function in the methodology of science; it is inspired by the statement *ignoramus et ignorabimus* – we do not know and will not know the truth. In *Logical-philosophical letters* of 1933, Alfred Tarski argues that "for infinite languages the problem of the definition of truth has a negative solution"[372]. He puts forward the thesis that it is not possible to build a correct definition of a true statement for the vernacular or for infinite languages

---

370 S. Hawking, *Wszechświat w skorupce orzecha* (The Universe in a Nutshell), trans. P. Amsterdamski, Poznań, 2018, p. 125
371 Leonard Mlodinow, *Stephen Hawking: A Memoir of Friendship and Physics,* Knopf Doubleday Publishing Group, 2021, pp. 111–112.
372 A. Tarski, *Pisma logiczno-filozoficzne. Prawda* (Logical-philosophical letters. Truth) vol. I, trans. J. Zygmunt, Warszawa, 1995, p. 11.

of formalised deductive sciences, it can only be defined for an infinite language in a deductive system. He believes that "in reference to colloquial language it is impossible (…) not only to define the notion of truth" but even to use this concept consistently and in accordance with the laws of logic[373].

The language of science, which is an infinite language, cannot define a true statement. In justifying the impossibility of constructing a definition of truth for infinite languages, Tarski references the thesis of Kurt Gödel (1906–1978), who showed that in arithmetic, being an infinite language, it is not possible to construct a strict definition of truth. Gödel's 1931 incompleteness theorem demonstrated that any coherent mathematical system powerful enough to perform what we know as elementary arithmetic must be subject to the constraint that its own coherence can never be proved[374].

Two conclusions can be drawn from the above deliberations – firstly, objective truth about reality eludes science, philosophy and mathematics, and secondly, there exists an abyss between that which can be proven and that which is true. Both theories, Tarski's and Gödel's, reveal a fundamental truth concerning the classically understood truth of statements, representing empirical proof that it is impossible to outline an unambiguous criterion defining the objective truthfulness of a hypothesis or theory. In the context of Smoluchowski's usage principle, this is an argument of key importance as it justifies the thesis that in verifying a scientific theory, the utility criterion is more effective than the truth criterion.

According to Józef Życiński, a breakdown occurred in the hope held for twenty-five centuries of achieving certain knowledge of reality and its truths[375], which is well illustrated by the statement that, "*Epistēmē* shrinks, giving way to the omnipotent *doxa*"[376].

The aim of science is knowledge, not truth. On that principle, science strives in the first instance not for truthful beliefs as such but for beliefs accepted as true, in relation to which we have the appropriate proofs to believe them. That we are engaged in broadening our knowledge, and that we then accept the best available explanation of all mysteries, only proves that we are trying to explain them.

Though Peirce, as well as Gregory W. Dawes and Gilbert Harman, use the same logical thought as Smoluchowski, they do not get to the heart of the matter,

---

373   Ibidem, p. 15.
374   R. Smullyan, *Na zawsze nierozstrzygnięte. Zagadkowy przewodnik po twierdzeniach Gödla* (Forever Undecided: Puzzle Guide to Godel), trans. J. Pogonowski, Warsaw 2007, p. 7.
375   J. Życiński, *Język i metoda* (Language and method), Kraków 1983, p. 102.
376   Ibidem, p. 109.

to the criterion that specifies which hypothesis, being more probable, provides a better explanation. Peirce did not take the next step and did not refer to his own "defective" theory of utility, stating that the truthfulness of the judgements and concepts of a given theory decide its usefulness. Similarly, Dawes does not define the phrase "satisfactory explanation of a puzzling fact" and does not define the principles on the basis of which we accept the best available explanation. Nor does not he give reasons why we should accept the best available explanations. Harman believes that certain data, despite the best explanation and despite their falsification, are among those things not fully verified, however science, in his view, demands theories that can be trusted without undue risk.

A question arises, however, as to how we should verify these best explanations. A common feature of these concepts is the lack of a category according to which the verification and acceptance of theories or hypotheses is conducted. A theory which is the most useful at a given level of science is that which enables the best possible progress of technology and at the same time stimulates the further development of science by inspiring the creation of new theories. Hypotheses are created based on facts accumulated through experiments and observations and the main driver of their creation is the pressure of possible changes and improvements to the utility of previously functioning theories through their alteration or exchange, falsification and approval as more useful. In the face of such unequivocal demands and at the same time through the impossibility of satisfying them, the utility criterion represents the pillar on which the credibility of science rests. Thomas Kuhn poses the questions:

> Why would this evolutionary process have occurred at all? What must nature, together with man, be like for science to be at all possible? Why should the academic community be capable of achieving a lasting consensus unachievable in other fields? Why can this agreement survive subsequent paradigm shifts? And why should paradigm shifts always lead to the emergence of tools more perfect than those known previously? Not only the scientific community, but the world of which that community is a part, must be characterized by special features; our deliberations have not brought us any closer to answering the question of what these properties should be. The question: how must the world be for man to be able to know it? is as old as science – and still remains unanswered[377].

It is not difficult to note that introducing the utility criterion to the discourse answers many of the questions posed by Kuhn. Smoluchowski's conception

---

377 T. Kuhn, *Przewrót kopernikański. Astronomia planetarna w dziejach myśli*, (The Copernican Revolution. Planetary Astronomy in the Development of Western Thought), trans. S. Amsterdamski, Warszawa 1966, s.294.

answers the scholar's doubts, including the essential question on finding a criterion initiating the evolutionary process of the development of science. Decisions concerning the use of particular theories are not only the domain of scientists; it happens that the whole of society takes part in this process through everyday decisions and choices as to what is the most useful. However, due to where new theories arise, it is the choices of the scientific community that primarily determine their usefulness.

Thomas Kuhn (1922–1996)

There exist theories that in the future may be falsified and do not function in the scientific discourse, and conversely – there are also those that will probably never be falsified and which scientists are using. In the history of science we have already seen the example of medieval alchemy, which in modern science has been transformed into the serious discipline of chemistry (this breakthrough was made by Robert Boyle in *The Sceptical Chymist*, published in 1661).

The process of creating science is a consequence of the make-up of the human mind, for which learning is the essence of its functioning. The imperatives of competition and rivalry imposed on that prompt ceaseless revision and improvement. It forces actions aimed at the best and fullest use of everything a person can have at their disposal. The achievement of lasting consensus by the academic community is mostly temporary as the use of a theory sooner or later leads to being tested. Everyday practice forces acceptance of those theories that seem the most useful. Change generally occurs gradually but can also take place abruptly with the toppling of prevailing paradigms, as in the case of the theories of Copernicus, Kepler, Newton, Darwin or Einstein.

Hilary Putnam (1926–2016)

According to Poincaré, paradigms are adopted conventions that are valid as long as they are useful, but when their usefulness comes to an end, they are rejected and replaced by new, more perfect, more useful ones, which, however, are also mere conventions. This rule does not only apply to scientists – the whole of human society functions in this way.

Hilary Putnam saw the success of science in the constant discovery of truth. He wrote: "[T]he typical realist argument against Idealism is that it makes the success of science a *miracle*. Berkley needed God just to account for the success of beliefs about tables and chairs (and trees in the Quad)"[378]. The "No Miracles Argument" as Putnam's reasoning is known, was intended as a defence of scientific realism, a scientific philosophy that does not perceive the success of science as a miracle. According to proponents of this school of thought, only explaining the successes of science through the notion of the truthfulness of theories enables us to not perceive them as something otherworldly and inexplicable. The term "No Miracles" should not be understood theologically but concerns a philosophical stance that interprets the success of science. It is not created by an inexplicable miracle, but the philosophical view posed by scientific realism is the only clear and comprehensible explanation of it[379].

---

378 H. Putnam, *What is "Realism"?* Proceedings of the Aristotelian Society, vol. 76, 1975, pp. 177–94.
379 Idem, *Philosophical Papers*, vol. 2, *Mind, Language and Reality*, Cambridge 1975.

Putnam's assertions that older scientific theories are superseded by new, better ones and the laws of new theories are drawn from the laws of earlier theories, are important assumptions both of the theory of scientific realism and of the "No Miracles Argument". However, this concept clearly lacks an explanation of the cause bringing these changes about. Putnam does not consider it though it is, after all, a key issue for understanding the emergence of new hypotheses and theories and, as a result, the "No Miracles Argument". Smoluchowski's utility is an essential complement to the concept of the success of science. The idea that emerged several decades earlier, confirming that the most useful theory is the closest to the truth, fulfils the American philosopher's theory.

Would the success of science be real were it not for the effect on scientific research of the driver of utility? It is like the market mechanism in economics; the mobilising force of the introduction of changes, at the same time verifying them. In deliberating on problems going beyond the boundaries of the philosophy of science contemporary to him, Smoluchowski became something of a precursor of scientific realism. The scope of impact of the notion of utility that functions in our culture can be understood very broadly. It translates into the application of the utility criterion to the creation of theories and hypotheses in science, at the same time forcing a more effective knowledge and understanding of reality as well as the use of this criterion in everyday life in order to fully exploit knowledge in the development of new technology. The mechanism of utility enables the creation of further new theories through the application of this criterion as a natural factor imposing and forcing the dynamics of change. Utility is a natural creator of the success of science. The permanent verification of scientific theories in terms of their current usefulness is a test to which all scientific theories are subject. Such verification means the ceaseless testing of the correctness and effectiveness of prevailing scientific theories in an empirical way and within the field of the functioning scientific discourse. The theory of utility explains the simple mechanism present both in science and in everyday, practical life, which is the driver or progress. As already mentioned, it clearly explains the mechanisms operating in selected aspects of the success of science.

In the past it has happened that theories were so far ahead of their time that not only were they not accepted, they were often opposed and most often went unnoticed. Nicholas of Cusa, a medieval philosopher, theologian and mathematician, argued many years before Nicholas Copernicus[380] that the universe is not geocentric but that Earth moves, like other heavenly bodies, around the Sun. He

---

380   He died nine years before Copernicus was born.

went further in his hypotheses than Copernicus as he said the universe is made up of an infinite number of planets circling an infinite number of stars. He also acknowledged that the universe has no perimeter, or a so-called centre, as the centre is everywhere. He claimed that all planets revolve on their own axis and that rotational movement caused their flattening[381]. A hundred and fifty years before Kepler, using thought experiment, he posited the hypothesis that planets orbit the Sun not in circles but in ellipses. He created a theory describing the inertia of material bodies, which was only later given a real scientific basis by Newton. The use of a theory, and hence its acceptance, is subject to societal need. Many years had to pass for the utility of Cusa's theory to be accepted within the framework of the specified understanding of the laws of nature. The fact that Nicholas of Cusa's theories were closer to reality, more in line with the truth than the previously functioning views, was of no importance in the face of those theories' lack of utility value at the time.

The "No Miracles Argument", or scientific realism discussion taking place around the success of science was limited mainly to stating facts with no consideration within its framework as to seeking a mechanism driving that progress or finding a criterion according to which verification could be made of further steps in the investigation of the truths of nature.

The utility criterion makes the functioning in science possible of an old theory that has been replaced by a new one as long as it is useful. An example of this phenomenon, mentioned earlier, is the functioning of Newton's classical physics. Its use in everyday life is possible because its degree of imperfection is acceptable in many fields of science and technology. Czesław Białobrzeski notes that deviations from Newton's law of gravity and the principles of classical mechanics are barely perceptible, for example in the solar system they become visible only in the perihelion motion of Mercury[382]. We have the opposite situation in the case of Ptolemy's theory, which has been ousted completely from daily life as useless.

Verification of a theory's utility is not always as simple and obvious. The period in which we come to a conclusion on the greater utility of a new theory over an old one varies not only for different theories but also for the same theories in different societies. The usefulness of Darwin's theory of natural selection has been questioned constantly by various societies since the publication in 1859 of *On the Origin of Species*. Whether Darwinism will be accepted depends not only on

---

381 The diameter between the poles is smaller than the diameter than at the equator.
382 C. Białobrzeski, *Podstawy poznawcze fizyki świata atomowego* (The cognitive foundations of the physics of the atomic world), Warszawa 1984, s. 24.

changes to paradigms rooted deep within societies but is also conditioned by the fact that the utility of creationism in a given society is sometimes greater than the utility of evolutionary theory.

It happens that the utility of a theory may be perceived relatively because it is unverifiable in everyday life or because its usefulness does not depend on the category of truth for every observer. Then, in line with the thesis posited more than a hundred years ago by Poincaré, a key role is played by the decision of the scientist or scientific community introducing the new theory to science. However, the scientific community, as evidenced by the example of the theory of global warming, may be divided, and then the verifier of utility of a given theory becomes time. The falsehood of an adopted theory verifies the usefulness of the next new theory. It should be more true than its predecessor. Such was the case with the theory of Ptolemy, which was refuted by Copernicus's theory positing circular orbits of planets around the Sun, changed by Kepler's theory describing orbits in the form of ellipses, which was in turn superseded by Newton's theory introducing gravitational dependence in planetary motion, finally corrected by Einstein's theory combining gravity and space. A theory positively verified by its utility naturally replaces the previous one.

For a clearer understanding of Smoluchowski's proposition, we can imagine an abstract equilateral triangle, which we can give the working name of the triangle of science. On its base we place a well-founded set of scientific views and theories the utility of which is rarely undermined. The higher up, towards the apex, the less certain the theories' utility, often temporary, impermanent by definition. The movement of theories that takes place within the triangle runs from top to bottom and bottom to top, the lower, the more significant and certain they are, building the foundation of the field of science. Mostly based on this, with the help of induction, new ideas and new theories are introduced, which from the top, from the apex of the triangle, are included in the field of science. The more their utility is positively verified, the more certain their position in the triangle is. Theories whose utility is poorly verified are located at the peripheries of the triangle's apex. Outside the triangle, in the vicinity of the apex, are theories whose utility has not been established, which does not mean they are entirely discarded. An example here might be atomic theory, which functioned outside of the "triangle of science" until the end of the 19th century and whose utility was not confirmed until the 20th century, only then being included within the triangle. The triangle of science built into the framework of popular culture is not the only option for understanding nature. Other triangles are possible, created based on different, alternatively understood utility. However, the triangle created within the framework of our culture dominates, in which a key verifier and driver of expansion has been and remains the verified utility of a given theory, which is supported by the principle of the economy of thought and falsifiability.

What is the relation of the triangle of science to the truth of reality? In simple terms, the same as the relation of a geometric triangle to the infinite surface of the plane on which it stands. This is the relation of the epistemic triangle of science, which happens to adjoin to the infinite ontological surface of the truths of nature.

In the article *A Material Theory of Induction*, John Norton claims that in inductive reasoning we assume so-called inductive risk, which consists of the danger of adopting a high factor of faith in a fixed outcome, which ultimately transpires to be false. In science we strive to reduce inductive risk to the greatest degree possible. Norton's diagnosis is correct concerning the model of deductive logic seeking relations to induction based on universal schemas, which for centuries imposed a way of understanding nature, thus blocking the possible development of a system of induction without universal schemas. However, in the perspective of time, this is *post factum* wisdom, not answering the question of to what extent deductive reasoning was an essential intellectual step towards the conception of induction. Historical examples of the more or less conscious use of induction may testify to the fact that the idea of applying it matured slowly.

Competition within systems of inductive reasoning and in the results achieved from this reasoning is constant and permanent, forced by utility. Less useful theories and hypotheses, arising as a result of the operation of adopted inductive systems, are constantly pushed out of the triangle of science by other, more useful ones. Hence the extraordinary success of science in learning about the world, achieved through competition between inductive inquiries. Nothing motivates thought like competition.

Today, nobody questions a methodology based on induction. It should be noted, however, that it need not represent the crowning methodology of scientific thought. If we thought so, it would mean that the history of science has not taught us much. The requirements for a theory's compliance with truth are not fixed and uniform; they are different in the case of facts we achieve 'locally', for example in relation to the actions of colloids or the operating principles of a diesel engine, and different again in the analysis of cosmogonic theories. Inductive reasoning cannot always be applied to processes inaccessible to us, the best examples of which are quantum physics, superstring theory or, in cosmology, M-theory and brane cosmology. Some hypotheses of theories created in these fields go beyond the framework of induction as understood to date. It is probable that the accusation of holding back the development of induction, which Norton makes against deductive logic, will one day be by made against inductive thought, which holds back another way of creating theories, more effective in these fields of science and already appearing in some deliberations. This is, simply put, positing theories based on created mathematical constructs. Although in this case, the creation of theories is not accompanied by evidence or repeatability, nonetheless their credibility and usefulness are determined by the utility expected and made use of. Theories are accepted that cannot be falsified but which are sufficiently useful that they become a discursive part of science. A visible *novum* is the fact that the creation of hypotheses in these areas is not accompanied by classical methods. The methodology applied goes beyond the deductive and inductive construction of hypotheses. Analysing the problem from the perspective of utility theory, the assertion can be risked that use of the inductive method is insufficient in the case of this type of knowledge and we are witnesses to the creation of a more useful new methodology of creating theories.

In the article *Hume, Norton, and Induction without Rules*, Thomas Kelly writes that the reliability of the inductive rule, in contrast to the credibility of the deductive rule, seems to vary depending on the nature of the world in which it is used. Even if a given inductive rule is trustworthy in our world or in the world which in reality, in our opinion, is our world, there are probably other possible worlds in which the same rule is not reliable. Moreover, it is a natural supposition that that there exist not only possible worlds in which the rule is unreliable but also those in which proof can be obtained that a given rule is untrustworthy. It seems that this way of thinking runs counter to the idea that there may exist some truly universal inductive principles, or rules that it would be reasonable to apply without regard to the results of 'material facts'[383].

Norton claims that the basic principle is a theory's ability to bring with it evidence of truth in terms of its hypothetical confirmation. According to him, evidence that a triangle exists in which the angles add up to 180° is not only proof of Euclidean geometry but brings with it the whole baggage of the science that contains that geometry and the whole of modern physics, chemistry, and astrophysics. We can therefore note that everything – apart from an entirely banal version of hypothetical- deductive confirmation – demands additional limits based on facts[384].

---

[383] T. Kelly, *Hume, Norton, and Induction without Rules*, "Philosophy of Science" 2010, vol. 77, no. 5, pp. 754–764.

[384] J.D. Norton, *A Material Theory of Induction*, "Philosophy of Science" 2003, vol. 70, no.4, p. 653.

The example of the triangle given by Norton is a specific case of *differentia specifica* geometric theories, in which utility plays an important role. Let's look at Friedman's model. A triangle with the sum of angles of 180° in the flat model of the universe that occurs in Euclidean geometry can be replaced by a triangle with a sum of angles less than 180° appearing in the hyperbolic model. A triangle with the sum of angles greater than 180° occurs in the spherical model of the universe. Today we do not know which of these geometries is true in the sense that we do not know which is crucial to our reality. Hawking writes in the book *The Universe in a Nutshell* that it may be Riemann's saddle-shaped geometry[385].

However, in everyday life we use Euclidean geometry and even if Hawking was right, it would not change anything for millions of people. We therefore ask what the quantifier is of the geometry we use. The answer is unambiguous – this determinant is its utility. The aims towards which a geometry is used determine which geometry we apply.

Utility is a universal category referring to all aspects of man's functioning in the natural world. It determines the value of a theory or concept. As a result of its action, we come closer to objective knowledge of the truth of reality. Utility tests every theory, hypothesis or premise. It works on a similar principle to the "invisible hand of the market" introduced to economics in the 18th century by Adam Smith[386] with the difference that rather than in the free market of economics it functions in the non- ideologised space of knowledge. To paraphrase Smith's idea, it can be said that the driving force is that everyone thinks about achieving success in the creation of something new, unique, whether in science, business, production or even everyday life, that will be accepted, and maybe even admired, by society[387]. We act in this way to strive for the creation of something that would be more useful than that which functions now. The usefulness of every idea, theory or hypothesis, every thought, is determined by the criterion of utility and society constantly inspires and activates those actions. The principle of the free competition of theories holds true here. Every theory functioning in science, in everyday life, even that which has not been sufficiently falsified, is recognised and used until it transpires that there exists a theory whose utility is greater, which provides greater possibilities and – very importantly – enables

---

385   S. Hawking, *Wszechświat w skorupce orzecha* (The Universe in a Nutshell), op. cit., p. 19.
386   A. Smith, *Badania nad naturą i przyczynami bogactwa narodów* (An Inquiry into the Nature and Causes of the Wealth of Nations), trans. A. Prejbisz, vol. II, Warsaw 1954, p. 46.
387   Ibidem, p. 42.

more effective further development. The endless decisions taken concerning the choice of verifying utilised theories result in some theories being constantly used in science and technology, others being abandoned, and others still being consigned to history. The utility of theories is inherently related to their use, which determines the usefulness of a theory. This idea of Smoluchowski's, banal in content, brilliant in its simplicity, has driven our epistemology for over a century. On the basis of the above considerations, in the context of many dissertations of thinkers involved in the theory, philosophy or methodology of science, it can be seen that Smoluchowski's utility criterion introduces its own kind of order to the understanding of that timeless stimulus to which we owe the ceaseless pace of the development of science.

The principle of utility has been applied in quite a radical form both to Smoluchowski himself and to his scientific achievements and philosophical reflections. When the political need arose, with no resistance and contrary to the facts, the scientific worldview of the physicist was reformulated ascribing him materialistic views. This was not in line with the truth, but that was of no importance at the time.

Chapter XVI Materialism

In studies published particularly after 1952, suggestions appeared, supported by various arguments, of Smoluchowski's materialistic worldview. Are they really supported by source materials? If so, what justification do we find in the scientist's work to categorise his philosophy in this way? If not, what is the origin of such a belief about him?

Materialism argued that the only intrinsically existing entity is matter. This view – known in an older version as mechanistic materialism – assumed that everything that happens in nature can be explained and reduced to causes. In epistemological terms, it was assumed that only material reality, existing objectively and independently of the subject perceiving it, is an object of human cognition and consciousness and mental experience are secondary to matter. In this point of view, thoughts, feelings, emotions, the whole of human culture and social relations were to be the result of material and economic conditions.

The dialectical materialism key to the current considerations was initiated by two 19th century German thinkers – Karl Marx (1818–1883) and Friedrich Engels (1820– 1895). Applying Hegel's dialectic principles, they assumed that the whole of reality was material and that the characteristic differentiating dialectic materialism from mechanistic or ontological materialism was recognition that one of the traits of matter is its dialectic nature. According to this position, matter is not only subject to the laws of the natural sciences, but also to the general laws of dialectics[388].

It can be debated whether Smoluchowski can be attributed certain formulations related to elements of materialistic philosophy, however the assertion that he was a materialistic philosopher is an over-interpretation and attempts to ascribe dialectic materialism views to him are just absurd. The author's materialistic views cannot be determined by a suggestion taken out of context, since on the basis of the same statement, Smoluchowski can often be attributed views typical of philosophical pragmatism, positivism, conventionalism or empiriocriticism.

In the 1950s, Marxism was an ideology imposing political, economic, social and philosophical views based on certain works by Marx, Engels and Lenin, although the most troubling was the Stalinist version. In the period of that ideology being in force in the Soviet Union and the so-called socialist countries, including Poland, there was no way of separating philosophy from ideology in the publications that appeared. Officially, Stalinism ended in 1956 with the 20th congress of the Communist Party of the Soviet Union, although both earlier and

---

[388] J. Turek, *Materializm* (Materialism), w: *Powszechna encyklopedia filozofii* (The universal encyclopedia of philosophy), vol. VI, Lublin 2005, pp. 913–917.

later, communist ideologues formulated the principles of new theories, creating a scientific worldview based on dialectical materialism.

On May 29, 1952, the daily *Trybuna Ludu* (The People's Tribune), published a picture of Marian Smoluchowski on the front page with a caption recalling the 80th anniversary of the scientist's birth. The text below the picture directed the reader to the paper's back page, which bore an article by Władysław Krajewski (1919–2006) titled '*The great physicist and materialist philosopher (On the 80th anniversary of Marian Smoluchowski's birth)*'. The text, covering about a third of a column, described the basic events from Smoluchowski's biography, introducing him and his scientific achievements. A reader in the second decade of the 21st century would get the impression that the eightieth anniversary of the birth of a great scientist (falling on May 28, 1952) had become primarily an opportunity to recall the Polish physicist's work, but that is an illusory belief. For a party daily of the time, the article was surprisingly extensive and was reprinted multiple times in various magazines, which could not have been a coincidence. The publication of Krajewski's article was an important event – it started a series of publications intended to present the scholar's views as being of materialistic philosophy.

In the early 1950, in the daily party papers, the so-called organs of the Polish United Workers' Party, there are no politically neutral topics. Every newspaper of the time, but especially a party paper, was above all a propaganda vehicle rather than an informational one, and articles had ideological references or were tools of clear indoctrination, hence it is impossible to believe in the author's pure intentions. This conviction is reinforced by the history of this intentionally written article. The same article about Smoluchowski was reprinted in the following days in six other papers, which cannot be an act of chance. Knowing the reality of the time, the assertion can be hazarded that it was a thought-through political decision of the party authorities. What were the intentions of the decision-makers and consequently of the article's author?

In 1948, the first edition of the magazine '*Zagadnienia Filozofii*' ('Issues of Philosophy'), published under the patronage of the presidium of the Academy of Sciences of the Soviet Union, the article *A discussion on the nature of physical knowledge* appeared in which Ł. Storczak, as part of a debate on the nature of physical knowledge and referring to a previous statement by Andrey Markov, wrote: "It transpired that Smoluchowski's extensive investigations were necessary to finally establish that statistical regularity is an entirely new type of regularity, strictly determined by physical conditions (…). Unfortunately we still propagate Smoluchowski's wonderful materialistic ideas very little"[389].

---

389 Ł.I. Storczak, *Diskussija o prirodie fiziczeskogo znania* (Discussion on the origin of physical knowledge), "Woprosy Fiłosofii" ("Problems of Philosophy"), 1948, no 1, p. 206.

Ł. Storczak's article in 'Issues of Philosophy' magazine

In 1951, the third, May-June, edition of a bimonthly for teachers, '*Fizyka i Chemia*' ('Physics and Chemistry') carried a short article by Armin Teske on Smoluchowski.

In the final part of this text, Teske included information about Storczak's article and his praise for Smoluchowski. It cannot be ruled out that this is how Krajewski found out about him and used it as inspiration to propagate the thesis of the Polish physicist's alleged materialistic worldview. In the quarterly '*Myśl Filozoficzna*' ('Philosophical Thought') Krajewski wrote that "Smoluchowski, as one of the creators of statistical physics, and at the same time a confirmed materialist, took up the task of demonstrating the objective nature of the concepts of chance and probability, in order to provide solid methodological foundations for their application in science"[390]. This claim is not supported by any argument and a few pages earlier Krajewski even contradicts his later thesis, writing that Smoluchowski "does not call himself a materialist"[391]. In another place the assertion appears that the researcher did not realise himself that he was a materialist[392]. It is hard to imagine how the outstanding scientist could have had a "decidedly materialistic" worldview without being aware of it.

The anniversary in round numbers of the great physicist's birth was a pretext to use Smoluchowski's authority to propagate socialist ideology. Four years later, in 1956, in the book *Światopogląd Mariana Smoluchowskiego* (Marian Smoluchowski's Worldview), Krajewski developed his thesis, trying to annex Smoluchowski's scientific and philosophical legacy to the needs of party ideology. The communist system most clearly needed authorities it could invoke. Krajewski not only presented Smoluchowski as a philosopher with materialistic views but also ascribed to him inclinations towards dialectical materialism, suggesting that the Polish physicist's beliefs were consistent with Marxist philosophy. Such a reading of Smoluchowski's philosophy was unfounded and gave the impression of being prepared for a predetermined thesis. Władysław Krajewski's book fits in with the overall picture of publications of the time. It is no accident that it bears the title *Marian Smoluchowski's Worldview*, though it was more the worldview of citizens that was to be reshaped. Many books published at the time were intended to make the creation of a new society credible. There were

---

390 W. Krajewski, *Marian Smoluchowski jako filozof i materialista* (Marian Smoluchowski as a philosopher and materialist), "Myśl Filozoficzna" ("Philosophical Thought), 1952, no. 4 (6), pp. 243–244.
391 Ibidem, p. 241.
392 Ibidem, p. 126.

attempts to show that more enlightened scientists had already – consciously or unconsciously – proposed a new, fundamentally materialistic approach to science. They were not concerned by whether the contents found in the works of their predecessors were in line with their intentions, progressing in the spirit of Hegel's statement, "If the facts contradict the theory, so much worse for the facts". Because interpretations often related to the works of authors no longer living, various manipulations were permitted.

In his work, Krajewski used the method of dialectical materialism in the field of the philosophy of physics and other natural sciences[393]. The dialectic method in the communist system was a propaganda tool enabling a change in the meaning of words and concepts, thereby facilitating manipulation and imposing a defined interpretation on the author's intention. In *The Main Currents of Marxism*, Leszek Kołakowski describes this approach:

> Diamat [dialectical materialism – Ed.] is made up of theorems of various kinds. Some of them are banal common sense and contain nothing specifically Marxist. Others are philosophical confessions of faith, unprovable and unresolvable with the aid of scientific means. Still others are simply nonsense. To the fourth category belong claims that can be interpreted in various ways and, depending on the interpretation, belong to the first, second or third earlier mentioned[394].

Krajewski's reflections at the end of the book are an example of a dialectical treatment of Smoluchowski's views:

> In summary, we can say that in the field of the methodology of physics, Smoluchowski took a decidedly materialistic and often exuberantly dialectical position, while in his general philosophical deliberations – the materialistic trend stemming from his whole attitude was covered in numerous positivistic layers resulting from succumbing to fashionable bourgeois philosophy[395].

Kołakowski cites the Hungarian philosopher György Lukács who, by using the word "dialectical", invalidates all empirical circumstances, proving that they can look different but dialectically look quite the opposite. In his book about Lenin, Lukács accuses the reformists of having "a non-dialectical notion of majority". It transpires that a majority can be understood not only in the usual sense but

---

393 See W. Słomski, *Władysław Krajewski*, in: *Polska filozofia powojenna* (Post-war Polish philosophy), ed. W. Mackiewicz, Warsaw 2001, pp. 537–554.
394 L. Kołakowski, *Główne nurty marksizmu* (Main Currents of Marxism) vol. 3, Warsaw 2009, p. 157.
395 W. Krajewski, *Światopogląd Mariana Smoluchowski* (Marian Smoluchowski's worldview), Warsaw 1956, pp. 165–166.

also in the dialectical sense, as the opposite of a normal majority. Communists in Polish society did not have an ordinary majority but did have a majority in the deeper "dialectical" sense[396].

Krajewski assumed that Smoluchowski was an atheist because he never made any statements on the subject of God and so surely did not believe in Him. This is a good example of the dialectical method of drawing conclusions. Smoluchowski's views cannot be reduced to the level at which the scientist's worldview is based on issues from the realm of faith, which for Smoluchowski was an entirely different sphere of knowledge to science. The Polish scholar would not only have disagreed with being described as a materialist, but would most probably have disagreed with such a definition of materialism as Marxist ideologues proposed.

Smoluchowski's philosophy stands in opposition to Krajewski's thesis at least on the issue of the possibility of clear and exhaustive knowledge of the world of available phenomena. The scientist was close to pragmatism on this issue, believing that the scientific view of the world is not in line with the true state of things. He therefore asked the question: are we allowed to believe in this view? Answering this, he argued that this is a personal matter having nothing to do with physics; that it is a question of faith. "People generally do not like to cease doubting, they do not like the constant mental effort that critical thinking requires; they prefer to rely on certain belief in reality"[397]. Everyday existence inclines us to create certain cognitive frameworks in which our mind must function. Is this a though that a materialist could formulate?

In his book, Krajewski includes many themes referring to social or economic relationships, which were probably intended to evoke specific associations, to persuade the reader of the author's arguments, proffering ideas, thoughts and attitudes that the reader should in time consider his or her own. This work appeared during the most restrictive period of Marxism when indoctrination affected almost every aspect of life and science, especially philosophy. This was the end of the Stalinist era – in fact just after Stalin's death but before the so-called thaw, which would not reach Poland until after October 1956. Krajewski's book can therefore not be read today as an objective work of science as it is laden with the Marxist ideology of the Stalinist period, marked by a lack of impartiality and integrity on the part of the author. Inference based on quotations from

---

396 L. Kołakowski, *Główne nurty marksizmu* (Main Currents of Marxism), op. cit., p. 308.

397 M. Smoluchowski, *Przedmiot, zadanie, metoda oraz podział fizyki* (Subject, task, method and the division of physics), op. cit. pp. 165–166.

Smoluchowski's work, intended to prove his materialism, is neither objective nor trustworthy. The quotes cited and analysed were over-interpreted and sometimes even manipulated.

Krajewski's observations at the semantic level bear the stamp of the times in which the text was written. The understanding discussed earlier of the issues of chance and probability is summarised by the author in the following way:

> A consistent scientific solution to issues of chance and probability, as well as other philosophical matters, is possible only from the perspective of dialectical materialism. The foundations of such a solution are to be found in the works of classical Marxism. Using and critically processing the valuable contents of Hegel's dialectics, Marx and Engels raised the whole of philosophical materialism – including determinism – to a new, higher developmental stage[398].

Smoluchowski would not have participated in such an unscientific discourse, full of ideological references and dogmatic arguments. Continuing his train of thought, Krajewski confirms his discourse's lack of substantive value:

> As for the interpretation of the concept of probability and statistical regularity, the classics of Marxism did not directly address these issues. However, we find in these works certain "points of support" for the solution and these problems. Apart from the concept of chance already discussed, attention should be drawn to Marx's account in *Das Kapital* of the economic laws operating in the capitalist market, in particular the law of value[399].

The conviction that a scientific solution to the issue of chance and probability is possible only from the point of view of dialectical materialism is groundless, especially since Krajewski goes on shortly to state that the classics of Marxism did not directly address the interpretation of those issues and that in Marx's *Das Kapital* some undefined closer references to the analysed issues might be sought. Evidence developed by Smoluchowski of relationships between chance and probability and dialectical materialism is therefore very unconvincing.

Krajewski repeatedly includes comments and comparisons in the text which in the field of physics and natural philosophy are very unpersuasive. For example, he writes: "We know that Marxism understands from widespread social practice primarily production, industrial activity. Such an approach to the practice is, as the classics of Marxism show, the foundation and goal of theoretical knowledge and the highest criterion of truth"[400]. It is doubtful that Smoluchowski would

---

398 W. Krajewski, *Światopogląd Mariana Smoluchowski* (Marian Smoluchowski's worldview), op. cit., p. 108.
399 Ibidem, p. 111.
400 Ibidem, pp. 231–232.

have agreed with the thesis that the practice of production is the basis of the development of physics. "A Marxist analysis of the laws of economics can be of some help in the philosophical analysis of the nature of statistical laws studied by physics"[401], Krajewski claims. It is worth emphasising that the classics of Marxism were not authorities on physics and that economic and social issues were not a path leading to the truths of this field.

A characteristic opinion of Krajewski's concerns Bohr and Heisenberg: "So, for example, today we negatively evaluate the views of the outstanding physicists ⁻ Bohr and Heisenberg primarily because they occupy the opposite position in the methodology of physics, which inhibits the development of science (indeterminism, subjectivism, the absolutisation of quantum mechanics)"[402]. The cited example illustrates how much ideology and political attitude influenced Krajewski's scientific beliefs.

In studies published in the 1950s and 1960s, the authors often invoked the classics of Marxism in a way completely divorced from the subject of the publication. This was necessary and in some cases even essential. Every article and book was subject to a censor's assessment and citing the appropriate names could decide whether a work was published. Most frequently, the authors' intentions were quite apparent; it was mostly about meeting the censor's requirements. Unfortunately, Krajewski goes further in his efforts – his book is a serious and deliberate attempt to distort Smoluchowski's philosophical thoughts for the needs of the political system of the time.

It is not possible to fit Smoluchowski into a narrow system of perceiving the world, all the more so if it were to be a system of materialistic views. It cannot be claimed that he was a materialist if one remembers that he relativized every theory and definition. He did not refer directly to materialism because that issue was not a scientific problem to him. The mention in which he indirectly expresses an opinion on this school of thought is unambiguous. He writes that it is dangerous to believe in the reality of physical hypotheses; he also supposes that those who believe only in the real existence of matter (due to the immutability of mass) may one day seem as naïve as those who worship energy in a similar way.[403] He treated the theoretical position according to which matter is the only

---

401 Ibidem, p. 112.
402 Ibidem, p. 235.
403 M. Smoluchowski, *Kilka uwag oanalogiach fizycznych, zwłaszcza w teoriach prądówelektrycznych, prądów cieplnych i zjawiskach dyfuzji* (Several observations on physical analogies, especially in theories of electrical currents, heat currents and diffusion phenomena) „Wiadomości Matematyczne" ("Mathematical News") 1918,

material of reality[404] not as a scientific approach but more as an ideological one and from this point of view – naïve.

Smoluchowski was a physicist so it is impossible not to notice certain elements of materialistic thinking in his scientific methodology or epistemological treatment of reality. However, attempts at an ontological perception are devoid of aspects of materialistic philosophy. It is hard to find statements in the physicist's publications referring to his religious views. There are no clear, straightforward declarations or reflections on the subject of the existence of God, or thoughts about faith. From some philosophical queries the nature of Smoluchowski's religious views can be surmised, or – more generally – attempts can be made to gauge the direction of his thoughts. Hence the suggestion of his atheistic worldview should be considered misguided. Just as it is hard to find religious declarations in the Polish scientist's writings, so it is also impossible to find proof of his unequivocal atheist attitude. His agnostic theses in the context of deliberations on the existence of the external world, employing materialistic arguments, are made from the position of a naturalist, have no metaphysical overtones and do not refer to the problem of God's existence.

Teske's comment on the subject is restrained but unambiguous: "To prevent false conclusions, it should be highlighted that Smoluchowski was far removed from any mystical institutions. He was also indifferent in relation to religious beliefs; to quote his family's words, 'He did not believe in life after the grave' "[405]. Indifference towards religious beliefs and a disbelief in life after death prove that religious thought was not very close to him at the exoteric level. However, this does not imply a lack of this type of sensitivity and certainly does not constitute evidence of his atheistic attitude. From his philosophical considerations it can be surmised that he did not believe in a personal God. Goetel stated that death held no fear for Smoluchowski:

> Smoluchowski was not afraid of death, just as he was never afraid of anything in life. In the moments in which that romantic of knowledge and life thought about death, he wrote about it like this: 'Raise yourself up above human vanities and take a look at life from the position of eternal Nature. After all, we all travel that road that trillions have walked before us and trillions will walk after us. If we play for a moment shorter or longer at this stop, what does it mean!'[406].

---

404   vol. XXII, cit. after: Pisma Mariana Smoluchowskiego (The Writings of Marian Smoluchowski), vol. 3, op. cit., p. 242.
405   A. Teske, *Marian Smoluchowski: życie i twórczość* (Marian Smoluchowski: life and works), op. cit., p. 35.
406   W. Goetel, *Ze wspomnień osobistych o Maryanie Smoluchowskim* (From personal recollections of Marian Smoluchowski), op. cit., p. 229–230.

His philosophy alluded neither to theology nor religion but had a secular aspect based on paradigms of science. His love of beauty and nature suggested that he loved life the way it presents itself to us: "When, however, at other times he became aware of the immeasurable expanse of Beauty, Truth and Sublimity that life sowed around him, another death-defying statement sprang from his chest: 'After all, life is worth living'!"[407].

Smoluchowski was a passionate advocate of Darwin's theory, writing about it many times in various places. He maintained that it was chiefly due to Darwin that science had departed from the concept of purposefulness in nature, which had been a misleading *a priori* assumption, and that he had shown in various examples how apparently purposeful adaptations occur which are *de facto* "a natural result of natural causes"[408]. If we remove associations such as striving or action that arise in us when we use them to conceptualise strength – Smoluchowski argued – then we will understand that such a thought is an anthropomorphic element alien to inanimate nature. This means that if we claim that a general force of gravity exists, we conclude the regular existence of a certain property of the movements of all bodies[409]. Another of his statements is significant:

> No naturalist today accepts purposefulness in nature because nature cannot be personified as if it were a thinking and planning being. Purposefulness would be understood only from the point of view of those who adopt the hypothesis that nature was created by an intelligent being, with a plan and aim set in advance. We perceive no traces of purposefulness in inanimate nature, and we have also learned from Darwin to dispense with this notion in the field of living nature[410].

Smoluchowski did not believe that the sensory world was created by a personal God with some predetermined plan and his approach to religion was very sceptical, although we also do not find any statements from him questioning the existence of God. It is impossible to claim that he had no understanding of an esoteric concept of spirituality or that he questioned the existence of any relation between man and God since in this realm of human cognition the Polish scientist remained silent. This lack of any statement is not evidence of an atheistic attitude but is more a manifestation of scientific agnosticism. As an open-minded person sensitive to spiritual feelings and reaching beyond the discourse of the exact sciences, Smoluchowski did not restrict the known world to scientific reality, as he

---

407 Ibidem, p. 230.
408 M. Smoluchowski, *Poradnik dla samouków* (The Self-Study Handbook), vol. I, op. cit., p. 18.
409 Ibidem, p. 21.
410 Ibidem, p. 18.

often said. Most probably he remained silent, wishing to be honest in his statements. As another philosopher wrote: "Whereof one cannot speak, thereof one must be silent"[411]. Smoluchowski did not find in himself that undisputable truth to which he could relate and which would convince him that science requires truth and religion requires faith.

Faith and science are two different orders of knowledge that function in different cognitive spaces. Smoluchowski focused attention entirely on research of the physical reality available to him. There are no deliberations in his works going beyond the sphere of science, which, however, cannot constitute proof of an atheistic worldview.

The above considerations are aimed at revising the perception of the worldview and various aspects of Marian Smoluchowski's philosophy which have unjustly been labelled as materialistic. A scientist, a physicist, is for obvious reasons always close to a materialistic treatment of reality, especially in the epistemological sense, as the material he or she works on primarily manifests such a nature. Unfortunately, although the era of elevating materialism to the status of an idea binding in science is far behind us, the materialist and atheist label has stuck to Smoluchowski and this thesis is often reiterated in publications devoted to him. How unbefitting that label is to our professor is evidenced by the whole of his scientific life. For Smoluchowski, the world was a far more complex potentiality for which science, its hypotheses and theories were merely accepted conventions enabling the physicist to function in the scientific space when it was impossible to ultimately determine how reality actually looks. In this context, ascribing him a materialistic worldview is an unjustified narrowing of the horizons of thought about the reality surrounding him. Zygmunt Klemensiewicz puts Smoluchowski among the scientists for whom materialism would be a depreciation of their own personality:

> "Lutosławski taught us to differentiate between two types of activists: social activists and pillarets. The first work on the masses, the second on the person themselves; these first wish to cover as wide an area as possible, the second to penetrate as deeply as possible. As a scientist and as a mountaineer, Smoluchowski was a pillarlet in the purest meaning of the word"[412].

We are therefore dealing with a lover and propagator of science for whom the material taught to students obviously had to be in line with his deepest convictions.

---

411 L. Wittgenstein, *Tractatus logico-philosophicus* (Routledge & Kegan Paul, ed. 1922), Section 7, Page 189
412 Z. Klemensiewicz, *Marian Smoluchowski*, op. cit., ps. 3.

Chapter XVII Professor

Smoluchowski returned from his tour of European universities and in 1898 received the distinction *Veniam legendi*, or the right to lecture, and at the same time the position of Privatdozent at the University of Vienna. The following year he moved to the University of Lviv at the instigation of Kazimierz Twardowski, also for a Privatdozent role. He accepted the invitation to the University of Lviv in the hope of a professorship.

Kazimier Twardowski (1866–1938)

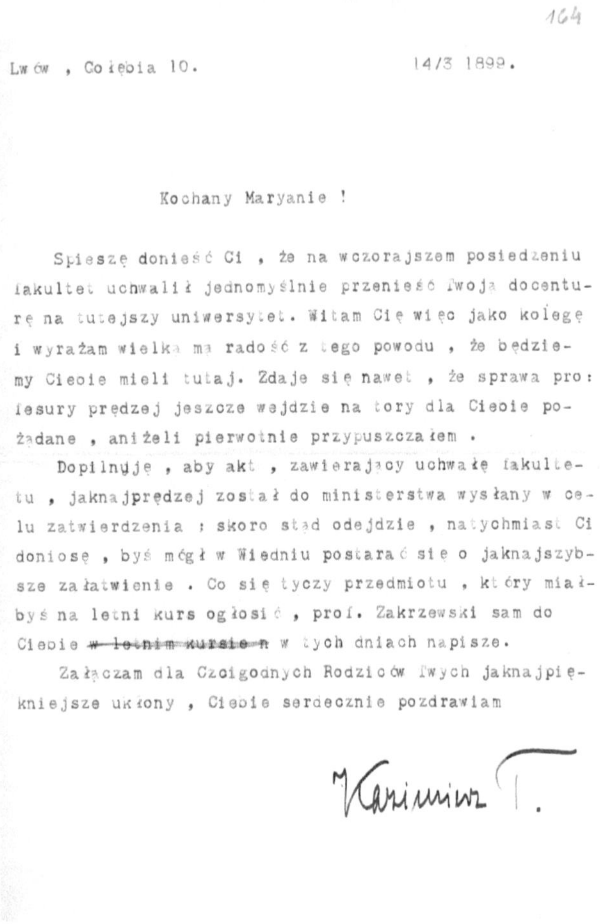

164

Lwów , Gołębia 10.   14/3 1899.

Kochany Maryanie !

    Spieszę donieść Ci , że na wczorajszem posiedzeniu fakultet uchwalił jednomyślnie przenieść Twoją docenturę na tutejszy uniwersytet. Witam Cię więc jako kolegę i wyrażam wielką mą radość z tego powodu , że będziemy Ciebie mieli tutaj. Zdaje się nawet , że sprawa profesury prędzej jeszcze wejdzie na tory dla Ciebie pożądane , aniżeli pierwotnie przypuszczałem .
    Dopilnuję , aby akt , zawierający uchwałę fakultetu , jaknajprędzej został do ministerstwa wysłany w celu zatwierdzenia : skoro stąd odejdzie , natychmiast Ci doniosę , byś mógł w Wiedniu postarać się o jaknajszybsze załatwienie . Co się tyczy przedmiotu , który miałbyś na letni kurs ogłosić , prof. Zakrzewski sam do Ciebie ~~w letnim kursie n~~ w tych dniach napisze.
    Załączam dla Czcigodnych Rodziców Twych jaknajpiękniejsze ukłony , Ciebie serdecznie pozdrawiam

Kazimierz T.

A letter from Kazimierz Twardowski

Undoubtedly a professorship in the faculty of theoretical physics at the University of Lviv was a challenge for the physicist of only 28 years of age, who thus became the youngest professor in the whole of the vast Austro-Hungarian Empire.

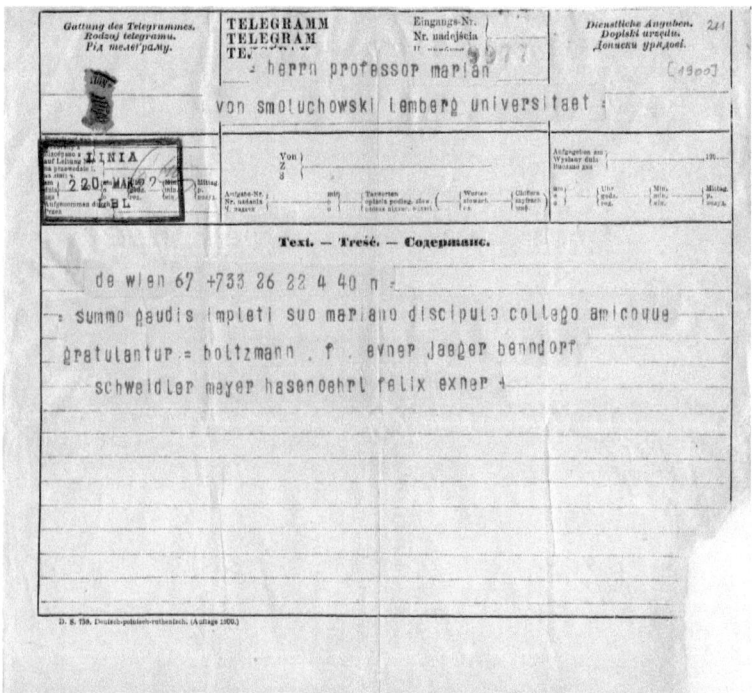

A telegram with congratulations on receiving the title of professor

The young, extremely gifted scientist doubtless received propositions from other European universities at the time, however, as he stated in one letter: "I would certainly very willingly take up the mathematical physics chair in Lviv (...), because although the scope (...) of activity would not be as great as in many other universities, I would rather my work bring direct benefits to my compatriots than to foreigners"[413]. Shortly after, in 1900, Smoluchowski received the title of extraordinary professor, and after another three years – professor of physics. At the university he lectured in mechanics, potential theory, optics, electricity, thermodynamics, the kinetic theory of gases, differential equations, and mathematical physics.

---

413 A. Teske, *Marian Smoluchowski: życie i twórczość* (Marian Smoluchowski: life and works), op. cit., p. 129.

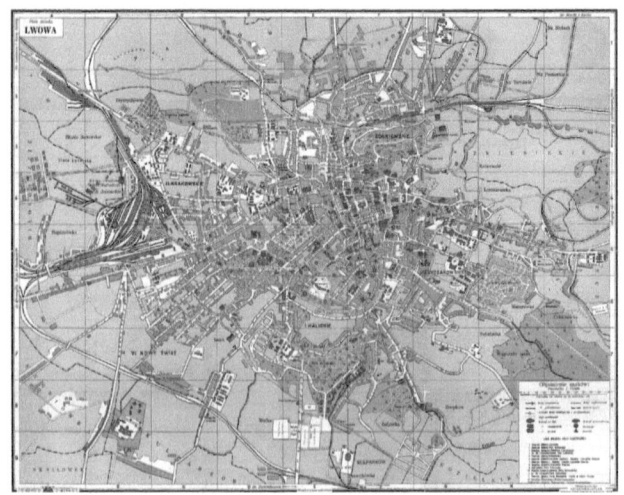

In Lviv

From August 1905 until April 1906, he took a sabbatical and went to Cambridge, where he worked at the famous Cavendish Laboratory. He made contact at that time with other scientists working on similar problems. Among them were mathematicians Robert Bill and Ernest William Hobson and physicists Joseph J. Thomson and Ernest Rutherford[414].

Smoluchowski during the birthday celebrations of Prof. Viktor von Lang (Vienna 1908)

An anecdote remains from the voyage around European universities. In the Polish language, the name Marian has its female equivalent – Maria. However, in German and French, both forms are identical. Due to this, various amusing events occurred. His Viennese friends sometimes jokingly called Smoluchowski 'Marianne'.

However, in France, while discussing his work, everyone was convinced it had been written by a woman – Madame Marie Smoluchowski. Maria Curie-Skłodowska had clearly made the French accustomed to the thought that Polish physicists were female and had the name Maria[415].

---

414 S. Ulam, *Marian Smoluchowski and the Theory of Probabilities in Physics*, op. cit., p. 477.
415 After: http://fizyka.net.pl/ciekawostki/ciekawostki_aou.html (access: 5.08.2020).

Professors and students of the Faculty of Philosophy at the University of Lviv (June 1904)

Honourable distinctions befell Smoluchowski – he received an honorary doctorate from the University of Glasgow and the Haitinger Prize. After returning to Lviv in the 1906–1907 academic year, he was elected as dean of the Faculty of Philosophy at the University of Lviv.

An important character trait of the scholar was his unambiguous attitude to both his career and social activities, consisting primarily in building his own position. Living and working in Lviv, he criticised the excessive politicisation of the local society. He wrote unenthusiastically: "I have not yet become infected with the Galician disease: politicking"[416]. He also noted that Poles are not a nation of high social awareness, well-developed and capable of common action:

> The whole of society seems somehow powerless, anaemic and partly resigned. Like an inanimate machine, rusted, lacking the moving force of a spring. I believe a contribution to this state of affairs is made by the second special Galician disease of officiating, as a result of which everyone would rather be a wheel or cog in this machine than a spring[417].

---

416 A. Teske, *Marian Smoluchowski: życie i twórczość* (Marian Smoluchowski: life and works), op. cit., p. 136.
417 Ibidem, p. 137.

In 1913, Smoluchowski moved to Kraków where, thanks to a recommendation by the outstanding physicist and former rector of the Jagiellonian University Professor August Witkowski, he took up the chair of experimental physics.

August Witkowski (1854–1913)

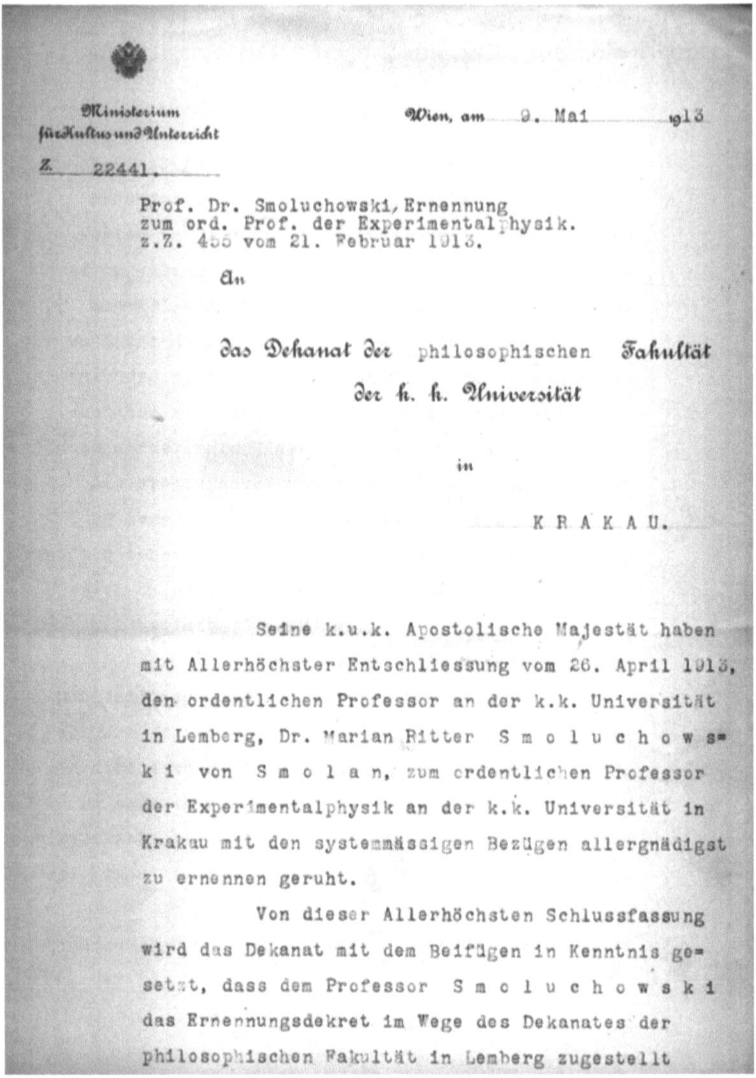

> Ministerium
> für Kultus und Unterricht
>
> Z. 22441.
>
> Wien, am 9. Mai 1913
>
> Prof. Dr. Smoluchowski, Ernennung
> zum ord. Prof. der Experimentalphysik.
> z.Z. 455 vom 21. Februar 1913.
>
> An
>
> das Dekanat der philosophischen Fakultät
> der k. k. Universität
>
> in
>
> KRAKAU.
>
> Seine k.u.k. Apostolische Majestät haben mit Allerhöchster Entschliessung vom 26. April 1913, den ordentlichen Professor an der k.k. Universität in Lemberg, Dr. Marian Ritter Smoluchowski von Smolan, zum ordentlichen Professor der Experimentalphysik an der k.k. Universität in Krakau mit den systemmässigen Bezügen allergnädigst zu ernennen geruht.
>
> Von dieser Allerhöchsten Schlussfassung wird das Dekanat mit dem Beifügen in Kenntnis gesetzt, dass dem Professor Smoluchowski das Ernennungsdekret im Wege des Dekanates der philosophischen Fakultät in Lemberg zugestellt

The naming of Marian Smoluchowski as professor of experimental physics in Kraków

Witkowski College (around 1912), ul. Gołebia 13

Smoluchowski went to Kraków as a known physicist. He had fundamental work to his credit in the field of the kinetic theory of matter as well as papers on density fluctuations in gas, gas opalescence in the vicinity of the critical point, the blueness of the sky as a consequence of the scattering of light on fluctuations in the atmosphere, he had worked on an explanation of Brownian motion and already formulated the equations named after him today[418].

---

418   A. Strzałkowski, *Wprowadzenie* (Introduction), op. cit., pp. 11–12.

Witkowski College, ul. Gołębia 13 (current view)

Following the death of Friedrich Hasenöhrl (1874–1915), a committee was established at the University of Vienna to select candidates to take up the chair in theoretical physics. In 1907, Hasenöhrl had become Boltzmann's deputy. On behalf of the committee, Stefan Meyer approached Smoluchowski to ask him if he would be interested in the position. At the same time, Höfler, who was also a member of the committee, urged Smoluchowski in a letter to take on the challenge.

The Collegium Physicum building, ul. Św. Anny 6, Kraków

Marian Smoluchowski and Tadeusz Zakrzewski (brothers-in-law) at the exit of Wiślna Street (Olszewski College in the background)

Smoluchowski would thus indirectly become Boltzmann's deputy at the University of Vienna and would surely have had better conditions there to conduct research and scientific work. The offer did not particularly come as a surprise since he was widely considered within the community as the best candidate for the position. The scientist considered the proposal seriously and after a sleepless night sent a telegram of acceptance. He wrote to Alois Höfler about his dilemma:

> As you know, I am a Pole and will remain a Pole. But I am not so chauvinistically blind as to feel unhappy leaving my narrower linguistic area. In fact, I feel as at home in Vienna as in Kraków or Lviv and the atmosphere of a big city suits me rather better than the narrower attitudes of a provincial town… So what makes it so difficult for me to leave Kraków is not some special attachment to that city but the feeling that I am needed here, the sense of a certain duty towards the University. I feel that many of my colleagues will take my decision not only with regret but with disapproval. I would be very sorry to have to admit they were right. I also do not know whether I would have made the decision I did were it not for the fact that I am essentially in the wrong place here. Although previously working as a theoretician, I have allowed myself to be persuaded to take up experimental physics and now feel the burden of that position acutely. It causes someone who has become used to lecturing graduate students in theoretical physics no minor unhappiness to recite every year in the same way elementary physics aimed at beginners in experimental physics, in addition to which (due to the poverty of my institute) with no possibility of showing really beautiful demonstrations… There I could live solely for science and would even have to exert all effort to live up to the hopes placed in me as a successor to Stefan, Boltzmann and Hasenöhrl. I realise that in the post in Vienna I would have more favourable conditions for my own scientific development – than the conditions currently prevailing – in Kraków, and I know that I could do more there and that is decisive. Actually, that is also basically egotism but a little different from what is usually meant by it. After all, I long ago gave up all social, political and national activities: when I decided after my matriculation exams to adopt science as my guiding star. This is the fundamental principle to which I will remain loyal and to which all other considerations must yield[419].

The above words evidence a progressive, quite pioneering attitude to science for those times, not burdened with the nationalistic phobias nascent in Europe at the turn of the century. Today, almost 100 years later, such views are widespread and do not arouse great controversy. The vast majority of scientists, without cutting themselves off from their roots and their nation, work in various laboratories and at various universities the world over. They go where they can count on the best conditions for conducting research and for their own development. Today, universities and laboratories strive to invite research scientists with the keenest

---

419 Extract of a letter from M. Smoluchowski to Höfler, 1915, Jagiellonian Library, BJ. Rkp. 9412, k. 53-59.

minds to their facilities, most often with no regard to their origins. At the same time, for every true scientist, the possibility of influencing the development of science is of decisive importance.

Unfortunately, it was not merit that decided the choice of a new head of department, but nationalistic factors. Although initially the committee unanimously supported Smoluchowski's candidacy, later, under external pressure, some of its members withdrew their support and in the end the majority voted against the Polish physicist taking over the chair on national grounds.

Marian Smoluchowski (right) in front of Collegium Novum; to the left Witkowski College, currently the seat of the Faculty of Physics of the Jagiellonian University

After a 10-year break, in the 1916–1917 academic year, Smoluchowski again took up the position of dean of the Faculty of Philosophy, this time at the Jagiellonian University. The growing scientific importance of the researcher's work led to many institutes and universities trying to tempt him to leave Kraków. In 1917, the Warsaw Scientific Society offered him a private laboratory[420]. He seriously

---

420 A. Teske, *Marian Smoluchowski*, "Fizyka i Chemia: czasopismo dla nauczycieli" ("Physics and Chemistry: a magazine for teachers") 1951, no 3 (17).

considered this offer as he had very warm memories both of being in Warsaw and of contact with Warsaw university circles. During his work at the University of Lviv, he was invited many times to take part in conferences organised by the University of Warsaw and he recalled those sojourns fondly: "They know how to work there and they know how to play"[421] – he wrote. But Smoluchowski stayed in Kraków. His scientific achievements and devotion to work were finally appreciated and in 1917 he was elected rector of the Jagiellonian University.

The protocol of Prof. Marian Smoluchowski's election as rector of the Jagiellonian University for the 1917–1918 academic year

---

421 A. Teske, *Marian Smoluchowski: życie i twórczość* (Marian Smoluchowski: life and works), op. cit., p. 137.

Unfortunately, he did not manage to take up the new tasks as in the summer of 1917 he contracted dysentery (known at the time as shigellosis) as a result of which disease he died on September 5 at the age of 45.

Posthumous recollection of Marian Smoluchowski in '*Physikalische Zeitschrift*' (Physical Journal)

М. ф. Смолуховскій.

(Marian v.-Smoluchowski).

(НЕКРОЛОГЪ).

Пятаго сентября 1917 года во время эпидеміи дизентеріи въ Краковѣ умеръ 45-ти лѣтъ знаменитый физикъ Смолуховскій. Онъ былъ одинъ изъ тѣхъ современныхъ выдающихся физиковъ—Ланжевенъ (Langevin), Перренъ (Perrin), Смолуховскій (Smoluchowski) и Эйнштейнъ (Einstein),—принадлежащихъ одному и тому же поколѣнію, которые предприняли переработку всѣхъ основныхъ принциповъ физики и примкнули непосредственно къ предшествующему поколѣнію великихъ физиковъ (Boltzmann, H. A. Lorentz, Planck. Poincaré, Rayleigh и J. J. Thomson).

XIX-ое столѣтіе можно назвать столѣтіемъ торжества термодинамики; все міровоззрѣніе, всѣ объясненія явленій природы стараются свести къ основнымъ принципамъ сохраненія энергіи и увеличенія энтропіи; подраздѣляютъ всѣ явленія природы на обратимыя и необратимыя, къ первымъ непосредственно приложимы принципы термодинамики, которые признаются непогрѣшимыми, напр., считаютъ, что всякое явленіе, протекающее въ природѣ, сопровождается всегда увеличеніемъ энтропіи и только въ крайнемъ случаѣ постоянствомъ ея. Работа термодинамиковъ въ концѣ XIX столѣтія направляется на изученіе необратимыхъ процессовъ и въ этомъ направленіи особенно важными являются изслѣдованія недавно умершаго физика Дюгема (Duhem).

Но термодинамика, которая имѣла такой огромный успѣхъ при изученіи физической и физикохимической статики, натолкнулась на непреодолимыя затрудненія при изученіи кинетики явленій природы. Кромѣ того, цѣлый рядъ данныхъ относительно диффузіи газовъ, внутренняго тренія и теплопроводности газовъ, вызывали уже въ концѣ XIX столѣтія постановку общихъ вопросовъ и давала массу новыхъ фактовъ, которые никакими способами не могли быть объяснены одними принципами термодинамики. Для этихъ явленій молекулярная физика давала простыя количественныя объясненія. Первая работа Смолуховскаго въ 1898 году и относится къ этимъ вопросамъ; онъ изучаетъ скачекъ температуры при переходѣ отъ теплой стѣнки къ газу, скачекъ этотъ является тѣмъ болѣе сильнымъ, чѣмъ давленіе газа меньше. Этотъ скачекъ температуры объясняется кинетической теоріей газовъ; точныя измѣренія распредѣленія температуры въ зависимости отъ разстоянія отъ нагрѣтой стѣнки, которыя были произведены въ 1910 году Лазаревымъ, позволяютъ непосредственно вычислить свободный путь молекулъ газа; термодинамика же не въ состояніи объяснить этого явленія.

Article in the magazine '*Postępy w Naukach Fizycznych*' (Advances in the Physical Sciences')

The extraordinary character of Smoluchowski, his progressive attitude towards many religious and social problems, must also have been visible in the issue of the perception of the position of women in science. The turn of the 19th and 20th centuries was characterised by an intensification in emancipation movements. During this period, from the perspective of a time termed the first wave of feminism, women – in addition to electoral rights – were also fighting for the right to study and work in science at universities. In the area of the former Polish Republic, attitudes towards allowing women into universities were highly critical, so the affirmative attitude of a professor of physics, a special science practised almost exclusively by men, was extremely important.

> Smoluchowski, a luminary intellectualist shaping education in the 20th century, could not fail to take a position on the issue of women's participation in science. He personally knew outstanding female physicists: Maria Skłodowska-Curie, Tatiana Ehrenfest, and Lisa Meitner. He was a passionate advocate of allowing girls into high schools and the broad opening of universities to girls[422].

He talked plainly about it during a speech given at the Lviv Scientific and Literary Society, which he titled *Women in the exact sciences*:

> Women who embark on the road of science should be facilitated in their vocation; all external barriers, peculiar superstitions, outdated views should cease which close off women's access to some scientific institutions, which hamper their education, scientific work, access to university chairs. Let the principle of free competition prevail here (as in every other field). May that competition be as lively as possible[423].

The psychologist Philip G. Zimbardo emphasises that in 2018, in the United States, women achieved more scientific titles than men in every discipline, even in engineering. They are smarter, work harder, exchange information, and create interest groups[424]. Over a hundred year ago, Professor Smoluchowski saw this dormant potential in girls wanting to learn.

> Female graduates of our high schools know about sines, cosines and logarithms no better or worse than boys; at university they attend lectures in higher mathematics, physics, chemistry with equal effect. Professors who have experience in this regard say that female student may even surpass male students in acumen, diligence and ease of assimilating the material[425].

422　Z. Gołąb-Meyer, *Poglądy Mariana Smoluchowskiego na nauczanie fizyki z perspektywy stulecia* (Marian Smoluchowski's views on the teaching of physics from a century's perspective), op. cit., p. 45.
423　M. Smoluchowski, *Kobiety w naukach ścisłych* (Women in the exact sciences), speech given at the Lviv Scientific and Literary Society in 1912, citation after: *Pisma Mariana Smoluchowskiego* (The Writings of Marian Smoluchowski), vol. 3, op. cit., p. 152.
424　A. Niezgoda, *Philip Zimbardo: Młodzi mężczyźni wycofują się z życia społecznego* (Philip Zimbardo: Young men withdraw from social life), Focus.pl, https://www.focus.pl/artykul/wylogowani-z-zycia (access: 10.01.2022).
425　M. Smoluchowski, *Kobiety w naukach ścisłych* (Women in the exact sciences), op. cit. pp. 138–139.

He perceived a difference in the treatment of men and women working in science by the conservative Polish scientific community:

> A male scientist appears almost impersonally as the author of certain scientific works; we judge his importance according to the value of these works without regard to any aspect of his private life. But when it comes to a woman scientist, everyone is interested above all in her private life which, after all, is irrelevant in assessing scientific merits; how easy it is to cast suspicions which have nothing to do with the matter and are harmful in the eyes of the general public[426].

Among the few women active in science that Smoluchowski mentioned in his speech, he singled out one person in particular: "In terms of general scientific importance, the activity of our compatriot, Mrs Curie-Skłodowska (…), is undoubtedly the most significant. This name has gained renown such as no other woman scientist has ever enjoyed; doubtless it will also be written permanently in a prominent place in the history of physics and chemistry[427]. Like many other scientists – physicists and chemists – he was very impressed with Skłodowska's work and fully appreciated her achievements:

> It is hard to give a brief, even cursory outline of Mrs Curie- Skłodowska's activities; it is a field of new phenomena; unusual and yet the material gathered in this field is so extensive, as evidenced by the size of Mrs Curies' work, *Traité de Radioactivité*; in two volumes, almost 1,000 pages in total. The author provides a laconic, concise description of her (and other scientists') work in the field of radioactive phenomena[428].

As a lecturer, Smoluchowski had great contact with young people. Walery Goetel mentions that:

> Being with youth brought Smoluchowski the greatest pleasure and joy. His attitude towards young people was something so extraordinary that his loss in this field is an irreparable catastrophe. Smoluchowski loved young people and understood them. The kindness and sincere friendliness with which he related to them, the understanding for all the flaws and shortcomings of youth, avoiding moralising and being aloof merely because of his seniority, the warm enthusiasm for the exuberance and vigour of youth, these were qualities that earned him the love and general veneration of the young. The healthy instinct of youth felt him as a friend, a guardian, an advocate, but above all as a person who himself was eternally young. Young people felt this particularly strongly during the expeditions which (…) he took almost constantly in their company. Here the main message he presented before us was the obligation to forget about his professorial

---

426   Ibidem, p. 145.
427   Ibidem, p. 146.
428   Ibidem.

dignity, demanding to be considered a companion equal to all those on the trip, and he put that principle into action[429].

To this same recollection Goetel adds:

> Professor Smoluchowski did not lead young people with his position but with his spirit, with his whole wonderful being. Wherever he appeared, the hearts of the young clung to him with irrepressible strength, they worshiped him as their master and leader, loved him as their dearest friend. His influence on youth was also unlimited and he needed few words to direct them to where he wished. (…) He did not hesitate as dean and rector of the Jagiellonian University to don the red cap and outfit of the student rowers on the Vistula and sit at the oars in a boat with students. On happy Vistula excursions with the University Sports Association, he played with the young with all his heart. When we wanted to do something for him he turned us down gently but firmly, repeating that he did not want to lead his sports companions due to his university status[430].

Walery Goetel with his daughter, Wanda (1917)

---

429 W. Goetel, *Ze wspomnień osobistych o Maryanie Smoluchowskim* (From personal recollections of Marian Smoluchowski), op. cit., p. 227.
430 Ibidem, p. 288. See also *Marian Smoluchowski (1872-1917). Fizyk, taternik, romantyk nauki* (Marian Smoluchowski (1872-1917), Physicist, mountaineer, romantic of science), op. cit., p. 42.

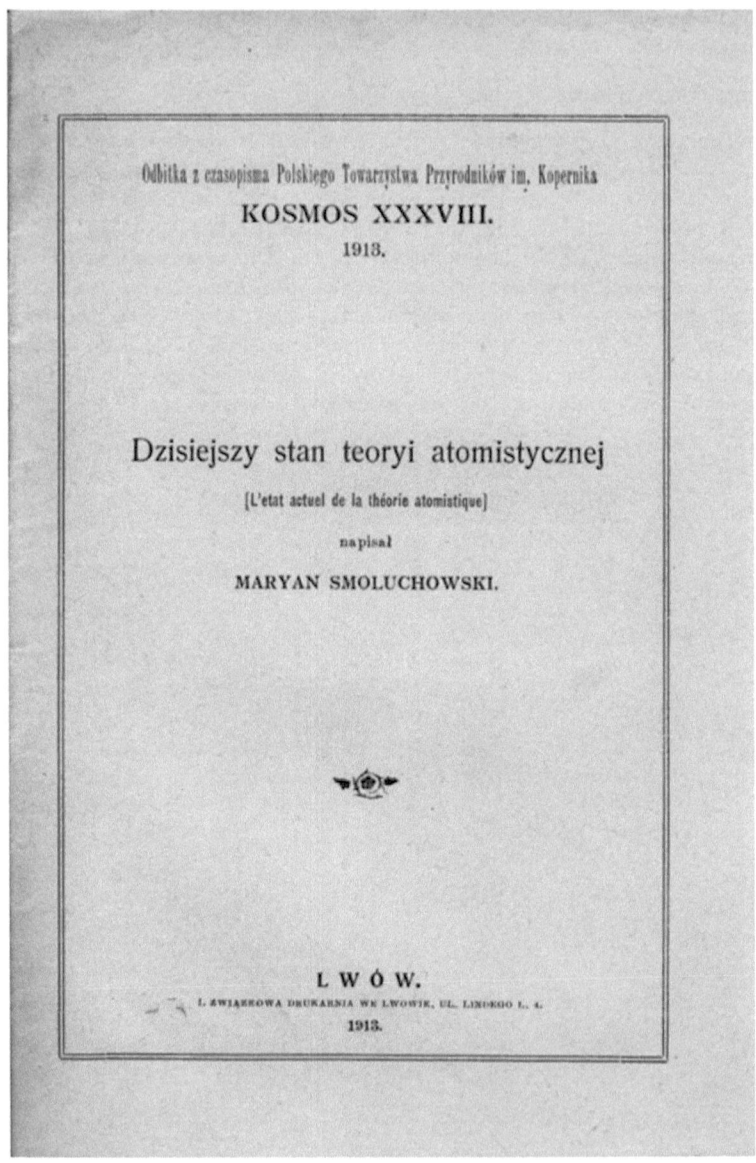

Members of the Polish Copernicus Society of Naturalists at the turn of the 19th and 20th centuries

This was not the usual behaviour of a rector, dean, or even professor at the time as university staff maintained much greater distance from students than today. Since his early youth, Smoluchowski had taken long highland hiking journeys as well as practising mountaineering and skiing. Now he was a professor, he did not give the impression of a frail scientist working from dawn to dusk in the privacy of a laboratory; quite the opposite – he was in great shape, was very athletic and had boundless energy, which he retained in his adult years.

An issue of '*Kosmos*' magazine with an article on Smoluchowski

Election of the Board of the Tourist Section of the Tatra Society, of which Smoluchowski was appointed Chairman

Elsewhere, in recalling a mountain excursion with academic youth, Goetel presents the extraordinary attitude of a professor who was already famous in the world of science; an attitude demonstrating incredible humility and exceptional selflessness.

I will never forget a certain episode from one of the Tatra expeditions. Smoluchowski had come with our group, all young people, for a skiing trip to the Western Tatras. We spent several frosty winter nights in a ski lodge in the Starorobociańska valley and from there we took ourselves off for trips to the nearby peaks. In the evening, as is usual when staying in a lodge, we took it in turns to take care of the kitchen, bringing water from a nearby stream etc. By an unspoken rule, the professor was of course exempt from this rota. But one evening, when my turn had come to go for water, Smoluchowski disappeared somewhere. When I got to work and started to look for buckets, a cheerful voice called me from the brook. I ran out of the door of the lodge and in the gloom I saw the figure of the professor laden with buckets full of water and trudging with them through the deep snow towards the lodge. When we wanted to take the buckets he scolded us and reminded us that he was not on the trip in order to break the rules[431].

Goetel also describes how after a long hike, at around noon and in not the best weather, the expedition ran into a problem at Mount Hoverla.

Smoluchowski suddenly remembered something and looking at his watch, he told everybody that on that day, in the evening, he had an important meeting he had to attend in Lviv. In order to make it to the train in Vorokhta in time he had calculated that he had to walk a kilometre in 8–9 minutes. So he set off immediately along with the whole expedition of more than 30 people. As Smoluchowski strode forth with steely determination, ever more people fell by the wayside, unable to keep up with the pace. Only four people arrived at the station in Vorokhta with Smoluchowski. Only Smoluchowski was not tired, but fresh and full of satisfaction that the sporting feat had been successful and thanks to that he would be able to take part in the evening's meeting[432].

Among Smoluchowski's uncommon personality traits, his ethical attitude also stands out, which was mentioned not only by relatives but also by all those who met him through work, during mountain expeditions or even just socially. Goetel noted that the high ethical standards the scholar set for himself never affected his judgement of other people. He tried not to demand too much of them and though he himself led a Spartan life, he did not demand the same of people around him. He was a believer in extending the utmost tolerance but could not stand a weak character or superficial disposition[433].

---

431 W. Goetel, *Ze wspomnień osobistych o Maryanie Smoluchowskim* (From personal recollections of Marian Smoluchowski), op. cit., p. 227–228. See also *Marian Smoluchowski (1872–1917). Fizyk, taternik, romantyk nauki* (Marian Smoluchowski (1872–1917), Physicist, mountaineer, romantic of science), op. cit., p. 36–38.
432 W. Goetel, *Marian Smoluchowski – człowiek gór* (Marian Smoluchowski – man of the mountains), op. cit., p. 92.
433 Idem, *Ze wspomnień osobistych o Maryanie Smoluchowskim* (From personal recollections of Marian Smoluchowski), op. cit. p. 219. See also *Marian Smoluchowski (1872–1917). Fizyk, taternik, romantyk nauki* (Marian Smoluchowski (1872–1917), Physicist, mountaineer, romantic of science), op. cit., p. 40–41.

Not only skis but oars (Marian Smoluchowski at the middle oars)

That these were not empty declarations is evidenced by Smoluchowski's effective opposition as a member of the Honorary Doctors Committee against an attempt in 1911 to grant an honorary doctorate of the University of Lviv's Faculty of Philosophy to Archbishop Józef Bilczewski and Michał Bobrzyński. In the years 1908 to 1913, Bobrzyński was the governor of Galicia and a member of the conservative 'Stańczyk' party. An outraged Smoluchowski wrote a letter to the rector of the university in which he stated:

> I do not know who is really responsible in this matter and actually I am not at all interested in making personal accusations against anybody. I merely regret that the whole matter has been settled in an informal manner and a candidate has been imposed on the Faculty of Philosophy who should naturally be assigned to the Faculty of Theology [this concerns Archbishop Bilczewski – J. G.], and above all a certain other personality has been imposed on the whole university who, according to the opinion of probably a large number of university colleagues, is decidedly unworthy of this distinction, although the mood in opposition to his candidature was well known [here he means Bobrzyński – J. G.].[434]

---

434 A. Teske, *Marian Smoluchowski: życie i twórczość* (Marian Smoluchowski: life and works), op. cit., p. 247–248.

Without doubt, the archbishop's later career adds spice to the whole issue. "In 1893, he received the title of full professor. In the 1896–97 academic year, he was dean of the Faculty of Philosophy and in 1900 rector of the Jan Kazimierz University. On June 26, 2001, he was beatified by Pope John Paul II during his visit to Lviv. The canonisation of Józef Bilczewski by Pope Benedict XVI occurred on October 23, 2005, in Rome"[435]. Also in the case of his successful support of the candidacy for rector of an atheist and advocate of Darwin's theories, Benedykt Dybowski, (N.B. he did not accept the election)[436], Smoluchowski opposed conservative university circles. His position on both issues had no religious grounds, he expressed an opinion in line with his own conviction and attitude towards those people[437]. Leaving aside the Polish scholar's personal view on the matter (which he nowhere expressed publicly), any attempt to reconstruct some of his statements from a religious point of view is an abuse[438].

---

435 *Józef Bilczewski*, timenote.pl, https://timenote.info/pl/Jozef-Bilczewski (access: 1.03.2022).
436 P. Polak, *Byłem pana przeciwnikiem [profesorze Einstein]...* (I was your opponent [Professor Einstein]...), Kraków 2012, p. 87.
437 W. Krajewski, *Światopogląd Mariana Smoluchowskiego* (Marian Smoluchowski's worldview), op. cit., p. 238.
438 M. Stawarz, *Rekonstrukcja i krytyczna analiza poglądów filozoficznych Mariana Smoluchowskiego* (Reconstruction and critical analysis of Marian Smoluchowski's philosophical views), doctoral thesis, promoter: dr hab. P. Polak, Kraków 2016, pp. 42–43.

Ernst Mach, as secretary of the Imperial Academy of Sciences, informs Marian Smoluchowski of the sending to print of the dissertation *Über den Temperatursprung bei Wärmeleitung in Gasen* (On the temperature jump in heat conduction in gases) with a request to make final amendments (1898)

Smoluchowski was a visionary, he considered research problems that went on to form the focal point of dynamically developing fields of science. He had no doubt that in the 20th century the exact sciences would dominate, as he wrote in the article *The importance of the exact sciences in general education*: "The 19th century has commonly been called 'the age of steam and electricity'. The 20th century has not yet been Christened but there can be no doubt that it will again receive a nickname from the field of the exact sciences or technology"[439].

Today we know that the 20th century was the age of the atom and again – like Smoluchowski a hundred years ago – we face a great unknown. It is difficult to predict today what discovery will dominate the 21st century and we can no longer be so sure that it will come from the field of the exact sciences. Since the 1970s, in the opinion of many physicists, we have been dealing with a crisis in physics. The similarity of these situations, separated by 120 years, prompts us to re-examine the discoveries made at the turn of the 19th and 20th centuries. We will do this from a philosophical point of view as that plane of perception is most conducive to a comparison of the position and state of physics now and a hundred years ago.

---

[439] M. Smoluchowski, *Znaczenie nauk ścisłych w wykształceniu ogólnym* (The importance of the exact sciences in general education), op. cit., pp. 124–125.

Chapter XVIII The philosophy of physics

In the exact sciences, as opposed to the humanities, time tests theories, changing the value of the discoveries made from the perspective of a new era. This is clearly visible in physics at the turn of the 19th and 20th centuries. In 1910, German physicist Felix Auerbach (1856–1933) published the book *Historical tables of physics* (*Geschichtstafeln der Physik*)[440] while in 1983 Jurij Chramow (born 1933), a Ukrainian science historian, published *A Biography of Physics*[441]. The authors of both works evaluate the discoveries made in physics in the years 1899–1900. The works are separated by more than 70 years of development in the field and it is worth noting how much the perspectives of the 1910s and 1980s influence the assessments of these discoveries' value.

According to Auerbach in 1910, 44 discoveries made in 1899 and 69 made in 1900 were considered important in physics. In 1983, Jurij Chramow repeated Auerbach's study and in the perspective of the late 20th century claimed that in 1899, 17 important discoveries were made in physics and in 1900 only 16.

From a list of 113 important discoveries, after 73 years only seven were recognised while as many as 26 entries on Chramow's list of discoveries made in 1899–1900 and considered significant in 1983 did not appear on Auerbach's list. These include: (1) Planck's law of radiation from a black body (the theory was not listed at all by Auerbach); (2) experimental confirmation of this law by Heinrich Rubens (1865–1922) and Ferdinand Kurlbaum (1857–1927); (3) the discovery of gamma rays by Paul Ulrich Villard (1860–1934); (4) the discovery of beta wave deviations in an electric field (Friedrich Ernst Dorn, 1848–1916, Antoine Henri Becquerel, 1852–1908); (5) the discovery that beta rays are negatively charged particles (Pierre Curie and Maria Skłodowska- Curie); (6) measurement of the e/m ratio for beta rays, yielding a result very close to the result for cathode rays (Becquerel); 7) the discovery by Pyotr Lebedev (1866–1912) of light pressure predicted by the Maxwell theory.

The peculiar state of physics in the late 19th century is evidenced not only by the non-appreciation of some important discoveries, which happens in every period because as Yogi Berra stated: "It is tough to make predictions, especially

---

440   F. Auerbach, *Geschichtstafeln der Physik* (Historical tables of physics), Leipzig 1910.
441   J.A. Chramow, *A Biography of Physics*, Kyiv 1983.

about the future"[442], but also by the prevailing conviction that physics had reached its peak.

During a talk given at the University of Munich on December 1, 1924, Max Planck recalled the views dominating in exact science circles at the end of the 19th century. It was widely believed that after the 19th century being full of discoveries, humanity would enter the 20th century with the awareness that exact sciences like physics and chemistry had already achieved a very high level of development and were as mature as geometry, which had been developing for centuries. Planck recalled:

> When I was beginning to study physics and asked my venerable teacher Philipp von Jolly for his opinion concerning the conditions and perspectives of my studies, he presented physics as a highly developed and almost fully mature science, which after its achievements had been crowned by the principle of conservation of energy, was shortly to assume its final form. True enough, in some corners there remained a speck or little bubble to be studied and removed, but the system itself was quite safe and theoretical physics was approaching a perfection which since centuries was the attribute of geometry[443].

Pieter Zeeman (1865–1943), who went on to win the Nobel Prize in 1902, made a similar statement, saying that in 1883 he was warned not to study physics as it was not a promising field – there was no room in it for anything new and important as it was an almost closed field of study[444]. Marcellin Berthelot (1827– 1907), the brilliant French chemist, stated in 1885 that *"Le monde est aujourd'hui sans mystère"* (The world is without mysteries today), while Albert A. Michelson (1852–1931), the American physicist of Polish-Jewish origin, winner of the Nobel Prize for physics in 1907, wrote in 1894: "The more important fundamental laws and facts of physical science have all been discovered, and these are now so firmly established that the possibility of their ever being supplanted in consequence of new discoveries is exceedingly remote. (…) future discoveries must be looked for in the sixth place of decimals"[445].

---

442 See N.N. Taleb, *The Black Swan: second edition*, New York, 2010, p. 136.
443 M. Planck, *Vom Relativen zum Absoluten* (From the relative to the absolute), lecture in Munich on December 1, 1924, in: M. Planck, *Jedność fizycznego obrazu świata. Wybór pism filozoficznych* (The Unity of the Physical Pattern of the World. A selection of philosophical writings), trans. R. i S. Kernerowie, Warsaw 1970, pp. 122–123.
444 See H. Casimir, *Haphazard Reality: Half a Century of Science*, New York 1984, p. 27.
445 Speech by A. A. Michelson at University of Chicago, after de Swart, *Philosophical and Mathematical Logic*, Cham, Switzerland, 2018, p. 497.

The successes of physics in the second half of the 19th century paradoxically inclined many physicists towards scepticism as to its future. The cause of this lack of faith in the future of physics was no simple mistake. Too many outstanding people of science had drawn similar conclusions. It is characteristic that, from their point of view, they were right in a sense. The era of mechanical physics and thermodynamics was coming to an end and both these concepts were actually exhausted. The pendulum of cultivating knowledge was swinging towards rationalistic science, which began with the Enlightenment. The hitherto philosophy of nature was being systematically eliminated from science in favour of specific sciences, assuming that the world could be interpreted in terms of Euclidean geometry. People distanced themselves from the Aristotelian method of perceiving natural phenomena, which Smoluchowski called "madness hampering the development (...) of the sciences, especially due to Aristotle's authority throughout the whole course of antiquity and the Middle Ages"[446]. It was argued that science needed objective criteria to assess the correctness of explanations concerning data as well as defining the premises for obtaining them. Taking the above in a wider perspective, it could be claimed that it was to be a science accepting only external facts, which – as Comte demonstrated – were held to be the only subject of true knowledge.

The consequences of the new philosophy were extremely significant – classical mechanics and the mechanistic conception of perceiving physical reality that started with Galileo and Newton, developed. Thermodynamics and electrodynamics, which worked perfectly in the macro-world, flourished. Intensive progress in both these fields led to natural philosophy becoming an anachronism in the view of many scientists. This resulted in various perspectives of philosophers' and naturalists' perceptions of reality as well as a different way of them treating the essence of things[447].

An important issue in the perception of perspectives of physics in the second half of the 19th century was the conventional attitude of academic circles, which had a conservative approach to the changes described in the works of Maxwell, Kelvin, Poincaré and Boltzmann that signalled the complexity of space, time and matter. We can see this conservativeness even in Lord Kelvin, who said in 1895 that "heavier-than-air flying machines are impossible"[448], and in 1897 that "radio

---

446 M. Smoluchowski, *Przedmiot, zadanie, metoda oraz podział fizyki* (Subject, task, method and the division of physics), op. cit., p. 12.
447 See idem, *Dwie książki z dziedziny „filozofii przyrody"* (Two books from the field of "the philosophy of nature"), op. cit. p. 293.
448 G. Dryden, J. Vos, *The Learning Revolution*, Stafford, 2005, s. 232.

has no future"[449]. He is also credited with a statement made in a speech to the British Association for the Advancement of Science in 1900: "There is nothing new to be discovered in physics now. All that remains is more and more precise measurement"[450]. In 1901, he apparently added that "X-rays will prove to be a hoax"[451].

Kelvin's statements are shocking; made 20 years earlier they would have been no surprise, but coming at the turn of the century they were a manifestation of intellectual conservatism. However, in the interests of complete honesty, it should be added that according to Davies and Brown, the aforementioned quote about the end of physics was wrongly attributed to Kelvin. There is no evidence that he ever said such a thing (or even that he spoke at that time at a meeting of the British Association for the Advancement of Science) and according to those authors, the quote is a paraphrase of the already-cited statement by Albert A, Michelson. Perhaps it concerned the 'Two Clouds' speech, given at the Royal Institution of Great Britain in 1900 in which he highlighted areas in which research later transpired to be revolutionary[452]. Regardless of to whom the quote is attributed, it illustrates the discourse ongoing among physicists at the time.

We owe progress in science, especially in physics and chemistry, in large measure to young minds. Both the *annus mirabilis* and the "miraculous year" have in common the young age of the researchers – Newton in 1666 was 24 years old and a student, and in 1905 Einstein was 26 and worked at a patent office. Both stood at the beginning of their scientific path, though the works they published were at the same time the culmination of that path. Many important discoveries have been made by scientists only in their twenties. Their success is often due to the fact that – as the saying goes – they did not know that what they were doing was impossible. Gustav Kirchhoff announced his laws when he was just 21. James Clerk Maxwell published the work On Faraday's Lines of Force when he was 24. Maria and Pierre Curie discovered two radioactive elements at the age of 29. Niels Bohr was 28 when he published *On the Constitution of Atoms and Molecules*. There are many similar examples.

---

449 S.J. Marshall, *Shaping the University of the Future*, Wellington 2018, p. 156.
450 W. Isaacson, *Einstein: His Life and Universe*, New York, 2017, p. 90.
451 See https://en.wikiquote.org/wiki/William_Thomson
452 *Superstrings: A Theory of Everything?*, eds. P. Davies, J. Brown, Cambridge 1988, p. 575.

"The owl of Minerva begins its flight only with the onset of the dusk," not at dawn, claims Georg Hegel in the preface to *Elements of the Philosophy of Right*[453]. Could he have been wrong? However, it is not only the two cases of Newton and Einstein that are evidence of his error. Discoveries in physics constitute the antinomy of the quoted thought as the genius of the two young physicists revealed itself "at dawn". Hegel adds:

> philosophy, at any rate, always comes too late (…). As the thought of the world, it appears only at a time when actuality has gone through its formative process and attained its completed state. (…) When philosophy paints its grey in grey, a shape of life has grown old, and it cannot be rejuvenated, but only recognized by the grey in grey of philosophy; the owl of Minerva begins its flight only with the onset of dusk[454].

We can look at Hegel's metaphor from many angles and each time understand it differently. The Owl of Minerva is philosophy itself which, in contrast to the exact sciences, always comes, as Hegel writes, too late. That philosophical thought "takes flight at dusk" is the culmination of the philosopher's deliberations.

The sentence could also refer to an understanding of nature's complexity and progress in science. Wisdom comes at dusk, at a time when the hitherto achievements of science are already in their closing stage, when the process of forming the paradigms of a given era has ended. Reflection summarising a period in science is possible when its dusk has already fallen, when we summarise from the perspective of new times, enabling real knowledge of the achievements of the bygone era.

Such an understanding of Hegel's thought, supported by the research of Auerbach and Chramow, prompts and attempt to understand a paradox – predictions of the end of physics at a time when its dynamic development started. The Owl of Minerva flies at the twilight of an era ending, heralding a new age in science. Foretelling the end of physics was not such a great mistake, as we know from today's perspective, as it was not about the whole of physics but about the era of mechanistic physics and thermodynamics. Michelson was partly right in saying that "[t]here is nothing new to be discovered in physics now. All that remains is more and more precise measurement"[455]. He lived at a time when

---

453 G.W.F. Hegel, *Elements of the Philosophy of Right*, Ed. Allen W. Wood, Cambridge, 2003, p. 23.
454 Ibidem.
455 *Speech by A.A. Michelsona at the University of Chicago*, after: A.K. Wróblewski, *Historia fizyki: od czasów najdawniejszych do współczesności* (The history of physics: from antiquity to modernity) Warsaw 2006, pp. 395–396.

Newtonian physics and thermodynamics reigned and it was this physics that he had in mind. Planck and Einstein ended that era with their papers. A similar situation occurred at the end of the Middle Ages, when few people realised they were dealing with the decline of old paradigms. In the 19th century, the changes were heralded by the development of statistical physics by Boltzmann and Smoluchowski, while the beginning of the revolution in the worldview of the 16th century was *De revolutionibus orbium coelestium*.

However, the early years of the 20th century was a time when physics was building its position as the dominant science. This fact reveals a change in the order of scientific problems. The centre of interest shifted in the direction of specific sciences, especially physics, astrophysics, cosmology, information technology and biology, in which new hypotheses, theories and constructs were emerging that changed the perception of reality. It is characteristic that in each era there exists an unjustified faith in the stability of the prevailing science, which often does not stand the test of time in the next era. An *ex post* philosophical reflection enables a more realistic assessment, as in the case of Hegel's Owl of Minerva. Which discoveries of the early 21st century are therefore really important? A current 'Auerbach list' checked 80 years later would as usual certainly be a shock, all the more so as, in the words of Berra: "The future ain't what it used to be"[456].

The changes that occurred in physics in the 19th and early 20th centuries, from a philosophical standpoint, prompt reflections on the direction in which thinking about reality was going. In the 19th century, philosophy was successively ousted from the pole position in science it had occupied for the previous 2,500 years. The year 1900, due to Planck, and 1905 through Einstein, were recognised as breakthrough years in the history of physics and became moments in which the scales began to tip in physics' favour. This does not mean that philosophy ceased to play an important role in science as it was still needed.

Throughout the whole of the 20th century, as philosopher Tadeusz Gadacz claims, more than a hundred outstanding philosophers can be counted who created their own new philosophical systems, leaving their mark. At the end of the 20th century and beginning of the 21st, this process halted, philosophy no longer being a magnet for great minds. Physics, astronomy, cosmology, information technology and biology became such magnets. In science, the specific sciences have developed and it is on these that modern geniuses focus.

A selective chronological-philosophical overview of changes presenting the history of discoveries in physics and chemistry in the 19th and early 20th

---

456   See N.N. Taleb, *The Black Swan: second edition*, op. cit., p. 136.

centuries draws attention to how perception of reality has changed and the transition that has taken place from a philosophical discourse to the mathematical methodology of the exact sciences. In it, the changes can be seen that have occurred in the thinking of physicists and chemists on the meta-science level and this in a way answers the thesis of physics being exhausted that was put forward by scientists of the late 19th century.

In order to capture the characteristic moment of change in the understanding of science it would be necessary to go back to the year 1783, when Antoine Lavoisier (1743–1794) went beyond the traditions of perceiving natural philosophy. Demonstrating that matter is neither created nor destroyed in chemical reactions and introducing the principle of conservation of mass and Lavoisier's law, he created the foundations of modern chemistry. This raised the discipline to the research standards and causal explanations that can be found in modern experimental physics. A farewell to the phlogiston theory was a symbolic break away from the Greek philosophy of explaining phenomena.

In 1774, Joseph Priestly isolated oxygen, which enabled an explanation of both quantitative and qualitative changes that occur during combustion, respiration and calcination. This justified the thesis that combustion is a chemical reaction consisting of the combination of various substances with oxygen. The new theory of combustion started a campaign to build chemistry according to its principles. In 1787, Lavoisier and three colleagues published a new chemical nomenclature which soon became widely accepted. The provisions of this work are currently widely used. Lavoisier showed that there are three states of matter: solid, liquid and gas and that any element can occur in any one of them.

An important date is 1827, when Robert Brown (1773– 1858) discovered Brownian motion (in an experiment using pollen grains and dye in water). This discovery had profound consequences for physics and 78 years later formed the basis for the formulation of essential proof of kinetic-atomic theory.

In 1843, James Prescott Joule (1818–1889) experimentally determined the mechanical correlate of heat by defining the amount of work required to produce a unit of heat, known as the mechanical equivalent of heat. He proved that heat is a form of energy independent of the substance heated. He thereby undermined the philosophical conviction of scholars at the turn of the 18th and 19th centuries that there existed an indestructible and immeasurable heat fluid known as caloric. In 1847, Hermann von Helmholtz proved that work cannot be produced from nothing and that the energy that appears lost is in reality transformed into heat energy, formulating the principle of the conservation of kinetic energy. Thanks to the work of such scientists as Helmholtz, Thomson and Clausius, the principle of the conservation of caloric was replaced by the principle of the conservation of energy. Helmholtz believed that the whole of knowledge comes from the senses and that science should be reduced to laws of classical mechanics encompassing matter, force and energy as the whole of reality.

Michael Faraday (1791–1867) was a scientist considered the greatest experimenter of all time. In his work *On the Various Forces of Nature and Their Relations to Each Other*[457], he wrote:

> I was formerly a bookseller and binder, but am now turned philosopher, which happened thus: Whilst an apprentice, I, for amusement, learnt a little of chemistry and other parts of philosophy, and felt an eager desire to proceed in that way further. After being a journeyman for six months, under a disagreeable master, I gave up my business and, by the interest of Sir H. Davy, filled the situation of chemical assistant to the Royal Institution of Great Britain, in which office I now remain; and where I am constantly employed in observing the works on nature, and tracing the manner in which she directs the order and arrangement of the world[458].

The change in the perception of knowledge in the 19th century also related to the practice of it as a subject. Faraday – referring to his life path – claimed that he had been a bookbinder and seller but had become a philosopher, by which he meant his research in the field of physics. No notable physicist after Faraday claimed that the crowning field of his research was philosophy, despite the fact that it achieved significant success in the second half of the 19th century and in the 20th.

The scientist, fascinated by the world of nature, asked:

> For what study is there more fitted to the mind of man than that of physical sciences? And what is there more capable of giving him an insight into the actions of those laws, a knowledge of which gives interest to the most trifling phenomenon of nature, and make

---

457 The work *On the Various Forces of Nature and Their Relations to Each Other* was created as a record of lectures given by Faraday at the Royal Institution in the 1860s aimed at a young audience. The publisher, physicist William Crookes, declared: "The lectures were published as they were given, *verbatim et literatim*. A careful and competent rapporteur noted their content, and the manuscript that was produced after deciphering the notes was then developed by the publisher with the utmost care". Quote after: M. Litwinowicz-Droździel, *Indukcje i przepływy. Michael Faraday – mikrostudium o romantycznej nauce* (Inductions and Flows. Michael Faraday – a Microstudy of Romantic Science), „Wiek XIX. Rocznik Towarzystwa Literackiego im. Adama Mickiewicza" ("The Nineteenth Century. Annual of the Adam Mickiewicz Literary Society"), 2015, vol. VIII (L), p. 99.

458 *The Philosopher's Tree. Michael Faraday's Life and Work in His Own Words*, ed. P. Day, London 1999, p. 255, quote after: M. Litwinowicz-Droździel, *Indukcje i przepływy* (Inductions and Flows), op. cit., pp. 93–94.

the observing student find – tongues in trees, books in the running brooks, sermons in stones and good – in everything[459].

This statement of Faraday's constitutes a programme of change in the shaping of knowledge. Methodological studies of nature became a key human endeavour for the next 200 years. In 1831, Faraday described the phenomenon of electromagnetic induction, self- induction. In 1834, Faraday discovered electrolysis and introduced the nomenclature to describe it. He formulated Faraday's first law concerning the process of electrolysis and the second law on the electrochemical coefficient – they are of fundamental significance to the theory of electromagnetism. In 1845, he discovered paramagnetism and diamagnetism, taking a big step in the direction of the emergence of thermodynamics. He built the first generator and the first model of an electric motor. He introduced the concept of field lines of force and advanced the theorem that electric charges act upon each other through such a field as well as discovering the magneto-optic phenomenon. Citing the concept of point atoms and the forces of their infinite fields, he suggested that the lines of electric and magnetic force related to these atoms could in fact act as a medium with the help of which light waves are propagated. On the basis of this same concept, Maxwell later developed the theory of the electromagnetic field and in 1881, Helmholtz posited the hypothesis that electricity, both positive and negative, is divided into elementary portions that behave like electricity atoms[460].

Faraday was an experimental genius – not having a full mathematical education, he illustrated observed phenomena such as the lines of force of a magnetic field. His lack of deep mathematical knowledge was not a hurdle for him, but was more of supporting factor. He discovered laws of physics directly, through observations and experiments, not guided by the theories or beliefs of others. He was the initiator of many fundamental concepts, such as electrolysis, the electrolyte, the cation, anion, cathode, and anode. "I need new names," he wrote, "to express my discoveries in the science of electricity, however without involving an excessive theory which I won't know how to master"[461].

---

459   Ibidem, p. 1710, quote after: M. Litwinowicz-Droździel, *Indukcje i przepływy* (Inductions and flows), op. cit., p. 89.
460   Ibidem, p. 1573, quote after: M. Litwinowicz-Droździel, *Indukcje i przepływy* (Inductions and flows), op. cit., p. .
461   Ibidem, p. 1049, quote after: M. Litwinowicz-Droździel, *Indukcje i przepływy* (Inductions and flows), op. cit., p. 93.

A characteristic feature of Faraday's personality is his attitude towards lectures, for which he prepared with extraordinary care. He believed that the items needed to conduct a lecture and its accompanying demonstrations (and only they) should be on a table; all had to be in order and visible. A lack of transparency raises suspicions – the scientist appears like a suspicious schemer and a cheat. The language of the lecture also has to be appropriate: flowing, harmonious, simple, accessible. The experiments he conducted for public display, Faraday did not call 'experiments' but 'illustrations'. They should be characterised by minimalism – everyday items should be sufficient to conduct them (paper, a candle, water, a cup, a pencil, a piece of string; they could also be glass, a child's dummy or raisins). Failure is acceptable as only conjurers and jugglers are always successful in their arts. Illustrations are by no means intended to portray the horror or majesty of nature, nor the infallibility of the science that claims to represent it. They are intended help show that the learning process starts with being mindful of the world and observing the natural phenomena that reveal themselves constantly and spontaneously. He considered the most important aim of his lectures to be kindling the flame of knowledge; it should appear in the audience when they participate in a spectacle of science and they should leave the lecture hall in a state of mental exaltation[462].

Faraday was very devout and saw science and the exploration of nature as an extension of faith; he was a life-long active member of a small protestant church, of the benign Sandemanian Christian sect. Despite this, he was primarily not a philosopher but above all a brilliant classical physicist-experimenter, an adept of a science in which a mechanistic view of the world reigned.

In 1845, Kirchhoff announced laws enabling the calculation of the currents, voltages and resistances of electrical networks. In collaboration with Robert Bunsen (1811–1899), he developed chemical spectroscopy, which divides light into different wavelengths, enabling the determination of the chemical composition of object such as the Sun and other stars. They demonstrated that every element, after being heated to an incandescent state, emits a characteristic coloured light, which when refracted by a prism has an individual wavelength pattern specific to each element. Using this new research apparatus, they discovered two new elements. Kirchhoff established that when light passes through gas, it absorbs those wavelengths that it would emit after heating. This was the dawn of a new era, introducing a different way of searching for undiscovered elements.

---

462   M. Litwinowicz-Droździel, *Indukcje i przepływy* (Inductions and flow), op. cit., pp. 99–100.

Kirchhoff used this principle to explain numerous dark lines (Fraunhofer lines) in the Sun's spectrum. In 1959, he proved that the energy emission from a perfectly black body is only a function of temperature and frequency.

The discoveries of Lavoisier, Faraday and Kirchhoff proved that reality can be explained on the basis of known and accepted laws of physics and that even if new laws are discovered, they fit into the conventions of the philosophy of rationalism and empiricism. At the turn of 1855 and 1856, James C. Maxwell published his work *On Faraday's Lines of Force*, referring to Faraday's experimental research[463]. His aim in researching electricity and magnetism was to create mathematical frameworks underpinning Faraday's experimental results. The four Maxwell equations are recognised along with Newton's laws of motion and Einstein's theory of relativity as the most fundamental contribution to physics. Of all the discoveries of the 19th century, Maxwell's work had the most profound effect on the physics of the next century. His electromagnetic theory and the field equations related to it paved the way for Einstein's special theory of relativity. It was Maxwell who initiated the idea that in the 20th century took the form of quantum theory.

In the years 1861–1865, Maxwell formulated a theory of electromagnetism, unifying the interactions of electricity and magnetism and proving that electricity and magnetism are two aspects of the same phenomenon. The Maxwell equation introduced in 1861 showed that electrical and magnetic fields propagate in a vacuum in the form of a wave at the speed of light. Maxwell relied on mechanistic analogies and although his work ultimately contributed to the invalidation of mechanicism, it must be said that the way of perceiving nature that he presented was still mechanistic. Although the mechanical image of the world was destroyed by the 'great revolution', which will always be associated with Faraday, Maxwell and Hertz, and Maxwell must be credited with an essential contribution to achieving that revolution, the interpretation of the shift was made only in the 20th century, after Planck and Einstein. Although since the time of Maxwell we have believed that physical reality is described by continuous fields, and we treat that change in the conception of reality as the most profound and fruitful change in physics since the days of Newton, its interpretation made by contemporary physicists was mechanistic and hence they did not see the prospect that it opened up to physics.

---

463   J.C. Maxwell, *On Faraday's Lines of Force*, "Transactions of the Cambridge Philosophical Society" 1864, vol. X, part I, no. III, pp. 27–83.

Maxwell's research on electromagnetism assured him a place in the history of science along with the greatest scientists. In an attempt to illustrate Faraday's law of induction, he constructed a mechanistic model. He claimed that it led to the emergence of the appropriate 'displacement current' in a dielectric medium that could be the seat of transverse waves. By calculating the velocity of these waves, he discovered that they are very close to the speed of light and argued that it was difficult to avoid the conclusion that light is made up of transverse waves of the same medium that gives rise to electrical and magnetic phenomena[464].

According to Roger Penrose, Maxwell and Faraday were the creators of the third revolution in our views of physical reality. The first occurred in ancient Greece with the emergence of Euclidean geometry and the statistics of rigid bodies. The second was in the 17th century and its pioneers were Galileo and Newton explaining the movements of heavy bodies with reference to forces acting between the particles of which they are composed. The third, in the 19th century, showed that the notion of particles is insufficient to describe nature and that the existence of continuous fields permeating space has to be taken into consideration[465]. Penrose is right, with the proviso that in the last case, the importance of Maxwell's and Faraday's findings was only appreciated 150 years later. At the time they made their discoveries, their usefulness was understood at the level of mechanistic physics.

The transition from the philosophical to the scientific level can also be seen in the example of William Thomson, who published papers devoted to theories of electrical and magnetic phenomena and referred to Faraday's method of analysing electrical phenomena. In 1857, he observed a change in electrical resistance occurring under the influence of a magnetic field. Conducting research into the propagation of electrical impulses along cables, he came to develop the theory of electrical vibrations. He conducted preliminary research into the relationship between magnetism and electricity which was later developed by Maxwell. He presented a theory of oscillating currents to explain the phenomenon of electricity.

Originally, Thomson's beliefs were based on the assumption that the phenomena that create force as a result of their action, i.e. electricity, magnetism and heat, are the result of the action of an invisible material in constant motion. With time he changed his mind and presented a mathematical structure that

---

464   Ibidem, p. 497.
465   A. Einstein, *5 prac, które zmieniły oblicze fizyki* (Five papers that changed the face of physics) op. cit., pp. 7–8.

underlies experimental results in various fields of physics, such as heat, mechanical movement, fluid movement (gas or liquid), electricity and magnetism. He thus became the first to suggest that mathematical analogies exist between types of energy. He was the first physicist to try to treat Faraday's lines of force concept mathematically, which led Maxwell to the problem of the electromagnetic field. In 1848, Thomson developed an absolute temperature scale. Absolute zero was to be the lowest possible temperature, which according to him was -273.15°C. This was later dubbed in his honour 0°K (zero on the Kelvin scale).

In 1851, he formulated one of the rules of the second law of thermodynamics, stating that heat cannot be totally converted into work. He recognised that a key issue in the interpretation of this law is the explanation of irreversible processes. Thomson also formulated the principle of the dissipation of energy in the paper *On the Dynamic Theory of Heat*, in which he wrote that if entropy always grows, the universe will ultimately achieve a state of uniform temperature and maximum entropy, from which it would be impossible to extract any work. He described his theory as *Heat Death of the Universe*, which contributed to a lively discussion between physicists on the subject.

In 1847, Thomson analysed Joule's argument on the mutual exchangeability of heat and mechanical work and their mechanical equivalence and established that different forms of mechanical, electrical and heat energy are essentially the same and can be exchanged. In 1856, he discovered the so-called Thomson effect, consisting of the transfer of heat when electric current passes through a circuit composed of a single material. Both issues fitted in with a mechanistic composition of the universe.

In 1865, Rudolf Clausius (1822–1888) introduced the modern, macroscopic concept of entropy. His most famous paper, entitled *On the moving force of heat and the laws regarding the nature of heat itself which are deducible therefrom*, concerned the laws governing the relation between heat and mechanical work. Clausius created two laws of thermodynamics. The first states that there exists a constant relationship between the work performed and the heat produced as a result, or conversely – through the absorption of heat we can create work.

The second is connected with the observation that whenever heat is transformed into work, it is always accompanied by a certain amount of heat flowing from a hotter body to a colder one.

In 1879, Slovenian physicist Jožef Stefan discovered the law according to which the entire radiation emitted from a black body is proportional to the fourth power of its absolute temperature, and in 1844, Ludwig Boltzmann deduced this fact from thermodynamics, taking into account the second law of thermodynamics and the kinetic theory of gases. This theory is known as the Stefan-Boltzmann law.

According to Boltzmann, apart from Stefan, only Helmholtz realised the importance of Maxwell's theory. Other physicists were sceptical of it or even rejected it, mostly due to the appearance of incomplete or poorly developed vector analysis. Boltzmann published a series of papers in which he showed that the second law of thermodynamics, concerning energy exchange, can be explained through the application of mechanical laws and the theory of probability in relation to the movements of atoms. In 1872, he derived a theorem for the temporal development of the distribution function in phase space.

Boltzmann developed a general law of energy distribution and a theorem of energy equipartition, known as the Maxwell-Boltzmann distribution law. He derived an equation for the change in energy distribution between atoms occurring as a result of atomic collisions and laid the foundations of statistical mechanics. He explained that the second law is essentially statistical and that a system approaches a state of thermodynamic equilibrium (a uniform distribution of energy in the whole system), because equilibrium is the most probable state of a material system. He was the first physicist to perceive a decline in the mechanical-thermodynamic understanding of physics.

Boltzmann's greatest achievement was developing statistical mechanics, which explains and predicts how the properties of atoms (such as mass, charge and structure) determine the visible properties of matter (such as viscosity, thermal conductivity, and diffusion). His work on statistical mechanics was fiercely attacked and for a long time not understood, but the conclusions stemming from it were finally supported by the discoveries of atomic physics. It was recognised that fluctuation phenomena such as Brownian motion can be explained only through statistical mechanics and that the randomness of the behaviour of atomic objects, changing position in an uncontrolled way, can be described only through probability.

Boltzmann's entropy is proportional to the logarithm of the (average) number of states available to the system in given external conditions, denoted by W. The inverse of W can be treated as the probability of filling each available state. Boltzmann showed, in the example a perfect gas, that for an isolated system, entropy so-defined grows, achieving an asymptotic maximum value. He thus challenged the foundations of thermodynamics, proposing the fact be accepted that not everything can be defined and calculated with 100-percent certainty and that some results can be given only with a certain probability. Boltzmann was a 20th-century physicist and the equation $S = k \times \log W$ became the symbol of the end of the thermodynamic era.

The idea that matter and all complex things – water, fire, life – are subject to entropy and probability prompted serious changes in the physics world, but not without resistance from some scientists, including Ostwald and Mach, two of Boltzmann's greatest opponents. At the end of the 19th century, phenomenological views on the composition of matter dominated the beliefs of European physicists. Atomic science was considered obsolete, a theory destined to be forgotten as an unscientific fantasy. Sceptics of the legitimacy of atomic theory held as their chief argument the incompatibility of the theory's main assumptions with everyday observations. Another hurdle hampering the acceptance of atomic theory was the reversibility of natural phenomena assumed in kinematics.

The dispute between energeticists, representing a phenomenological perception of matter, and advocates of the atomistic conception, was not superficial as it touched on both the understanding of the essence of science and the paradigms underpinning scientific thought. It was claimed that explaining physical events with the aid of probability calculus was contrary to the experience of everyday life, hence theories based on such mathematics constituted mere mental speculation. One burden was the Platonic understanding of reality, still present in science, and its tendency to "mistake the map is for the territory"[466], with philosophical speculations being more trusted than mathematics.

In 1900[467], Planck proposed quantum theory using Boltzmann's statistical theory but the latter had already used the notion of the quantisation of energy in his work in 1872, some 28 years before Planck's papers, dividing the energy of a system into very small and separate packets. Hence Boltzmann and Planck are known as the father and mother of the quantum. Boltzmann conceived of quantisation as a sort of mathematical trick enabling the application of combinatorial equations in probability calculations. Energy quanta did not appear further in the final equations, but there is no doubt that through his approach, Boltzmann helped pave the way for quantum theory[468].

---

466 N.N. Taleb, *The Black Swan*, op. cit. p xxix.
467 In the years 1897–1899, Max Planck presented his thesis related to quantum theory numerous times. In 1899, he presented the value of Planck's constant in 'Sitzungsberichte der Königlich Preussischen Akademie der Wissenschaften zu Berlin. Mitteilung' ('Session reports of the Royal Prussian Academy of Sciences at Berlin. Communication'). Historically, the date of quantum theory's introduction is considered to be December 14, 1900, the publication date of *Über irreversible Strahlungsvorgänge* (On Irreversible Radiation Processes), "Annalen der Physik" ("Annals of Physics"), 1900, vol. 306, No. 1, pp. 69–122.
468 See A. Eftekhari, *Ludwig Boltzmann (1844–1906)*, https://pdfs.semanticscholar.org/5c96/924ab515da7ebb6cb7601ec916099b03aed0.pd f (access: 14.01.2021), p. 21.

Non-equilibrium thermodynamics, or the thermodynamics of irreversible processes, also of huge importance to philosophy, concerns the thermodynamic process causing an increase in the total entropy of a system and its vicinity. The name suggests it is impossible to reverse an irreversible process, however, due to the statistical nature of thermodynamic phenomena, reversal is possible though the probability of it occurring is close to zero. A logical consequence of adopting the kinetic theory of matter was the acceptance of Boltzmann's hypothesis of the reversibility of phenomena. As a consequence, ice in a glass of water, hypothetically, according to Boltzmann's hypothesis and kinetic theory, can cool down on its own and simultaneously warm up the water in the glass though this is such an improbable phenomenon that in practice it is ignored[469].

Boltzmann had the greatest influence on research into non- equilibrium states, irreversibility and the irreversible process due to the combination of the kinetic theory of gases and thermodynamics. He built models based on the behaviour of atoms at a time when the existence of atoms had yet to be proven.

In 1896, Wilhelm Wien (1864–1928) published a formula to determine the composition of black-body radiation. Wien's work enabled Planck to solve the problem of radiation in thermal equilibrium using quantum physics.

A basic conclusion resulting from the equations formulated by Maxwell in 1864 was the existence of electromagnetic waves. In 1886, they were discovered by Heinrich Hertz and in the years 1887–1888, he proved that electrical energy moves in waves with a frequency that can be calculated. The German scientist was the first to produce waves (called Hertz waves) using an electrical oscillator. He was convinced that these waves would have no practical application and believed they were irreversibly dissipated in space. However, the experiments he conducted led to the invention of wireless telegraphy.

This was a great achievement of physics in serving technology and it fitted in perfectly with the 20th-century model of mechanistic physics.

In 1892, Hendrik Lorentz advanced the idea of introducing the discrete structure of electricity to the Maxwell equations. He assumed the existence of ether as an unchanging dielectric devoid of internal movements and not subject to mechanical forces as well as a substance composed of only elementary particles of positive or negative electricity. Ether constituted an element of building physical theories in a philosophical convention.

In 1896, a student of Lorentz's, Pieter Zeeman, discovered the phenomenon of light spectrum lines splitting under the influence of a magnetic field and Lorentz

---

469   Ibidem, pp. 19–20.

provided a theoretical explanation of the phenomenon within the framework of electron theory. For this explanation, Lorentz used the term "charged particles", changing it in 1895 to "ions" and only in 1899 started to speak of them as "electrons". In 1895, Lorentz introduced the concept of local time (different times for different locations), and discovered that bodies in motion approaching the speed of light contract in the direction of the movement. He found that if instead of Galilean transformations, describing the transition to a moving frame of reference, other transformations were to be used (named Lorentz transformations by Einstein in his honour), the Maxwell equations concerning the propagation of light remain unchanged. Lorentz transformations cause the equations of mechanics to undergo a change that seemed absurd at the time, but Einstein proposed a modification to the mechanical equations so that the Lorentz transformations did not change their form, so it could be said that to a certain degree, Lorentz was a forerunner of that theory.

On November 8, 1895, Wilhelm Röntgen (1845–1923) discovered so-called Rentgen (X-ray) radiation. In 1901, when the first Nobel Prize was awarded in the field of physics, it was given to Röntgen.

Joseph John Thomson (1856–1940) assumed that the atom is a sphere of positively charged matter in which electrons are suspended. However, it was not clear in what way these protons and electrons were arranged in the atom. Thomson chose the 'plum pudding' model in which protons and electrons are evenly distributed. In the book *Recollections and Reflections* he wrote:

> At first there were very few who believed in the existence of these bodies smaller than atoms. I was even told long afterwards by a distinguished physicist who had been present at my lecture at the Royal Institution that he thought I had been "pulling their legs". I was not surprised at this, as I had myself come to this explanation of my experiments with great reluctance, and it was only after I was convinced that the experiment left no escape from it that I published my belief in the existence of bodies smaller than atoms[470].

According to Thomson, all matter, regardless of its source, contains particles of the same type, which are much less massive than atoms and constitute their parts.

In 1896, Antoine Henri Becquerel discovered radioactivity. Between June and December 1896, Pierre Curie (1859–1906) and Maria Skłodowska-Curie discovered two radioactive elements: polonium and radium. During the Nobel Prize award ceremony in 1911, the president of the Royal Swedish Academy of Sciences said to Maria Skłodowska-Curie:

---

470   J.J. Thomson, *Recollections and Reflections*, London 1936, p. 341.

Madam. In 1903, the Swedish Academy of Sciences had the honour of conferring upon you the Nobel Prize for Physics for the part which you, together with your late husband, took in the momentous discovery of spontaneous radioactivity.

This year, the Academy has decided to award you the prize for Chemistry in recognition of the eminent services you have rendered to this science by your discovery of radium and polonium, by your description of the characteristics of radium and its isolation in the metallic state, and by your research into the compounds of this remarkable element. During the eleven years in which Nobel Prizes have been awarded, this is the first time that the distinction has been conferred upon a previous prizewinner. I beg you, Madam, to see in this circumstance a proof of the importance which our Academy attaches to your most recent discoveries, and I invite you, Madam, to receive the prize from His Majesty the King, who has graciously consented to present it to you[471].

On December 14, 1900, Max Planck published the paper *On the irreversible processes of radiation*, which changed the face of physics. In it, he described a revolutionary idea, according to which the energy emitted by a resonator can only take discrete values or quanta. Planck's new thesis saw the light of day following afternoon tea at the Planck household on October 7, 1900, during which Heinrich Rubens told Planck that the results of measurements showed a deviation from the predictions of Wien's second law, defining the radiation distribution of a perfectly black body. This led Planck to the thought of improving Wien's formula and consequently led to the formulation of a new theory of black body radiation. He presented the results of his deliberations on October 19, 1900, at a meeting of the German Physical Society in Berlin. This was a turning point in the history of physics. The significance of this discovery, with its far- reaching impact on classical physics, was not at first appreciated. However, proof of its validity gradually became overwhelming as the application of this theory explained many discrepancies between observed phenomena and classical theory.

Surprising in this context is the issue of Planck's attitude towards atomic theory, of which he was an opponent. He changed his mind only at the turn of the century while conducting research into black body radiation[472]. Planck had to give up one of his most cherished convictions, that the second law of thermodynamics was an absolute law of nature. In place of that, he was forced to accept Boltzmann's interpretation, that it is only a statistical law.

---

471 Speech by Erik Wilhelm Dahlgren, head librarian of the National Library and President of the Royal Swedish Academy of Sciences, on December 10, 1911, https://www.nobelprize.org/prizes/chemistry/1911/ceremony-speech/ (9.04.2020).

472 See J. Bernstein, *Einstein and the Existence of Atoms*, op. cit., p. 865.

Between March 18 and December 19, 1905, Albert Einstein wrote six papers which altered the nature of physics. They concerned the theory of relativity, the photoelectric effect and the theory of Brownian motion. According to Roger Penrose, by publishing the first five works, Einstein started the fourth revolution in physics, manifesting in the way of perceiving nature (there was discussion earlier of the three previous revolutions). John Stachel (born 1928), a physicist and director of the Boston University Centre for Einstein Studies, believes that these five papers caused the year 1905 to be known in the general discourse of physics as the 'miraculous year' analogous to 1666, known as the *annus mirabilis*. Einstein's papers laid the foundation of a revolution in the physics of the 20th century[473].

On April 30, 1905, Einstein finished his doctoral dissertation titled *Eine neue Bestimmung der Moleküldimensionen* (A New Determination of Molecular Dimensions), which went on to become one of his most cited works, although it does not concern the theory of relativity. On May 11, 1905, Einstein's work *On the Movement of Small Particles Suspended in Stationary Liquids Required by the Molecular-Kinetic Theory of Heat*[474] was published, explaining Brownian motion. It contained novel physical concepts arguing for the atomic structure of matter. Many years later, John Stachel, the publisher of a multi-volume collection of Einstein's writings, claimed that the research he conducted showed the visible influence on the above-mentioned Einstein paper of another paper: *On irregularities in the distribution of gas molecules and their effect on entropy and the equation of state*, written by Marian Smoluchowski in 1903. As mentioned earlier, Einstein had reviewed a book in which that paper was included. Smoluchowski's work contained suggestions related to research and conclusions on Brownian motion, which the Polish physicist had already fully developed in 1903. Smoluchowski did not publish his work until 1906, as a result of which all the glory for discovery of the phenomenon went to Einstein.

On June 30, 1905, Einstein's best-known work was published: *On the Electrodynamics of Moving Bodies*[475], which invalidated Newton's absolute time, simultaneously giving great importance to a physical constant – the speed of light, which initiated the era of relativistic physics.

---

473 Einstein's Miraculous Year: Five Papers That Changed the Face of Physics, op. cit. p 3.
474 Idem, *Über die von der molekularkinetischen Theorie der Wärme geforderte Bewegung von in ruhenden Flüssigkeiten suspendierten Teilchen*, op. cit.
475 Idem, *Zur Elektrodynamik bewegter Körper*, op. cit.

September 27, 1905, saw the publication of the fourth and shortest paper[476], titled *Does the Inertia of a Body Depend Upon its Energy Content?* It contained the derivation of the most famous formula in the world: $E = mc2$, although it did not appear in the article in this form as Einstein only performed the proof leading to the equation.

On March 18, 1905, the paper *On a Heuristic Point of View Concerning the Production and Transformation of Light*[477] appeared. The scientist believed that only this publication had truly revolutionary value. He showed in it that light can be treated not only as a wave but also as a collection of particles – quanta of energy. This paper is currently cited as the work that contained the explanation of measurement results of the photoelectric phenomenon, but its content is much broader and concerns in large measure dark body radiation, including analyses of radiation entropy. Einstein thus became one of the founders of quantum physics.

On December 19, 1905, Einstein finished and sent to the editorial office of *Annalen der Physik* a second paper as a complement to his earlier deliberations on Brownian motion: *On the theory of Brownian motion*[478].

On July 20, 1906, Marian Smoluchowski published a paper on Brownian motion in *Annalen der Physik* under the title *Zarys teorii kinetycznej ruchów Browna i roztworów mętnych* (An outline of the kinetic theory of Brownian motion and cloudy solutions). This paper, together with Einstein's earlier publications, constitutes a breakthrough in the perception of the atomic hypothesis proclaimed by some physicists and a resolution to the dispute ongoing at the time on the structure of matter.

Einstein, Smoluchowski and Sutherland discovered that the nature of Brownian motion is purely physical. Atoms and molecules, being in constant chaotic motion, surround a suspension particle and collide with it. As a result of the collisions of molecules, the particle makes uncontrolled movements of rapidly changing velocity and direction. A mathematical description of Brownian motion was one of the elements that contributed to the emergence of the theory of stochastic processes. Jean Perrin quantitatively tested the

---

476 Idem, *Ist die Trägheit eines Körpers von seinem Energieinhalt abhängig?*, „Annalen der Physik"1905, vol. 18, pp. 639–641.
477 Idem, *Über einen die Erzeugung und Verwandlung des Lichtes betreffenden heuristischen Gesichtspunkt*, „Annalen der Physik" 1905, vol. 17, pp. 132–148.
478 Idem, *Zur Theorieder Brownschen Bewegung*, op. cit.

Einstein-Smoluchowski formula, showing that the mean squared shift in a given direction is proportional to time.

In 1906, Walter Nernst posited the third law of thermodynamics, assuming it is impossible to achieve a temperature of absolute zero (zero Kelvin) in a finite number of steps if a non-zero absolute temperature is taken as a starting point.

In 1909, Greek mathematician Constantin Carathéodory (1873– 1950) developed an axiomatic system of thermodynamics. He presented in it all previously formulated laws of thermodynamics using a mathematical tool.

In April 1911, Ernest Rutherford performed an important experiment. Negative electrons, compensating for the positive charge of the nucleus, regarded as travelling in circular orbits around the nucleus, and the force of attraction between the electrons and nucleus Rutherford compared to the force of gravity between revolving planets and the Sun. Most of the atom was empty space which offered no resistance to the passage of the alpha particles. In this way, Rutherford confirmed experimentally the existence of the atomic nucleus.

July 1913 saw the publication of the first part of Niels Bohr's paper *On the Constitution of Atoms and Molecules*[479], and the other two parts were published in September and November. Bohr referred to Rutherford's model of the atom, which was supposed to consist of a heavy, positively charged nucleus with much lighter negatively charged electrons orbiting it at a considerable distance. Despite the fact that according to the laws of classical physics such a system should be unstable, Bohr adopted made assumptions concerning the electrons circling in the atom in defined circular orbits. According to the first assumption, electrons may not emit any radiation and their energy may be stable. The second assumption foresaw the possibility of an electron changing its orbit from one to another, whereby this jump was to be accompanied by the emission or absorption of radiation of a defined frequency. Bohr's paper was received with incredulity by proponents of classical physics, and two future Nobel Prize winners – Otto Stern (1888–1969) and Max von Laue (1879–1960) vowed that "If this nonsense of Bohr should prove to be right in the end, we will quit physics"[480].

The chronologically presented philosophical outline of the discoveries in physics and chemistry in the 19th and early 20th centuries shows the changes that occurred in the methodology of science over the space on more than a hundred years. The mechanistic- thermodynamic perception of nature was initially

---

479   N. Bohr, *On the Constitution of Atoms and Molecules*, "Philosophical Magazine and Journal of Science" 1913, vol. 26, no. 1, pp. 1–25.
480   A.K. Wróblewski, *Historia fizyki* (The History of Physics), op. cit., p. 453.

conducive to the development of physics, prompting a departure from the perception of nature in qualitative terms and a transition towards quantitative terms. Over time it became a limitation due to the extremely rationalistic construction of science, to the point of opposition to creating scientific hypotheses, which hampered its development. From the time of Maxwell, different attitudes are visible in scientists' perceptions of nature. The vast majority opt for a straightforward compatibility relationship between theory and observation and only a few treat observations as a contribution to the building theories and hypotheses. A breakthrough was the acceptance of atomic theory, quantum mechanics and the theory of relativity.

Writing in 1956 about Marian Smoluchowski's travels around Europe, Stanisław Ulam noted that

> experiences from travel provide very lively impressions of the genetic development of many ideas of contemporary physics in the germination phase. In physics, this period was perhaps one of the most interesting prior to the absolute violation of the theory of relativity and quantum theory. Suspicions can be sensed of the approaching revolution in the foundations of physics, and because perhaps the current time also gives a similar impression it may be instructive to look back and review the appearance at this period of the signs and portents of change[481].

Similarly to 120 years earlier, when the imminent decline of physics was predicted, at the end of the 20th century and the start of the 21st, many scientists foretell the end of philosophy, particularly of natural philosophy. Numerous physicists have referred to that field with some contempt, arguing that it is a science that essentially lacks a methodology of its own and its subject of activity is systematically shrinking and being taken over by the specific sciences.

However, the situation seems to be reversing somewhat since it is physics, supposedly being in bloom, that increasingly often needs the support of philosophy.

Sabina Hossenfelder's cited deliberations support this thesis, drawing conclusions concerning the verification of the theories emerging in modern physics[482]. Smoluchowski's utility criterion is more culturally burdened in the complex current conditions than would result from the rational functioning of science.

Nima Arkani-Hamed claims that naturalness is neither a principle nor a law and has been treated as a guide. According to some, naturalness is pure philosophy, although it has helped theoretical physicists achieve successes.

---

481 S. Ulam, *Marian Smoluchowski and the Theory of Probabilities in Physics*, op. cit., p. 476.
482 S. Hossenfelder, *Lost in Math*, op. cit.

Despite this, some have abandoned it in favour of another idea – so-called split supersymmetry[483].

Theoretical physicists have much cause for complaint about the laws of physics discovered so far. They especially dislike numbers bearing the hallmarks of unnaturalness, which have served in the process of developing natural theories since at least the 16th century[484]. In today's theoretical physics, practices of creating new natural laws have been established, whose laws will not be verifiable for a long time. Contact with philosophy could help physicists define what questions are worth asking, however currently such contact is sporadic. Basing theoretic physics mostly on aesthetic criteria and the resulting lack of progress are evidence of the inability within this field to self-correct[485].

Citing simplicity or beauty as a result of achieving confirmation in science is implied in the attitude towards the alleged facts prevailing in a given field. If our system represents any kind of cyclical process, we will quickly review the sine or cosine functions, the polynomial expansion of which has infinite possibilities, in order to be able to use them in a justification of the hypothesis at hand. In short, we have no universal schema or universal formal rules that let us define what is simpler or the simplest, which in essence deserves the term beauty. As much as possible, we choose variables and functions appropriate to the facts that, in our opinion, prevail. These facts are material postulates of inference to adjust the simplest curve[486]. Agreements concerning universal principles defining what is appropriately simple are important but secondary since, in contrast to what Norton asserts, we have an instrumental universal principle enabling definition not of what is simpler or beautiful but what is more useful. That which is more useful contains within itself that which is simpler because while agreement on what is simpler constitutes a postulative statement, discovering what may be more useful constitutes a substantive argument verified in daily life.

Despite the changes that have occurred in the development of physics and natural cosmology, one feature has remained constant. This is the quantitative approach to the world. It is believed tha mathematical models correspond objectively to a given order, which is more than the world taken perceptually. Starting from the times of Galileo, only quantitative aspects of the features of things have

---

483   W. Heinsenberg, *Część i całość. Rozmowy o fizyce atomu*, (The part and the whole. Talks in the vicinity of atomic physics), trans. K. Napiórkowski, Warszawa 1987, p. 119.
484   S. Hossenfelder, *Lost in Math*, op. cit., p. 128.
485   Ibidem, p. 316.
486   J.D. Norton, *A Material Theoryof Induction*, op. cit., pp. 655–657.

been taken into account. They are believed to be primary qualities and – being objective – are contrasted with subjective, sensually perceptible qualities[487].

The crisis in physics Hossenfelder writes about is qualitatively similar to the one 120 years ago. Then, the mechanistic perception of nature focused on a quantitative approach to the world convinced physicists of the end of the development of physics. A mechanistic and thermodynamic understanding of nature became a limitation to the understanding of reality and that had to be changed. Today, a similar limitation is the conception of the standard model, the limitations of which can be overcome by scientists similar to Planck and Einstein at the start of the 20th century. Recently, unusual signals have been recorded that might suggest the discovery of unknown properties of neutrinos or even the attainment of proof for the existence of the axion, a hypothetical particle beyond the standard model.

Kathryn Zurek, a physicist-theoretician at the California Institute of Technology, admits that if the signals come from axions, the chief candidates for the particles creating dark matter, or from non-standard neutrinos, the recorded signals might confirm that neutrinos have a strong magnetic moment. A neutrino with a magnetic moment, like the axion, does not fit into the standard model[488].

---

[487] Z. Hajduk, *Filozofia przyrody. Filozofia przyrodoznawstwa. Metakosmologia* (The Philosophy of nature. The Philosophy of natural science. Metacosmology), Lublin 2007, p. 58.

[488] K. Zurek, *High Energy Theory, Particle Astrophysics and Early Universe Cosmology*, http://www.kzurek.theory.caltech.edu/ (acess: 15.01.2022).

Chapter XIX Children

Marian Smoluchowski had two children with Zofia Baraniecka, daughter of Jagiellonian University mathematics professor Marian Baraniecki, whom he married in 1901 team. Aldona's mother, Zofia, was inconsolable that her daughter had fallen in love with a jockey. He was a rather unusual horseman though – during this visit he also negotiated a loan that his bank granted to Poland. Apart from horses, his second passion was aviation and he was one of the first Americans to gain a pilot's licence (Naval Aviator Force no 145) with so-called 'Gold Wings' for achievements in aviation. In addition, in the Second World War, Read served as the assistant chief of staff on Guam in the Pacific. His third passion was astronomy – he witnessed almost every solar eclipse in the 20th century (he died in 1998 at the age of 101 having been a widower for 14 years). Working in a bank was merely an unpleasant necessity for him and Aldona quite quickly took control of the domestic finances[489].

---

489 J. Zakrzewski, *Dzieci Mariana Smoluchowskiego* (Marian Smoluchowski's children), "Pauza Akademicka" (the weekly of the Polish Academy of Arts and Sciences) 2017, no 380–381, p. 11.

Aldona (Donia) Smoluchowska (1902–1984)

Children

Donia and Romek Smoluchowski (1910)

Zofia Smoluchowska, Romek, Teofila Smoluchowska and Donia (Aldona)

Aldona and her husband, Duncan H. Read, with their children

Roman Smoluchowski was born in Zakopane (reflecting the alpine interests of his father, Marian). He was left fatherless at the age of seven and apparently did not remember his father well. According to the family, he remembered his black beard soaked in tomato soup. He graduated from the V High School. A. Witkowski, opposite which the Smoluchowskis lived[490].

Zofia Smoluchowska with son Roman

---

490 Ibidem.

Following the outbreak of the Second World War, Roman tried several times to escape from occupied Poland – legend has it that among the detainees he was recognised by a Ukrainian member of a firing squad, also a physicist. The American Astronomical Society also writes in memory of Roman Smoluchowski about the help given to him by his sister and her friend – the Queen of Sweden. As a result, Roman made it to the United States where, thanks to the help of Eugene Wigner, with whom he worked during a post-doctoral internship at Princeton, he quickly became naturalised and worked as a physics instructor at Princeton and later at the General Electric Research Laboratory in Schenectady, NY. There he gained access to secret German documents in which he found his own name as a person earmarked for "liquidation". He worked at the Carnegie Institute of Technology in Pittsburgh and then at Princeton University to later find himself – in the footsteps of Steven Weinberg among others – in Austin, where there were no strict retirement rules in force[491].

The main achievements of Roman Smoluchowski's early research period concern solid-state physics. An article on symmetries in solid-state physics (in crystals)[492], written with Wigner and L.P. Bouckaert, and the most cited work on the anisotropic work function of the exit from metal[493], has achieved more than a thousand citations. To this day, after several decades, the second of them gets over 200 citations a year[494]!

After moving to Texas, Roman became increasingly interested in astrophysics and the physics of planets. In 1991, in recognition of his contribution among other things to research on the atmosphere of Jupiter, on his eightieth birthday, a planetoid was named after him[495].

Roman Smoluchowski also had many other interests (from activities in the American Physical Society, where he was a co- founder of the section on condensed phase physics, to painting – he took part in paining courses while already in his pre-retirement years). Apart from strictly scientific work, he also enthusiastically embraced the popularisation of physics[496].

---

491   Ibidem.
492   L.P. Bouckaert, R. Smoluchowski, E. Wigner, *Theory of Brillouin Zones and Symmetry Properties of Wave Functions*, "Crystals. Physical Review" 1936, no. 50, pp. 58–67.
493   R. Smoluchowski, *Anisotropy of the Electronic Work Function of Metals*, "Physical Review" 1941, no. 60, pp. 661–674.
494   J. Zakrzewski, *Dzieci Mariana Smoluchowskiego* (Marian Smoluchowski's children), op. cit., p. 11.
495   Ibidem.
496   Ibidem.

Children

Roman Smoluchowski

Roman Smoluchowski

In 1948, Roman met a staffer at the editorial office of *Physics Today*, Louise Riggs, who came from a naval family (which has given the United States many admirals, including Louise's father). In 1950, they got married and had two children: Peter and Irena.

Zofia Smoluchowska with son Roman, his wife Louise, and grandson Peter (1952)

Roman never returned to Poland. As he said, it would have been too difficult for him. However, he maintained contact with Poles, not only with his family. He died in January 1998[497].

Zofia Smoluchowska (circa 1952)

---

497 Ibidem.

Chapter XX Commemoration

 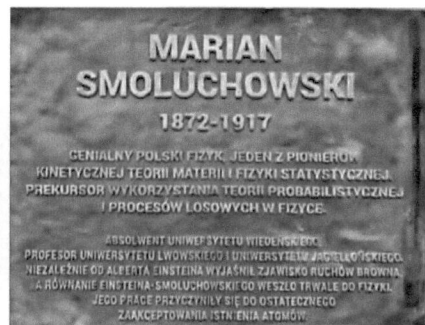

Marian Smoluchowski's presence in contemporary scientific life is represented by three prizes awarded to outstanding physicists and named after him.

The Marian Smoluchowski Medal, established in 1965, is the highest distinction conferred by the Polish Physical Society (PTF) for scientific achievement in the field of the physical sciences. It is awarded every two years during the Congress of Polish Physicists or at a special ceremony organised by the Board of the PTF. The winner gives a special lecture at the ceremony in memory of Marian Smoluchowski. Among its laureates are three Nobel Prize-winning physicists: Ben R. Mottelson (1975), Subrahmanyan Chandrasekhar (1983), and Vitaly Lazarevich Ginzburg (2003)[498]. It is worth emphasising that Chandrasekhar and Lazarevich received their medals before they won the Nobel, which demonstrates the accuracy of the Awards Committee's selections.

---

498 M. Kluza, *Pod przewodnią gwiazdą nauki: Marian Smoluchowski w stulecie śmierci* (Under the guiding star of science: Marian Smoluchowski on the centenary of his death) https://jbc.bj.uj.edu.pl/dlibra/publication/492587/edition/467084/con- tent (dostęp: 15.01.2022).

The Marian Smoluchowski Medal

List of Marian Smoluchowski Medal winners[499]

1965	Wojciech Rubinowicz	Warzaw
1968	Aleksander Jabłoński	Toruń
1969	Jerzy Pniewski	Warsaw
1969	Marian Danysz	Warsaw
1970	Marian Mięsowicz	Kraków
1970	Jerzy Gierula	Kraków
1972	Leonard Sosnowski	Warsaw
1973	Subrahmanyan Chandrasekhar	Chicago
1974	Georgij Nikołajewicz Florow	Dubna
1975	Gerald Leondus Pearson	Stanford
1976	Arkadiusz Henryk Piekara	Warsaw
1977	Victor Frederick Weisskopf	Cambridge
1979	Włodzimierz Trzebiatowski	Wrocław
1980	Ben R. Mottelson	Copenhagen
1981	Adriano Gozzini	Pisa
1982	Władysław Opęchowski	Vancouver
1983	Jan Rzewuski	Wrocław
1984	Witalij Łazarewicz Ginzburg	Moscow
1985	Joseph Henry Eberly	Rochester
1986	Andrzej Trautman	Warsaw

---

499 Ibidem.

1987	Wojciech Królikowski	Warsaw
1988	Andrzej Hrynkiewicz	Kraków
1989	Zdzisław Szymański	Warsaw
1990	Władysław Świątecki	Berkeley
1991	Jacek Prentki	Paris, Geneva
1992	Arnold Whittaker Wolfendale	Durham
1993	Stanisław Kielich	Poznań
1994	Ryszard Sosnowski	Warsaw
1997	Włodzimierz Zawadzki	Warsaw
1998	Kacper Zalewski	Kraków
1999	Andrzej Kajetan Wróblewski	Warsaw
2000	Bohdan Paczyński	Princeton
2001	Aleksander Wolszczan	Toruń Pennsylvania State University
2002	David Shugar	Warsaw
2003–2004	Stefan Pokorski	Warsaw
2003–2004	Andrzej Białas	Kraków
2005	Jan Żylicz	Warsaw
2007	Robert R. Gałązka	Warsaw
2008	Józef Barnaś	Poznań
2009	Wojciech Żurek	Los Alamos
2010	Tomasz Dietl	Warsaw
2011	Krzysztof Pomorski	Lublin
2012	Douglas Cline	Rochester
2013	Jan Misiewicz	Wrocław
2015	Henryk Szymczak	Warsaw
2017	Jerzy Lukierski	Wrocław

The Marian Smoluchowski–Emil Warburg Prize is awarded biannually for outstanding achievement in the field of pure or applied physics. Its winner can be a scientist whose work is related to a scientific centre in Poland or Germany. The prize's symbol is a medal made of silver and the financial reward accompanying it was funded by the Meyer-Viol Foundation. The Awards Ceremony is held either during the Congress of Polish Physicists or during the Spring meeting of the German Physical Society (*Deutsche Physikalische Gesellschaft*)[500].

---

500   Ibidem.

The Marian Smoluchowski-Emil Warburg Prize

List of Smoluchowski-Warburg Prize winners[501]

1998	Andrzej Białas	Kraków
1999	Ludger Wöste	Berlin
2001	Janusz Zakrzewski	Warsaw
2003	Fritz Haake	Essen
2005	Andrzej Warczak	Kraków
2007	Andrzej Buras	Munich
2009	Andrzej L. Sobolewski	Warsaw
2011	Peter Fulde	Dresden
2013	Krzysztof Redlich	Wrocław
2015	Werner Hofmann	Heidelberg
2017	Andrzej M. Oleś	Kraków

The Smoluchowski Award

The Smoluchowski Award is given annually by the international Association for Aerosol Research (*Gesellschaft für Aerosolforschung*, GAeF), headquartered in Germany. It is awarded to young scientists for outstanding achievement in aerosol research made within the previous three years. The distinction may be bestowed upon one or two laureates. A diploma and financial reward of EUR 2,000 is given during the annual European Aerosol Conference[502].

---

501 Ibidem.
502 Ibidem.

Commemoration

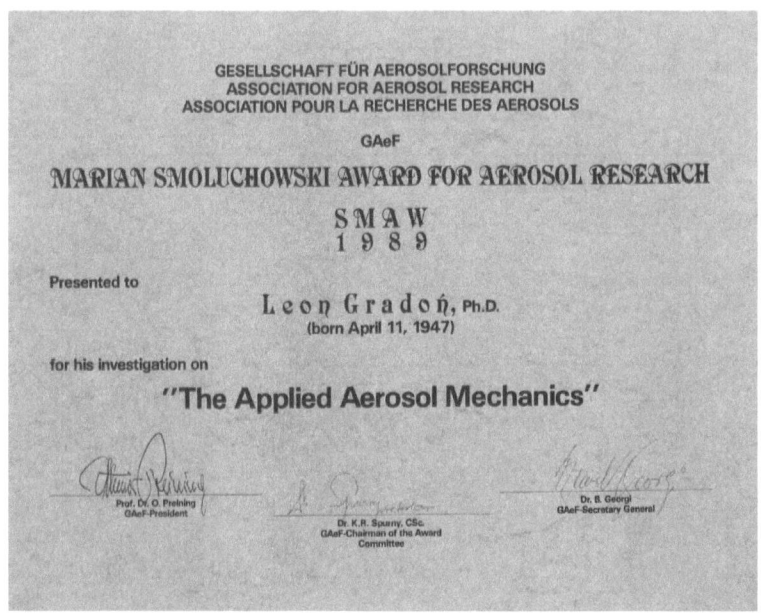

The Smoluchowski Award

List of Smoluchowski Award winners[503]

1986	Paul E. Wagner	Vienna
1987	Helmut Bunz	Karlsruhe
1988	Andreas Schmidt-Ott	Zurich
1989	Leon Gradoń	Warsaw
1990	Richard C. Flagan	Pasadena
1991	Reinhard Niessner	Munich
1992	David Y.H. Pui Andrey V. Filippov	Minneapolis Duisburg
1993	Vladek Szymanski	Vienna
1994	Heinz Burtscher Wolfgang Möller	Zurich Dresden
1995	Sotiris E. Pratsinis Michael Shapiro	Cincinnati Haifa
1996	Sergey Grinshpun	Cincinnati

---

503   Ibidem.

1997	Markku Kulmala	Helsinki
1998	Kimberly A. Prather Alfonso M. Gañán-Calvo	San Diego Seville
1999	Ian J. Ford	London
2000	Einar Kruis	Duisburg
2001	Alfred P. Weber D.A. Edwards	Clausthal Newark
2002	Da-Ren Chen	Richmond
2003	Ignacio González Loscertales	Malaga
2004	Colin O'Dowd	Galway
2005	Lutz Mädler	Zurich
2006	Jean-Pascal Borra	Paris
2007	Veli-Matti Kerminen	Helsinki
	Markus Kalberer	Bern
2009	Suwan Jayasinghe	London
	Wendelin J. Stark	Zurich
	Albert G. Nasibulin	Helsinki
2012	Not awarded	
2013	James D. Allan Christopher J. Hogan Jr. Maosheng Yao	Manchester Minneapolis Beijing
2014	Ilona Riipinen	Stockholm
2015	Andreas Kürten	Frankfurt
2016	Alexandra Teleki	Basel

Having such an outstanding figure as Smoluchowski in our science, we should cultivate memory of him. His achievements should be discussed at least briefly in physics textbooks so that teachers (to whose education in the field of the strict sciences he devoted much work and time) could tell their students about him. Smoluchowski wrote a textbook in which every contemporary high school teacher could find valuable tips on specific subjects as well as general comments on the subject of teaching. I believe that today too, many teachers of mathematics, physics or chemistry could gain a great deal by reading the *Self-Study Handbook*, written over a hundred years ago.

The Self-Study Handbook

Commemorative postcard

Marian Smoluchowski's grave

The figure of the great physicist Marian Smoluchowski should be brought back to Polish science and Polish culture – it would surely benefit everybody greatly.

Annex

List of Marian Smoluchowski's published works

The author of the chronological list of Smolochowski's published works presented below is Maciej Kluza[504]. It is not a complete list and does not cover the whole bibliography, however it presents the vast majority. The author accidentally omitted a few titles, but such situations are the exception. The list does not include handwritten manuscripts which can be found in digitised form, but there are also very few of those. Within the scope of this book's subject matter, the following list provides comprehensive information on the Polish physicist's scientific legacy.

Smoluchowski M., *On internal friction in non-aqueous solutions*, „Sitzungsberichte der Akademie der Wissenschaften mathematisch-naturwissenschaftliche Klasse" ("Meeting reports of the Academy of Sciences mathematical and scientific class") 1893, Division IIa, vol. 102, pp. 1136–1140; reprinted in: *Pisma Mariana Smoluchowskiego z polecenia Polskiej Akademii Umiejętności zgromadzone i wydane przez Władysława Natansona (The Writings of Marian Smoluchowski commissioned by the Polish Academy of Arts and Sciences collated and published by Władysław Natanson)*, vol. 1, Kraków 1924, pp. 1–4.

Smoluchowski M., *Akustische Untersuchungen über die Elastizität weicher Körper* (*Acoustic studies on the elasticity of soft bodies*) „Sitzungsberichte der Akademie der Wissenschaften mathematisch-naturwissenschaftliche Klasse" ("Meeting reports of the Academy of Sciences mathematical and scientific class") 1894, Division IIa, vol. 103, pp. 739–772; reprinted in: *Pisma Mariana Smoluchowskiego z polecenia Polskiej Akademii Umiejętności zgromadzone i wydane przez Władysława Natansona (The Writings of Marian Smoluchowski commissioned by the Polish Academy of Arts and Sciences collated and published by Władysław Natanson)*, vol. 1, Kraków 1924, pp. 5–35.

---

504 See. M. Kluza, *Pod przewodnią gwiazdą nauki* (Under the guiding star of science), op. cit.; (editor's note) notations have been adapted to the conventions of the book.

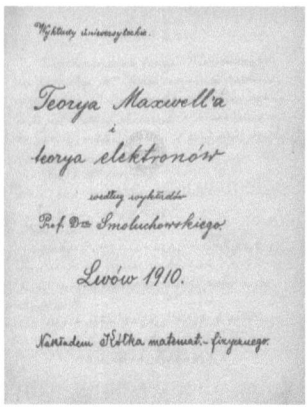

Maxwell's theory and the theory of electrons

Smoluchowski M., Lippmann M., *Recherches sur la dèpendance entre le rayonnement d'un corps et la nature du milieu environnant* (Research on the dependence between the radiation of a body and the nature of the surrounding medium), „Comptes Rendus Hebdomadaires des Séances de l'Académie des Sciences" ("Weekly Reports of the Sessions of the Academy of Sciences") 1896, vol. 123, no. 4, pp. 230–233; reprinted in: *Pisma Mariana Smoluchowskiego z polecenia Polskiej Akademii Umiejętności zgromadzone i wydane przez Władysława Natansona* (*The Writings of Marian Smoluchowski commissioned by the Polish Academy of Arts and Sciences collated and published by Władysław Natanson*), vol. 1, Kraków 1924, pp. 36–39.

Smoluchowski M., *Recherches sur une loi de clausius au point de vue d'une théorie générale de la radiation* (*Research on a Clausius law from the point of view of a general theory of radiation*), „Journal de Physique Théorique et Appliquée" ("Journal of Theoretical and Applied Physics"), 1896, series 3, vol. 5, pp. 488–499; reprinted in: *Pisma Mariana Smoluchowskiego z polecenia Polskiej Akademii Umiejętności zgromadzone i wydane przez Władysława Natansona* (*The Writings of Marian Smoluchowski commissioned by the Polish Academy of Arts and Sciences collated and published by Władysław Natanson*), vol. 1, Kraków 1924, pp. 40–51.

Kelvin W.T., Beattie J.C., Smoluchowski M., *Electrification of air by Röntgen rays*, "Nature: a weekly illustrated journal of science" 1897, vol. 55, no. 1418, pp. 199–200; reprinted in: *Pisma Mariana Smoluchowskiego z polecenia Polskiej Akademii Umiejętności zgromadzone i wydane przez Władysława Natansona* (*The Writings of Marian Smoluchowski commissioned by the Polish Academy of*

Arts and Sciences collated and published by Władysław Natanson), vol. 1, Kraków 1924, pp. 52–55.

Beattie J.C., Smoluchowski M., *Conductance produced in gases by Röntgen rays, by ultra-violet light, and by Uranium, and some consequences thereof*, "London, Edinburgh and Dublin Philosophical Magazine and Journal of Science" 1897, vol. 43, pp. 418–439 reprinted in: *Pisma Mariana Smoluchowskiego z polecenia Polskiej Akademii Umiejętności zgromadzone i wydane przez Władysława Natansona (The Writings of Marian Smoluchowski commissioned by the Polish Academy of Arts and Sciences collated and published by Władysław Natanson)*, vol. 1, Kraków 1924, pp. 56–79.

Kelvin W.T., Beattie J.C., Smoluchowski M., *Experiments on the electrical phenomena produced in gases by Röntgen rays, by ultraviolet light, and by uranium*, "Proceedings of the Royal Society of Edinburgh" 1897, vol. 21, pp. 393–428.

Kelvin W.T., Beattie J.C., Smoluchowski M., *Electrification of air by Röntgen rays*, "Nature" 1897, vol. 55, pp. 199–200.

Kelvin W.T., Beattie J.C., Smoluchowski M., *On the conductive effect produced in air by Röntgen rays and by ultra-violet light*, "Nature" 1897, vol. 55, pp. 343–347.

Kelvin W.T., Beattie J.C., Smoluchowski M., *On apparent and real diselectrification of solid dielectrics produced by Röntgen rays and by flame*, "Nature" 1897, vol. 55, pp. 472–474.

Kelvin W.T., Beattie J.C., Smoluchowski M., *On electric equilibrium between Uranium and an insulated metal in its neighbourhood*, "London, Edinburgh and Dublin Philosophical Magazine and Journal of Science" 1889, vol. 45, pp. 277–278; reprinted in: *Pisma Mariana Smoluchowskiego z polecenia Polskiej Akademii Umiejętności zgromadzone i wydane przez Władysława Natansona (The Writings of Marian Smoluchowski commissioned by the Polish Academy of Arts and Sciences collated and published by Władysław Natanson)*, vol. 1, Kraków 1924, pp. 80–82.

Kelvin W.T., Beattie J.C., Smoluchowski M., *On electric equilibrium between Uranium and an insulated metal in its neighbourhood*, "Nature" 1889, vol. 55, no. 1428, pp. 447–448.

Smoluchowski M., *Über Wärmeleitung in verdünnten Gasen (On conduction in rarefied gases)* „Annalen der Physik" ("Annals of Physics") 1898, vol. 300 (1), pp. 101–130;; reprinted in: *Pisma Mariana Smoluchowskiego z polecenia Polskiej Akademii Umiejętności zgromadzone i wydane przez Władysława Natansona (The Writings of Marian Smoluchowski commissioned by the Polish Academy of Arts and Sciences collated and published by Władysław Natanson)* vol. 1, Kraków 1924, pp. 83–112.

Smoluchowski M., *Über den Temperatursprung bei Wärmeleitung in Gasen (On the temperature jump in heat conduction in gases)* „Sitzungsberichte / Akademie der Wissenschaften in Wien" ("Reports of meetings / Academy of Sciences in Vienna") 1898, vol. 107, Dept. Ia., pp. 304-329; reprinted in: *Pisma Mariana Smoluchowskiego z polecenia Polskiej Akademii Umiejętności zgromadzone i wydane przez Władysława Natansona* (*The Writings of Marian Smoluchowski commissioned by the Polish Academy of Arts and Sciences collated and published by Władysław Natanson*), vol. 1, Kraków 1924, pp. 113-138.

Smoluchowski M., *On conduction of heat by rarefied gases*, "London, Edinburgh and Dublin Philosophical Magazine and Journal of Science" 1898, vol. 46, s. 192-206; reprinted in: *Pisma Mariana Smoluchowskiego z polecenia Polskiej Akademii Umiejętności zgromadzone i wydane przez Władysława Natansona* (*The Writings of Marian Smoluchowski commissioned by the Polish Academy of Arts and Sciences collated* and published by Władysław Natanson), vol. 1, Kraków 1924, pp. 139-155.

Smoluchowski M., *Neuere Untersuchungen über Wärmeleitung in Gasen* (*Recent investigations into heat conduction in gases*), „Oesterreichische Chemiker-Zeitung" (Austrian Chemists' Journal) 1898, vol. 2, pp. 385-392; reprinted in: *Pisma Mariana Smoluchowskiego z polecenia Polskiej Akademii Umiejętności zgromadzone i wydane przez Władysława Natansona* (*The Writings of Marian Smoluchowski commissioned by the Polish Academy of Arts and Sciences collated and published by* Władysław Natanson), vol. 1, Kraków 1924, pp. 156-164.

Smoluchowski M., *O przewodnictwie cieplnem gazów: według dotychczasowych teoryj i doświadczeń* (*On the thermal conductivity of gases: according to previous theories and experiments*). „Prace Matematyczno-Fizyczne" ("Mathematical-Physical papers") 1899, vol. 10, pp. 33-64; reprinted in: *Pisma Mariana Smoluchowskiego z polecenia Polskiej Akademii Umiejętności zgromadzone i wydane przez Władysława Natansona* (*The Writings of Marian Smoluchowski commissioned by the Polish Academy of Arts and Sciences collated* and published by Władysław Natanson), vol. 1, Kraków 1924, pp. 165-199. Smoluchowski M., *Etherion, a new gas?*, "Nature" 1899, vol. 59, no. 1539, pp. 223-224.

Smoluchowski M., *Weitere Studien über den Temperatursprung bei Wärmeleitung in Gasen* (*Further studies on the temperature jump in heat conduction in gases*), „Sitzungsberichte der Akademie der Wissenschaften mathematisch-naturwissenschaftliche Klasse" ("Meeting reports of the Academy of Sciences mathematical and scientific class") 1899, Department IIa, vol. 108, pp. 5-23; reprinted in: *Pisma Mariana Smoluchowskiego z polecenia Polskiej Akademii Umiejętności zgromadzone i wydane przez Władysława Natansona* (*The Writings of Marian Smoluchowski commissioned by the Polish Academy of Arts and Sciences collated and published by Władysław Natanson*), vol. 1, Kraków 1924, pp. 200-216.

Smoluchowski M., *O atmosferze Ziemi i planet* (*On the atmosphere of the Earth and planets*), in: *Commemorative book of the University of Lviv to celebrate the 500th anniversary of the Jagiellonian Foundation of the University of Krakow*, by the Senate of the University of Lviv (E. Winiarza printing house), Lviv 1900, pp. 1-28; reprinted in: Pisma *Mariana Smoluchowskiego z polecenia Polskiej Akademii Umiejętności zgromadzone i wydane przez Władysława Natansona* (*The Writings of Marian Smoluchowski commissioned by the Polish Academy of Arts and Sciences collated and published by Władysław Natanson*), vol. 1, Kraków 1924, pp. 217-247.

Smoluchowski M., *Über die Atmosphäre der Erde und der Planeten* (*On the atmosphere of the Earth and planets*) „Physikalische Zeitschrift" ("Physical Journal ") 1900, vol 2, No. 20, pp. 307-313; reprinted in: *Pisma Mariana Smoluchowskiego z polecenia Polskiej Akademii Umiejętności zgromadzone i wydane przez Władysława Natansona* (*The Writings of Marian Smoluchowski commissioned by the Polish Academy of Arts and Sciences collated and published by Władysław Natanson*), vol. 1, Kraków 1924, pp. 263-278.

Smoluchowski M., *O wynikach nowszych badań nad promieniowaniem* (*On the results of newer research on radiation*), „Kosmos: czasopismo Polskiego Towarzystwa Przyrodników im. Kopernika" ("Cosmos. Journal of the Polish Copernicus Society of Naturalists") 1900, vol. 25, pp. 74-87; reprinted in: *Pisma Mariana Smoluchowskiego z polecenia Polskiej Akademii Umiejętności zgromadzone i wydane przez Władysława Natansona* (*The Writings of Marian Smoluchowski commissioned by the Polish Academy of Arts and Sciences collated and published by Władysław Natanson*) vol. 1, Kraków 1924, pp. 248-262.

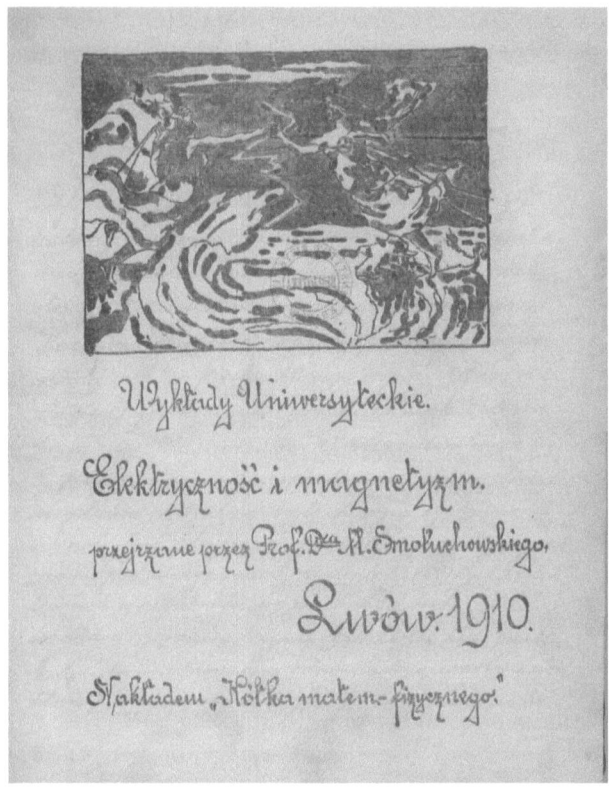

Electricity and magnetism

Smoluchowski M., *O nowszych postępach na polu kinetycznej teorii materii* (*On recent advances in the field of the kinetic theory of matter*), „Prace Matematyczno-Fizyczne" ("Mathematical-Physical Papers") 1901, no 12, pp. 112–135; reprinted in: *Pisma Mariana Smoluchowskiego z polecenia Polskiej Akademii Umiejętności zgromadzone i wydane przez Władysława Natansona* (*The Writings of Marian Smoluchowski commissioned by the Polish Academy of Arts and Sciences collated and published by Władysław Natanson*), vol. 1, Kraków 1924, pp. 279–305.

Smoluchowski M., *Kongres międzynarodowy fizyków odbyty w Paryżu od d. 6–12 sierpnia 1900 r.* (*International Congress of Physicists held in Paris from August 6–12, 1900*), „Wiadomości Matematyczne" ("Mathematical News") 1901, vol. 5, pp. 80–89.

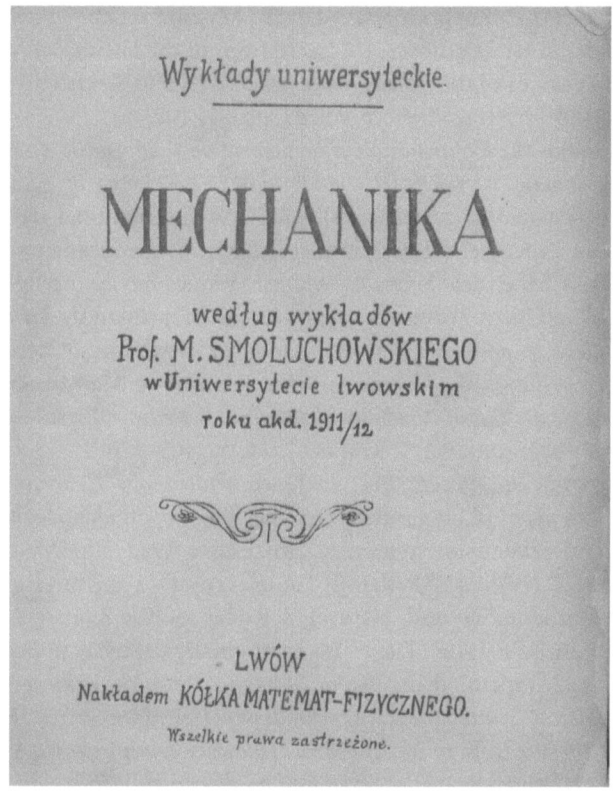

Mechanics (script from lectures)

Smoluchowski M., *Sur les phénomènes aérodynamiques et les effets thermiques qui les accompagnent* (On aerodynamic phenomena and the thermal effects that accompany them) „Bulletin International de L'Académie des Science de Cracovie" ("International Bulletin of the Academy of Sciences of Kraków") 1903, Classe des Sciences Mathématiques et Naturelles (Class of Mathematical and Natural Sciences), pp. 143–182; reprinted in: *Pisma Mariana Smoluchowskiego z polecenia Polskiej Akademii Umiejętności zgromadzone i wydane przez Władysława Natansona* (*The Writings of Marian Smoluchowski commissioned by the Polish Academy of Arts and Sciences collated and published by Władysław Natanson*), vol. 1, Kraków 1924, pp. 306–345.

Smoluchowski M., *O zjawiskach aerodynamicznych i połączonych z nimi objawach cieplnych* (On aerodynamic phenomena and the thermal effects that *accompany them),* „Rozprawy Wydziału Matematyczno- Przyrodniczego Polskiej Akademji Umiejętności" ("Dissertations of the Faculty of Mathematics and Natural Sciences of the Polish Academy of Arts and Sciences") 1903, Dept. A, Mathematical-Physical Sciences, vol. 43, pp. 71-109.

Smoluchowski M., *Contribution à la théorie de l'endosmose électrique et de quelques phénomènes corrlatifs (A contribution to the theory of electrical endosmosis and a few related phenomena),* „Bulletin International de L'Académie des Science de Cracovie" ("International Bulletin of the Academy of Sciences of Kraków"), Classe des Sciences Mathématiques et Naturelles (Class of Mathematical and Natural Sciences), pp. 182-199; reprinted in: *Pisma Mariana Smoluchowskiego z polecenia Polskiej Akademii Umiejętności zgromadzone i wydane przez Władysława Natansona (The Writings of Marian Smoluchowski commissioned by the Polish Academy of Arts and Sciences collated and published by Władysław Natanson),* vol. 1, Kraków 1924, pp. 403-420.

Smoluchowski M., *Przyczynek do teorii endosmozy elektrycznej i kilku zjawisk pokrewnych* (Contribution to the theory of electrical endosmosis and several related phenomena), „Rozprawy Wydziału Matematyczno-Przyrodniczego Polskiej Akademji Umiejętności" ("Dissertations of the Faculty of Mathematics and Natural Sciences of the Polish Academy of Arts and Sciences") 1903, Dept. A, Mathematical-Physical Sciences, vol. 43, pp. 110-127; reprinted in: *Pisma Mariana Smoluchowskiego z polecenia Polskiej Akademii Umiejętności zgromadzone i wydane przez Władysława Natansona (The Writings of Marian Smoluchowski commissioned by the Polish Academy of Arts and Sciences collated and published by Władysław Natanson),* vol. 1, Kraków 1924, pp. 384-402

Smoluchowski M., *O metodzie podobieństwa dynamicznego i jej zastosowaniach w mechanice cieczy i gazów (On the method of dynamic similarity and its application in the mechanics of liquids and gases),* „Prace Matematyczno-Fizyczne" ("Mathematical-Physical Papers") 1904, vol. 15, pp. 115-134; reprinted in: *Pisma Mariana Smoluchowskiego z polecenia Polskiej Akademii Umiejętności zgromadzone i wydane przez Władysława Natansona (The Writings of Marian Smoluchowski commissioned by the Polish Academy of Arts and Sciences collated and published by Władysław Natanson)* vol. 1, Kraków 1924, pp. 346-368.

Smoluchowski M., *On the principles of aerodynamics and their application by the method of dynamical similarity to some special problems,* "London, Edinburgh and Dublin Philosophical Magazine and Journal of Science" 1904, vol. 7, issue 42, pp. 667-681; reprinted in: *Pisma Mariana Smoluchowskiego z polecenia Polskiej*

Akademii Umiejętności zgromadzone i wydane przez Władysława Natansona (*The Writings of Marian Smoluchowski commissioned by the Polish Academy of Arts and Sciences collated and published by Władysław Natanson*) vol. 1, Kraków 1924, pp. 369-383.

Smoluchowski M., *Über Unregelmässigkeiten in der Verteilung von Gasmolekülen und deren Einfluss auf Entropie und Zustandsgleichung*, w: *Festschrift Ludwig Boltzmann gewidmet zum sechzigsten Geburtstage 20. Februar 1904* (*On irregularities in the distribution of Gas molecules and their influence on entropy and the equation of state*, in: Commemorative book dedicated to Ludwig Boltzmann on the occasion of his sixtieth birthday on February 20, 1904 (Leipzig 1904, s. 626-641; reprinted in: *Pisma Mariana Smoluchowskiego z polecenia Polskiej Akademii Umiejętności zgromadzone i wydane przez Władysława Natansona* (*The Writings of Marian Smoluchowski commissioned by the Polish Academy of Arts and Sciences collated and published by Władysław Natanson*), vol. 1, Kraków 1924, pp. 421-435.

Smoluchowski M., *Sur la formation des veines d'efflux dans les liquides* (*On the formation of the efflux veins in liquids*) „Bulletin International de L'Académie des Science de Cracovie" ("International Bulletin of the Academy of Sciences of Kraków") 1904, Classe des Sciences Mathématiques et Naturelles (Class of Mathematical and Natural Sciences), pp. 371-384; reprinted in: *Pisma Mariana Smoluchowskiego z polecenia Polskiej Akademii Umiejętności zgromadzone i wydane przez Władysława Natansona* (*The Writings of Marian Smoluchowski commissioned by the Polish Academy of Arts and Sciences collated and published by Władysław Natanson*), vol. 1, Kraków 1924, pp. 450-462. Smoluchowski M., *O powstaniu żył podczas wypływu cieczy* (*On the formation of veins during the outflow of liquid*), „Rozprawy Wydziału Matematyczno-Przyrodniczego Polskiej Akademji Umiejętności" ("Dissertations of the Faculty of Mathematics and Natural Sciences of the Polish Academy of Arts and Sciences") 1904, Dept. A, Mathematical- Physical Sciences, vol. 46, pp. 129-139; reprinted in: *Pisma Mariana Smoluchowskiego z polecenia Polskiej Akademii Umiejętności zgromadzone i wydane przez Władysława Natansona* (*The Writings of Marian Smoluchowski commissioned by the Polish Academy of Arts and Sciences collated and published by Władysław Natanson*), vol. 1, Kraków 1924, pp. 436-449.

Smoluchowski M. (ed.), *Sprawozdania z prac polskich na polu fizyki za lata 1901-1902* (*Reports on Polish papers in the field of physics for the years 1901-1902*), „Kosmos: czasopismo Polskiego Towarzystwa Przy rodników im. Kopernika" ("Cosmos. Journal of the Polish Copernicus Society of Naturalists") 1904, vol. 29, pp. 528-545.

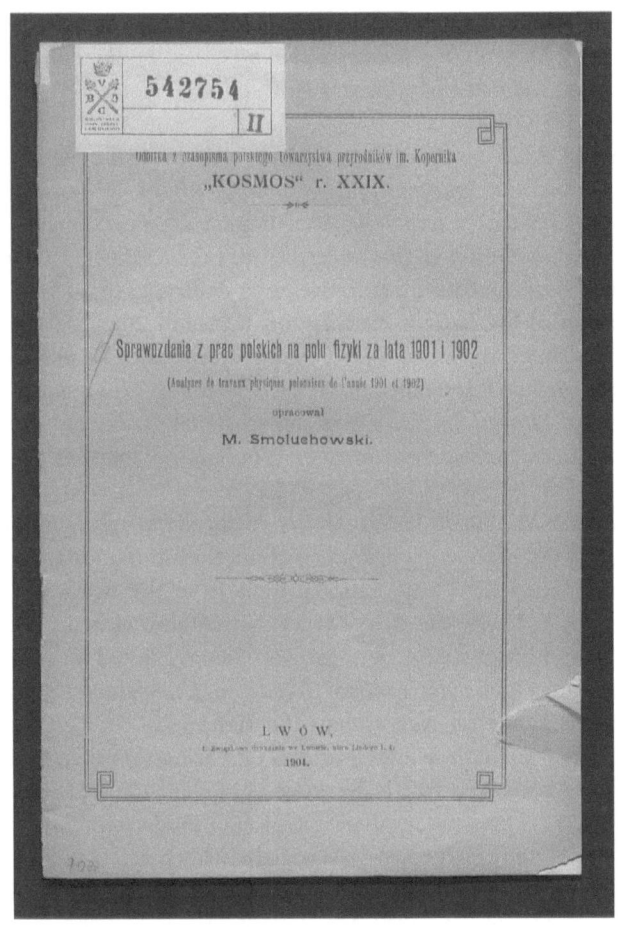

Reports on Polish papers in the field of physics for the years 1901–1902

Smoluchowski M., *Zur Theorie der elektrischen Kataphorese und der Oberflächenleitung* (*On the theory of electrical cataphoresis and surface conduction*), „Physikalische Zeitschrift" ("Physical Journal") 1905, vol. 6., No. 17, pp. 529–531; reprinted in: *Pisma Mariana Smoluchowskiego z polecenia Polskiej Akademii Umiejętności zgromadzone i wydane przez Władysława Natansona* (*The Writings of Marian Smoluchowski commissioned by the Polish Academy of Arts and Sciences collated and published by Władysław Natanson*), vol. 1, Kraków 1924, pp. 463–467.

Smoluchowski M., *Essai de théorie de la cataphorèse électrique et de la conduction superficielle* (Essay on Electrical cataphoresis and surface conduction) „Journal de Physique Théorique et Appliquée" ("Journal of Theoretical and Applied Physics") 1907, vol. 6, no. 1, pp. 659–660.

Smoluchowski M., *Sur le chemin moyen parcouru par les molecules d'un gaz et sur son rapport avec la théorie de la diffusion* (On the mean path of molecules of gas and its relationship to the theory of diffusion) „Bulletin International de L'Académie des Science de Cracovie" ("International Bulletin of the Academy of Sciences of Kraków") 1906, Classe des Sciences Mathématiques et Naturelles (Class of Mathematical and Natural Sciences), s. 202–213; reprinted in: *Pisma Mariana Smoluchowskiego z polecenia Polskiej Akademii Umiejętności zgromadzone i wydane przez Władysława Natansona* (The Writings of Marian Smoluchowski commissioned by the Polish Academy of Arts and Sciences collated and published by Władysław Natanson), vol. 1, Kraków 1924, pp. 479–489.

Smoluchowski M., *O drodze średniej cząsteczek gazu i o jej związku z teorią Dyfuzji* (On the mean path travelled by the molecules of a gas and its relation to the theory of diffusion) „Rozprawy Wydziału Matematyczno-Przyrodniczego Polskiej Akademji Umiejętności" ("Dissertations of the Faculty of Mathematics and Natural Sciences of the Polish Academy of Arts and Sciences") 1906, Dept. A, Mathematical-Physical Sciences, vol. 46, pp. 129–139.

Smoluchowski M., *Essai d'une théorie cinétique du mouvement Brownien et des milieux troubles* (An outline of the kinetic theory of Brownian motion and cloudy solutions) „Bulletin International de L'Académie des Science de Cracovie" ("International Bulletin of the Academy of Sciences of Kraków") 1906, Classe des Sciences Mathématiques et Naturelles (Class of Mathematical and Natural Sciences), pp. 577–602. Smoluchowski M., *Zarys teorii kinetycznej ruchów Browna i roztworów mętnych* (An outline of the kinetic theory of Brownian motion and cloudy solutions) „Rozprawy Wydziału Matematyczno-Przyrodniczego Polskiej Akademji Umiejętności" ("Dissertations of the Faculty of Mathematics and Natural Sciences of the Polish Academy of Arts and Sciences") 1906, Dept. A, Mathematical-Physical Sciences, vol. 46, pp. 257–281; reprinted in: *Wkład polskich uczonych do fizyki statystyczno- molekularnej* (The contribution of Polish scientists to statistical-molecular physics) ed. T. Piech, Wrocław 1962, pp. 129–151. Smoluchowski M., *Zur kinetischen Theorie der Brownschen Molekularbewegung und der Suspensionen* (On the kinetic theory of Brownian molecular motion and of suspensions) „Annalen der Physik" ("Annals in Physics") 1906, vol. 21, pp. 756–780.

Smoluchowski M., *Contribution à la théorie du mouvement des liquides visqueux; en particulier desproblems en deux dimensions* (Contribution to the theory of *movement of viscous liquids; in particular of two- dimensional problems*) „Bulletin International de L'Académie des Science de Cracovie" ("International Bulletin of the Academy of Sciences of Kraków") 1907, Classe des Sciences Mathématiques et Naturelles (Class of Mathematical and Natural Sciences), pp. 1–16; reprinted in: *Pisma Mariana Smoluchowskiego z polecenia Polskiej Akademii Umiejętności zgromadzone i wydane przez Władysława Natansona* (*The Writings of Marian Smoluchowski commissioned by the Polish Academy of Arts and Sciences collated and published by Władysław Natanson*), vol. 1, Kraków 1924, pp. 555–569.

Smoluchowski M., *Przyczynek do teorii ruchów cieczy lepkich, zwłaszcza zagadnień dwuwymiarowych* (*Contribution to the theory of movement of viscous liquids; in particular of two-dimensional problems*) „Rozprawy Wydziału Matematyczno-Przyrodniczego Polskiej Akademji Umiejętności" ("Dissertations of the Faculty of Mathematics and Natural Sciences of the Polish Academy of Arts and Sciences") 1907, Dept. A, Mathematical-Physical Sciences, vol. 47, pp. 1–16; reprinted in: *Pisma Mariana Smoluchowskiego z polecenia Polskiej Akademii Umiejętności zgromadzone i wydane przez Władysława Natansona* (*The Writings of Marian Smoluchowski commissioned by the Polish Academy of Arts and Sciences collated and published by Władysław Natanson*), vol. 1, Kraków 1924, pp. 539–554.

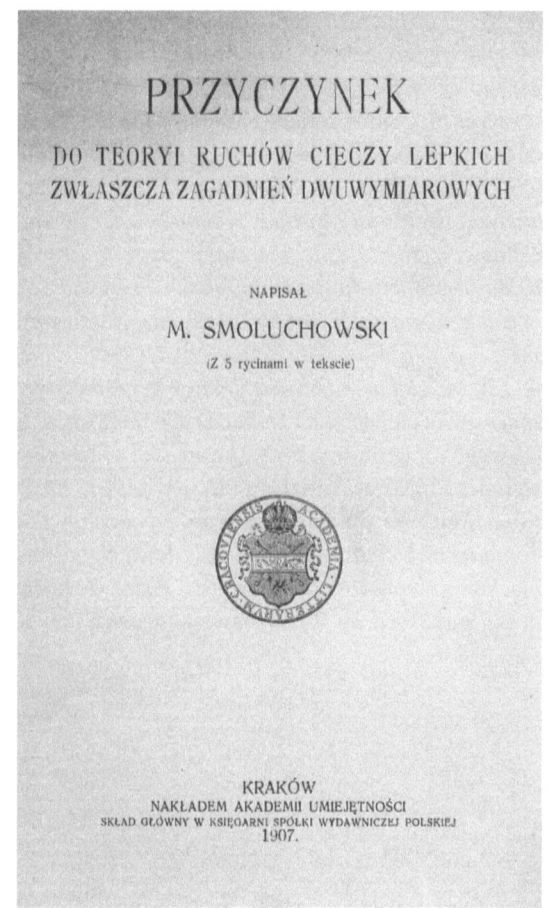

Contribution to the theory of movement of viscous liquids; in particular of two-dimensional problems

Smoluchowski M., *Théorie cinétique de l'opalescence des gaz a l'état critique et de certains phénomènes corrélatifs* (Molecular- kinetic theory of the opalescence of gases in the critical state and a few related phenomena) „Bulletin International de L'Académie des Science de Cracovie" ("International Bulletin of the Academy of Sciences of Kraków") 1907, Classe des Sciences Mathématiques et Naturelles (Class of Mathematical and Natural Sciences), pp. 1057–1075.

Smoluchowski M., *Teoria kinetyczna opalescencji gazów w stanie krytycznym oraz innych zjawisk pokrewnych* (Kinetic theory of the opalescence of gases in a critical state and of certain correlative phenomena). „Rozprawy Wydziału Matematyczno-Przyrodniczego Polskiej Akademji Umiejętności" ("Dissertations of the Faculty of Mathematics and Natural Sciences of the Polish Academy of Arts and Sciences") 1907, Dept. A, Mathematical-Physical Sciences, vol. 47, pp. 179–198; reprinted in: *Pisma Mariana Smoluchowskiego z polecenia Polskiej Akademii Umiejętności zgromadzone i wydane przez Władysława Natansona* (*The Writings of Marian Smoluchowski commissioned by the Polish Academy of Arts and Sciences collated and published by Władysław Natanson*), vol. 1, Kraków 1924, pp. 570–588.

Smoluchowski M., *Molekular-kinetische Theorie der Opaleszenz von Gasen im kritischen Zustande, sowie einiger verwandter Erscheinungen,* (Molecular-Kinetic theory of the opalescence of gases in a critical state and of certain correlative phenomena) „Annalen der Physik" ("Annals in Physics") 1907, vol. 25, pp. 205–226; reprinted in: *Pisma Mariana Smoluchowskiego z polecenia Polskiej Akademii Umiejętności zgromadzone i wydane przez Władysława Natansona* (*The Writings of Marian Smoluchowski commissioned by the Polish Academy of Arts and Sciences collated and published by Władysław Natanson*), vol. 1, Kraków 1924, pp. 589–609.

Przyczynek do teoryi opalescencyi w gazach w stanie krytycznym. — *Beitrag zur Theorie der Opaleszenz von Gasen im kritischen Zustande.*

Mémoire

de M. *MARYAN SMOLUCHOWSKI* m. c.,

présenté dans la séance du 17 Juillet 1911.

Vor einiger Zeit habe ich an dieser Stelle eine Theorie entwickelt[1]), der zufolge die in Gasen und Flüssigkeiten notwendigerweise vorhandenen und durch Anwendung von Wahrscheinlichkeitsbetrachtungen zu berechnenden Ungleichförmigkeiten der Verteilung der Moleküle sich durch das Tyndall'sche Phänomen bemerkbar machen müßten, und insbesondere die Opaleszenz von Gasen im kritischen Punkte hierauf zurückzuführen wäre. Seitdem ist diese Theorie mehrfach Gegenstand theoretischer Diskussion gewesen und es sind bemerkenswerte experimentelle Untersuchungen erschienen, welche die Möglichkeit einer allseitigen Verifizierung derselben durch Messung des Opaleszenzvermögens näher gerückt haben.

Deshalb möchte ich nun zu einigen hiebei aufgeworfenen Fragen Stellung nehmen und insbesondere einige die experimentelle Seite betreffenden Punkte näher ausführen.

Ein abfälliges Urteil über diese Theorie ist von Rothmund[2]) gefällt worden, anläßlich seiner Experimentaluntersuchungen über kritische Trübung von binären Flüssigkeitsgemischen, worin sich dieser Forscher folgendermaßen ausspricht: „Ich glaube, daß man von dem weiteren Ausbau der Hypothese von Donnan viel eher eine Klärung und Förderung der Frage erwarten darf, als von dem Hereinziehen des Wahrscheinlichkeitsbegriffes". Doch findet sich bei Rothmund kein tatsächlich gegen diese Theorie sprechendes

[1]) Bull. de l'Acad. d. Sc. Crac. Déc. 1907, S. 1057, Ann. d. Phys. **25**, S. 205 (1908).
[2]) Zeitschr. f. phys. Chem. **63**, S. 54 (1908).

Contribution to a theory of opalescence in gases in a critical state

Smoluchowski M., *Zarys najnowszych postępów fizyki* (*Outline of the latest advances in physics*) „Muzeum: czasopismo wydawane przez Towarzystwo Nauczycieli Szkół Wyższych" ("Museum: Magazine published by the Association of Higher School Teachers") 1907, vol. 1, book 1, pp. 43–60; book 2, pp. 144–165.

Smoluchowski M., *Uwagi o kilku zjawiskach drobinowych, związanych z przypadkowymi odchyleniami od stanu najprawdopodobniejszego* (*Observations on several particle phenomena related to random deviations from the most probable state*) in: *Sprawozdanie X Zjazdu lekarzy i przyrodników polskich* (Reports of the 10th Congress of Polish doctors and naturalists), Lviv 1908.

Smoluchowski M., *Lord Kelvin*, „Ateneum Polskie" ("Polish Athenaeum") 1908, vol. 1, pp. 212–228; reprinted in: *Pisma Mariana Smoluchowskiego z polecenia Polskiej Akademii Umiejętności zgromadzone i wydane przez Władysława Natansona* (*The Writings of Marian Smoluchowski commissioned by the Polish Academy of Arts and Sciences collated and published by Władysław Natanson*), vol. 3, Kraków 1928, pp. 1–15. Smoluchowski M., *Dr Władysław Natanson: odczyty i szkice* (*Dr. Władysław Natanson: readings and sketches*) „Ateneum Polskie" ("Polish Athenaeum") 1908, vol. 2, pp. 134–136.

Smoluchowski M., *Stanisław Kępiński*, „Ateneum Polskie" ("Polish Athenaeum") 1908, vol. 2, pp. 274–276.

Smoluchowski M., *Dwie książki z dziedziny filozofii przyrody* (Two books from the field of natural philosophy) „Ateneum Polskie" ("Polish Athenaeum") 1908, vol. 4, pp. 291–296.

Smoluchowski M., *Teorya kinetyczna gazów* (*A kinetic theory of gases*) Lviv 1908.

A kinetic theory of gases

Smoluchowski M., *Über ein gewisses Stabilitätsproblem der Elastizitätslehre und dessen Beziehung zur Entstehung von Faltengebirgen* (*On a certain stability problem in the theory of elasticity and its relationship to the formation of fold mountains*), „Bulletin International de L'Académie des Science de Cracovie" ("International Bulletin of the Academy of Sciences of Kraków") 1909, Classe des Sciences Mathématiques et Naturelles (Class of Mathematical and Natural Sciences), vol. 6, pp. 3–20; reprinted in: *Pisma Mariana Smoluchowskiego z polecenia Polskiej Akademii Umiejętności zgromadzone i wydane przez Władysława Natansona* (*The Writings of Marian Smoluchowski commissioned by the Polish Academy of Arts and Sciences collated and published by Władysław Natanson*), vol. 2, Kraków 1927, pp. 5–22.

Smoluchowski M., *O pewnym zagadnieniu z teorii sprężystości i o jego związku z wytworzeniem się gór fałdowych (On a certain problem in the theory of elasticity and its relationship to the formation of fold mountains),* „Rozprawy Wydziału Matematyczno- Przyrodniczego Polskiej Akademji Umiejętności" ("Dissertations of the Faculty of Mathematics and Natural Sciences of the Polish Academy of Arts and Sciences") 1909, Dept. A, Mathematical-Physical Sciences, vol. 49, pp. 223–226; reprinted in: *Pisma Mariana Smoluchowskiego z polecenia Polskiej Akademii Umiejętności zgromadzone i wydane przez Władysława Natansona (The Writings of Marian Smoluchowski commissioned by the Polish Academy of Arts and Sciences collated and published by Władysław Natanson),* vol. 2, Kraków 1927, pp. 1–4.

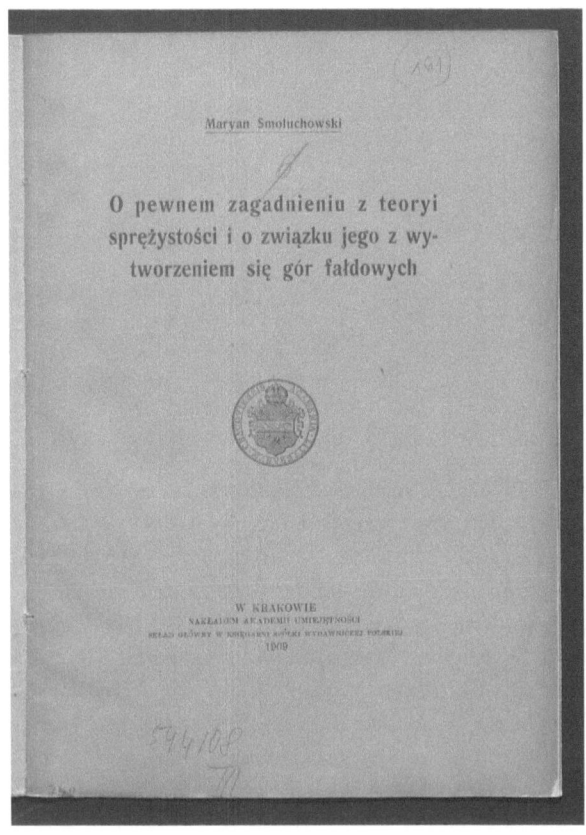

On a certain problem in the theory of elasticity and its relationship to the formation of fold mountains

Smoluchowski M., *Some remarks on the mechanics of overthrusts*, "Geological Magazine" 1909, vol. 6, 204–205; reprinted in: *Pisma Mariana Smoluchowskiego z polecenia Polskiej Akademii Umiejętności zgromadzone i wydane przez Władysława Natansona* (*The Writings of Marian Smoluchowski commissioned by the Polish Academy of Arts and Sciences collated and published by Władysław Natanson*), vol. 2, Kraków 1927, pp. 23–24.

Smoluchowski M., *Versuche über Faltungserscheinungen schwimmender elastischer Platten* (*Experiments on the folding phenomena of floating elastic plates*), „Bulletin International de L'Académie des Science de Cracovie" ("International Bulletin of the Academy of Sciences of Kraków") 1909, Classe des Sciences Mathématiques et Naturelles (Class of Mathematical and Natural Sciences), Series II, pp. 727–734; reprinted in: *Pisma Mariana Smoluchowskiego z polecenia Polskiej Akademii Umiejętności zgromadzone i wydane przez Władysława Natansona* (*The Writings of Marian Smoluchowski commissioned by the Polish Academy of Arts and Sciences collated and published by Władysław Natanson*), vol. 2, Kraków 1927, pp. 25–32.

Smoluchowski M., *Kilka uwag o fizycznych podstawach teorii górotwórczych* (A few remarks on the physical foundations of orogenic theories) „Kosmos: czasopismo Polskiego Towarzystwa Przyrodników im. Kopernika" ("Cosmos. Journal of the Polish Copernicus Society of Naturalists") 1909, yearbook 34, books. 7–9, pp. 547–579; reprinted in: *Pisma Mariana Smoluchowskiego z polecenia Polskiej Akademii Umiejętności zgromadzone i wydane przez Władysława Natansona* (*The Writings of Marian Smoluchowski commissioned by the Polish Academy of Arts and Sciences collated and published by Władysław Natanson*) vol. 2, Kraków 1927, pp. 33–62.

Smoluchowski M., *Sur la conductibilité calorifique des corps pulvérisés* (*On the heat conductivity of pulverized bodies*), „Bulletin International de L'Académie des Science de Cracovie" ("International Bulletin of the Academy of Sciences of Kraków") 1910, Classe des Sciences Mathématiques et Naturelles (Class of Mathematical and Natural Sciences). Series A: Mathematical Sciences, no. 5, pp. 129–153; reprinted in: *Pisma Mariana Smoluchowskiego z polecenia Polskiej Akademii Umiejętności zgromadzone i wydane przez Władysława Natansona* (*The Writings of Marian Smoluchowski commissioned by the Polish Academy of Arts and Sciences collated and published by Władysław Natanson*), vol. 2, Kraków 1927, pp. 78–101.

Smoluchowski M., *Sur la théorie mécanique de l'érosion glaciare* (*On the mechanical theory of glacial erosion*) „Comptes rendus hebdomadaires des séances de l'Académie des sciences" ("Weekly Reports of the Sessions of the Academy of Sciences") 1910, vol. 150, no. 21, pp. 1368–1371; reprinted in: *Pisma Mariana*

Smoluchowskiego z polecenia Polskiej Akademii Umiejętności zgromadzone i wydane przez Władysława Natansona (The Writings of Marian Smoluchowski commissioned by the Polish Academy of Arts and Sciences collated and published by Władysław Natanson), vol. 2, Kraków 1927, pp. 63–65.

Smoluchowski M., *Über Wärmeleitung pulverförmiger Körper und ein hierauf gegründetes neues Wärmeisolierungsverfahren*, w: Bericht über den II. Internationalen Kältekongress Wien 1910, 6–12 Oktober (*On thermal conduction of powdered bodies and a new thermal insulation process based on this, in: Report on the II International Congress of Refrigeration, Vienna 1910, October 6–12*), vol. 1, Vienna 1911, pp. 166– [172;] reprinted in: *Pisma Mariana Smoluchowskiego z polecenia Polskiej Akademii Umiejętności zgromadzone i wydane przez Władysława Natansona (The Writings of Marian Smoluchowski commissioned by the Polish Academy of Arts and Sciences collated and published by Władysław Natanson)*, vol. 2, Kraków 1927, pp. 112–120.

Smoluchowski M., *Contributions to the theory of transpiration, diffusion and thermal conduction in rarefied gases*, „Bulletin International de L'Académie des Science de Cracovie", ("International Bulletin of the Academy of Sciences of Kraków"), Classe des Sciences Mathématiques et Naturelles (Class of Mathematical and Natural Sciences), Series A: Mathematical Sciences, July 1910, pp. 295–312; reprinted in: *Pisma Mariana Smoluchowskiego z polecenia Polskiej Akademii Umiejętności zgromadzone i wydane przez Władysława Natansona (The Writings of Marian Smoluchowski commissioned by the Polish Academy of Arts and Sciences collated and published by Władysław Natanson)*, vol. 2, Kraków 1927, pp. 134–151.

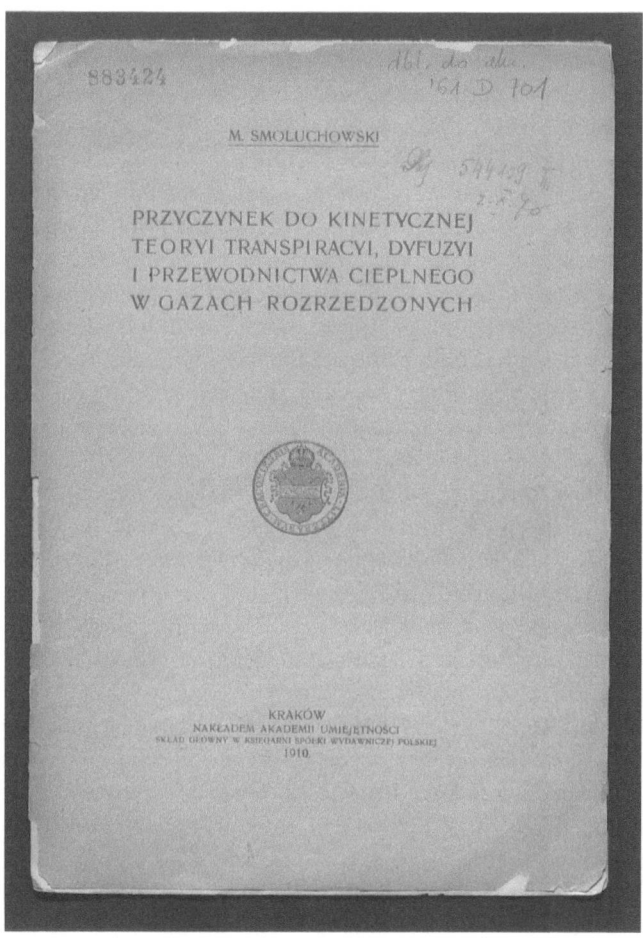

A contribution to the theory of transpiration, diffusion and thermal conduction in rarefied gases

Smoluchowski M., *Przyczynek do kinetycznej teorii transpiracji, dyfuzji i przewodnictwa cieplnego w gazach rozrzedzonych* (A contribution to the theory of transpiration, diffusion and thermal conduction in rarefied gases) „Rozprawy Wydziału Matematyczno- Przyrodniczego Polskiej Akademji Umiejętności" ("Dissertations of the Faculty of Mathematics and Natural Sciences of the Polish Academy of Arts and Sciences") 1910, Dział Adept A, Mathematical- Physical Sciences, vol. 50, pp. 209–214. reprinted in: *Pisma Mariana Smoluchowskiego* z polecenia Polskiej Akademii Umiejętności zgromadzone i wydane przez Władysława Natansona (*The Writings of Marian Smoluchowski commissioned by the Polish Academy of Arts and Sciences collated and published by Władysław Natanson*), vol. 2, Kraków 1927, pp. 128–133.

Smoluchowski M., *Zur kinetischen Theorie der Transpiration und Diffusion verdünnter Gase* (*On the kinetic theory of transpiration and diffusion of rarefied gases*), „Annalen der Physik" (Annals in Physics") 1910, vol. 33, pp. 1559–1570.

Smoluchowski M., *Van der Waalsa teoria stanu ciekłego a zjawiska lepkości* (*The Van der Waals theory of the liquid state and the phenomena of viscosity*) „Kosmos: czasopismo Polskiego Towarzystwa Przyrodników im. Kopernika" ("Cosmos. Journal of the Polish Copernicus Society of Naturalists") 1910, vol. 35, pp. 543–549; reprinted in: *Pisma Mariana Smoluchowskiego z polecenia Polskiej Akademii Umiejętności zgromadzone i wydane przez Władysława Natansona* (*The Writings of Marian Smoluchowski commissioned by the Polish Academy of Arts and Sciences collated and published by Władysław Natanson*), vol. 2, Kraków 1927, pp. 121–127.

Smoluchowski M., *Teorya Maxwella i teorya elektronów* (*Maxwell's theory and the theory of electrons*), Lviv 1910.

Smoluchowski M., *Some remarks on conduction of heat through rarefied gases*, "The London, Edinburgh, and Dublin Philosophical Magazine and Journal of Science" 1911, vol. 21, pp. 11–14.

Smoluchowski M., *Über die Wechselwirkung von Kugeln, die sich in einerzähen Flüssigkeit bewegen* (*On the interaction of spheres moving in a viscous liquid*) „Bulletin International de L'Académie des Science de Cracovie" ("International Bulletin of the Academy of Sciences of Kraków") 1911, Classe des Sciences Mathématiques et Naturelles (Class of Mathematical and Natural Sciences) Series A: Mathematical Sciences, January 1911, pp. 28–39.

Smoluchowski M., *O oddziaływaniu wzajemnym kul poruszających się w ośrodku lepkim* (*On the interaction of spheres moving in a viscous liquid*) „Rozprawy Wydziału Matematyczno-Przyrodniczego Polskiej Akademji Umiejętności" ("Dissertations of the Faculty of Mathematics and Natural Sciences of the Polish Academy of Arts and Sciences") 1911, Dept. A, Mathematical-Physical Sciences, vol. 51, pp. 1–3; reprinted in: *Pisma Mariana Smoluchowskiego z polecenia Polskiej Akademii Umiejętności zgromadzone i wydane przez Władysława Natansona* (*The Writings of Marian Smoluchowski commissioned by the Polish Academy of Arts and Sciences collated and published by Władysław Natanson*), vol. 2, Kraków 1927, pp. 179–181.

Smoluchowski M., *Zur Theorie der Wärmeleitung in verdünnten Gasen und der dabei auftretenden Druckkräfte* (*On the theory of heat conduction in rarefied gases and the pressure forces that occur*) „Bulletin de l'Académie des Sciences de Cracovie ("International Bulletin of the Academy of Sciences of Kraków") 1911, Classe des Sciences Mathématiques et Naturelles (Class of Mathematical and Natural Sciences). Series A: Mathematical Sciences, Julyt 1911, pp. 432– 453; reprinted in: „Annalen der Physik" ("Annals in Physics") vol. 340, pp. 983–1004 and *Pisma Mariana Smoluchowskiego z polecenia Polskiej Akademii Umiejętności zgromadzone i wydane przez Władysława Natansona* (*The Writings of Marian Smoluchowski commissioned by the Polish Academy of Arts and Sciences collated and published by Władysław Natanson*), vol. 2, Kraków 1927, pp. 155–176.

Smoluchowski M., *Beitrag zur Theorie der Opaleszenz von Gasen im kritischen Zustande* (*Contribution to the theory of the opalescence of gases in the critical state*), „Bulletin de l'Académie des Sciences de Cracovie"("International Bulletin of the Academy of Sciences of Kraków") 1911, Classe des Sciences Mathématiques et Naturelles (Class of Mathematical and Natural Sciences). Series A: Mathematical Sciences, October 1911, pp. 493–502; reprinted in: Pisma Mariana *Smoluchowskiego z polecenia Polskiej Akademii Umiejętności zgromadzone i wydane przez Władysława Natansona* (*The Writings of Marian Smoluchowski commissioned by the Polish Academy of Arts and Sciences collated and published by Władysław Natanson*), vol. 2, Kraków 1927, pp. 215–225.

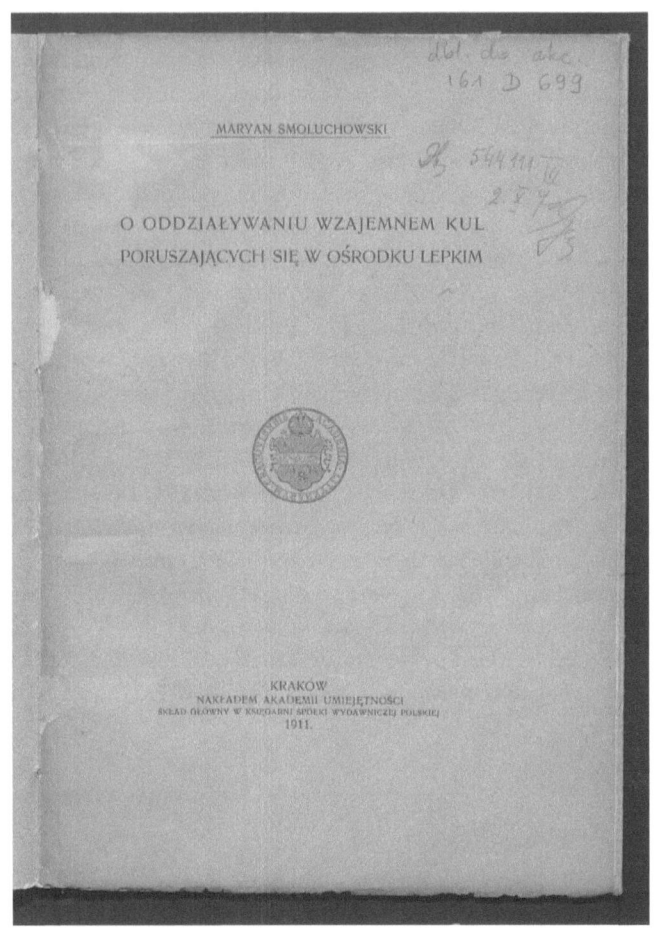

On the interaction of spheres moving in a viscous liquid

Smoluchowski M., *On opalescence of gases in the critical state*, "The London, Edinburgh, and Dublin Philosophical Magazine and Journal of Science" 1912, vol. 23, pp. 165–173.

Smoluchowski M., *Bemerkung zur Theorie des absoluten Manometers von Knudsen* (Comment on the theory of Knudsen's absolute manometer) „Annalen der Physik" ("Annals in Physics") 1911, vol. 339, pp. 182–184; reprinted in: *Pisma Mariana Smoluchowskiego z polecenia Polskiej Akademii Umiejętności zgromadzone i wydane przez Władysława Natansona* (*The Writings of Marian*

Smoluchowski commissioned by the Polish Academy of Arts and Sciences collated and published by Władysław Natanson), vol. 2, Kraków 1927, pp. 152–154.

Smoluchowski M., *Études sur la conductibilité calorifique des corps pulvérisés (suite)* (*Studies on the heat conductivity of pulverized bodies (continued)*), „Bulletin de l'Académie des Sciences de Cracovie, ("International Bulletin of the Academy of Sciences of Kraków") 1911, Classe des Sciences Mathématiques et Naturelles (Class of Mathematical and Natural Sciences), Series A: Mathematical Sciences, October 1911, pp. 550–557; reprinted in: *Pisma Mariana Smoluchowskiego z polecenia Polskiej Akademii Umiejętności zgromadzone i wydane przez Władysława Natansona* (*The Writings of Marian Smoluchowski commissioned by the Polish Academy of Arts and Sciences collated and published by Władysław Natanson*), vol. 2, Kraków 1927, pp. 102–111.

Smoluchowski M., *O pewnem zagadnieniu kinetycznej teoryi roztworów* (*On a certain problem of the kinetic theory of solutions*) in: *A commemorative book to celebrate the 250th anniversary founding of the University of Lviv by King Jan Kazimierz*, Lviv 1911, pp. 1–8; reprinted in: *Pisma Mariana Smoluchowskiego z polecenia Polskiej Akademii Umiejętności zgromadzone i wydane przez Władysława Natansona* (*The Writings of Marian Smoluchowski commissioned by the Polish Academy of Arts and Sciences collated and published by Władysław Natanson*), vol. 2, Kraków 1927, pp. 209–214. Smoluchowski M., *Ewolucja teorii atomistycznej* (*The evolution of atomic theory*) „Wiadomości Matematyczne" ("Mathematical news") 1911, vol. 15, books 5–6, pp. 201–216; reprinted in: „Rocznik Akademii Umiejętności w Krakowie"("The Yearbook of the Academy of Arts and Sciences in Kraków) year 1910–1911, pp. 131–154 and *Pisma Mariana Smoluchowskiego z polecenia Polskiej Akademii Umiejętności zgromadzone i wydane przez Władysława Natansona* (*The Writings of Marian Smoluchowski commissioned by the Polish Academy of Arts and Sciences collated and published by Władysław Natanson*), vol. 3, Kraków 1928, pp. 16–30.

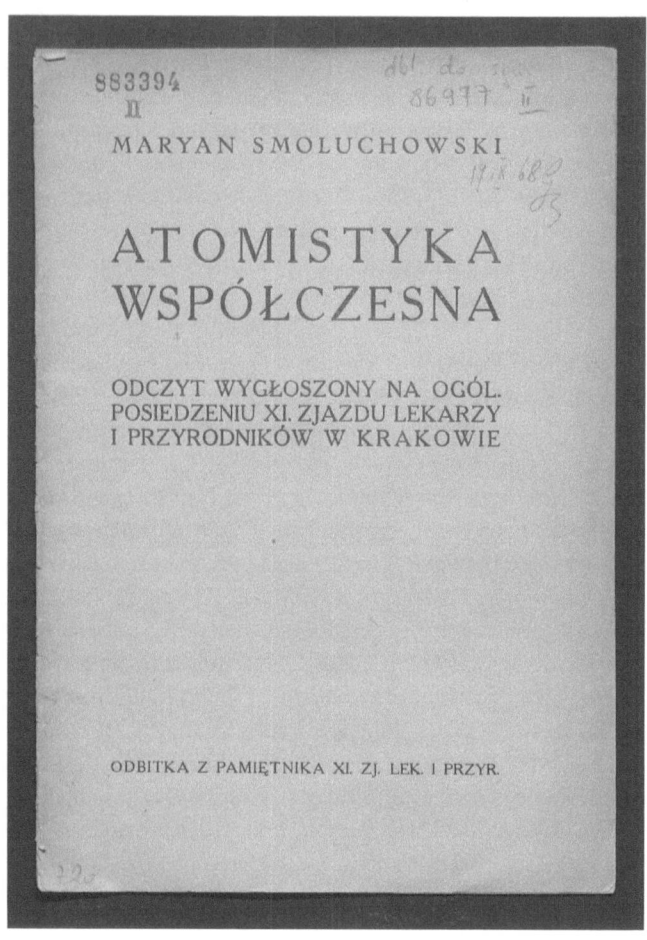

Contemporary atomic science

Smoluchowski M., *Atomistyka współczesna: odczyt wygłoszony na ogól. posiedzeniu XI. Zjazdu Lekarzy i Przyrodników w Krakowie* (Contemporary atomic science: reading delivered at the general sitting of the 11th Congress of Doctors and Naturalists in Kraków) in: *Księga pamiątkowa z IX Zjazdu Lekarzy i Przyrodników Polskich w Krakowie* (Commemorative book of the 11th Congress of Polish Doctors and Naturalists in Kraków) Kraków 1912, pp. 129–143; reprinted in: *Pisma Mariana Smoluchowskiego z polecenia Polskiej Akademii Umiejętności zgromadzone i wydane przez Władysława Natansona* (*The Writings of Marian*

Smoluchowski commissioned by the Polish Academy of Arts and Sciences collated and published by Władysław Natanson), vol. 3, Kraków 1928, pp. 31–44.

Smoluchowski M., *On the practical applicability of Strokes' lawof resistance, and the modifications of it required in certain cases*, in: *Proceedings of the Fifth International Congress of Mathematicians*, vol. 2, Cambridge 1913, pp. 192–201; reprinted in: *Pisma Mariana Smoluchowskiego z polecenia Polskiej Akademii Umiejętności zgromadzone i wydane przez Władysława Natansona (The Writings of Marian Smoluchowski commissioned by the Polish Academy of Arts and Sciences collated and published by Władysław Natanson)*, vol. 2, Kraków 1927, pp. 195–208.

Smoluchowski M., *Experimentell nachweisbare, derüblichten Thermodynamikwidersprechende Molekularphänomene (Experimentally verifiable molecular phenomena contradicting the usual thermodynamics)* „Physikalische Zeitschrift"("Physical Journal") vol. 13, 1912, pp. 1069–1080; reprinted in: *Pisma Mariana Smoluchowskiego z polecenia Polskiej Akademii Umiejętności zgromadzone i wydane przez Władysława Natansona (The Writings of Marian Smoluchowski commissioned by the Polish Academy of Arts and Sciences collated and published by Władysław Natanson)*, vol. 2, Kraków 1927, pp. 226–251.

Smoluchowski M., *Mechanika, (Mechanics)* Lviv 1912.

Smoluchowski M., *Einige Beispiele Brown'scher Molekularbewegung unter Einfluss äusserer Kräfte (Some examples of Brownian motion under the influence of external forces)* „Bulletin de l'Académie des Sciences de Cracovie"("International Bulletin of the Academy of Sciences of Kraków") 1911, Classe des Sciences Mathématiques et Naturelles (Class of Mathematical and Natural Sciences)· Series A: Mathematical Sciences, July 1913, pp. 418–434; reprinted in: *Pisma Mariana Smoluchowskiego z polecenia Polskiej Akademii Umiejętności zgromadzone i wydane przez Władysława Natansona (The Writings of Marian Smoluchowski commissioned by the Polish Academy of Arts and Sciences collated and published by Władysław Natanson)*, vol. 2, Kraków 1927, pp. 252–267.

Smoluchowski M., *Anzahl und Grösse der Moleküle und Atome (Number and size of molecules and atoms)* „Scientia" ("Science") 1913, vol. 13, Year 7., No. 27-1, pp. 27–44.

Smoluchowski M., *Liczba i wielkość cząsteczek i atomów (Number and size of molecules and atoms)* „Wiadomości Matematyczne" ("Mathematical News") 1913, vol. 17, pp. 315–329; reprinted in: *Pisma Mariana Smoluchowskiego z polecenia Polskiej Akademii Umiejętności zgromadzone i wydane przez Władysława Natansona (The Writings of Marian Smoluchowski commissioned by the Polish Academy of Arts and Sciences collated and published by Władysław Natanson)*, vol. 2, Kraków 1927, pp. 45–59.

Smoluchowski M., *Dzisiejszy stan teoryi atomistycznej* (*The current state of atomic theory*) „Kosmos: czasopismo Polskiego Towarzystwa Przyrodników im. Kopernika" ("Cosmos. Journal of the Polish Copernicus Society of Naturalists") 1913, vol. 38, books 4–6, pp. 355–373; reprinted in: *Pisma Mariana Smoluchowskiego z polecenia Polskiej Akademii Umiejętności zgromadzone i wydane przez Władysława Natansona* (*The Writings of Marian Smoluchowski commissioned by the Polish Academy of Arts and Sciences collated and published by Władysław Natanson*), vol. 2, Kraków 1927, pp. 60–73.

Smoluchowski M., *Mihailecul (1926 m) i Faracul (1961 m) w zimie* (*Mihailecul (1926 m) i Faracul (1961 m) in winter*) „Taternik: organ Sekcyi Turystycznej Towarzystwa Tatrzańskiego" ("Mountaineer: A unit of the Tourism Section of the Tatra Society") 1913, year 7, no 3, pp. 103– 107.

Smoluchowski M., *Gültigkeitsgrenzen des zweiten Hauptsatzes der Wärmetheorie* (*Validity limits of the second law of the theory of heat*), in: *Vorträge über die kinetische Theorie der Materieund der Elektrizität: gehalten in Göttingen auf einladung der Kommission der Wolfskehlstiftung*, Hrsg. M. Planck i in. (*Lectures on the kinetic theory of matter and electricity: held in Göttingen at the invitation of the Wolfskehl Foundation Commission*, ed. M. Planck), Leipzig, 1914, pp. 89–121; reprinted in: *Pisma Mariana Smoluchowskiego z polecenia Polskiej Akademii Umiejętności zgromadzone i wydane przez Władysława Natansona* (*The Writings of Marian Smoluchowski commissioned by the Polish Academy of Arts and Sciences collated and published by Władysław Natanson*), vol. 2, Kraków 1927, pp. 361–398.

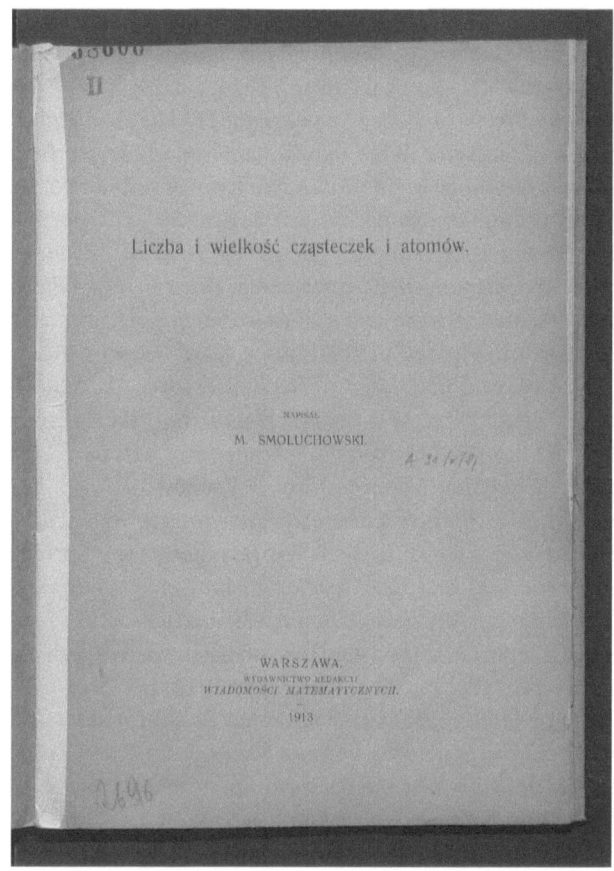

Number and size of molecules and atoms

Smoluchowski M., *Elektrische Endosmose und Strömungsströme* (*Electrical endosmosis and currents of flow*) in: *Handbuch der Elektrizität und des Magnetismus* (*Handbook of electricity and magnetism*), vol. 2, *Stationary Currents*, Edited by L. Graetz, Leipzig 1914, pp. 366–428; reprinted in: *Pisma Mariana Smoluchowskiego z polecenia Polskiej Akademii Umiejętności zgromadzone i wydane przez Władysława Natansona* (*The Writings of Marian Smoluchowski commissioned by the Polish Academy of Arts and Sciences collated and published by Władysław Natanson*), vol. 2, Kraków 1927, pp. 246–346.

Smoluchowski M., *O fluktuacjach termodynamicznych i ruchach Browna* (*On thermodynamic fluctuations and Brownian motion*) „Prace Matematyczno-Fizyczne" ("Mathematical-Physical Papers") 1914, vol. 25, pp. 187–263; reprinted in: *Pisma Mariana Smoluchowskiego z polecenia Polskiej Akademii Umiejętności zgromadzone i wydane przez Władysława Natansona* (*The Writings of Marian Smoluchowski commissioned by the Polish Academy of Arts and Sciences collated and published by Władysław Natanson*), vol. 2, Kraków 1927, pp. 354–360.

Smoluchowski M., *Bemerkung zu der Arbeit Hrn. B. Baule's: „Theoretische Behandlung der Erscheinungen in verdünnten Grasen"*, (*Commentary on Mr. B. Baule's work: 'Theoretical treatment of phenomena in rarefied grasses'*) „Annalen der Physik" ("Annals in Physics") 1914, Edition 4, vol. 45, pp. 623–624; reprinted in: *Pisma Mariana Smoluchowskiego z polecenia Polskiej Akademii Umiejętności zgromadzone i wydane przez Władysława Natansona* (*The Writings of Marian Smoluchowski commissioned by the Polish Academy of Arts and Sciences collated and published by Władysław Natanson*), vol. 2, Kraków 1927, pp. 177–178.

Smoluchowski M., *Studien über Molekularstatistik von Emulsionen und deren Zusammenhang mit der Brown'schen Bewegung* (*Studies on the molecular statistics of emulsions and their relation to Brownian motion*), „Sitzungsberichte der Akademie der Wissenschaften mathematisch-naturwissenschaftliche Klasse"("Meeting reports of the Academy of Sciences mathematical and scientific class") Section IIa, vol. 123, December 1914, pp. 2381–2405 reprinted in: *Pisma Mariana Smoluchowskiego z polecenia Polskiej Akademii Umiejętności zgromadzone i wydane przez Władysława Natansona* (*The Writings of Marian Smoluchowski commissioned by the Polish Academy of Arts and Sciences collated and published by Władysław Natanson*), vol. 2, Kraków 1927, pp. 399–421.

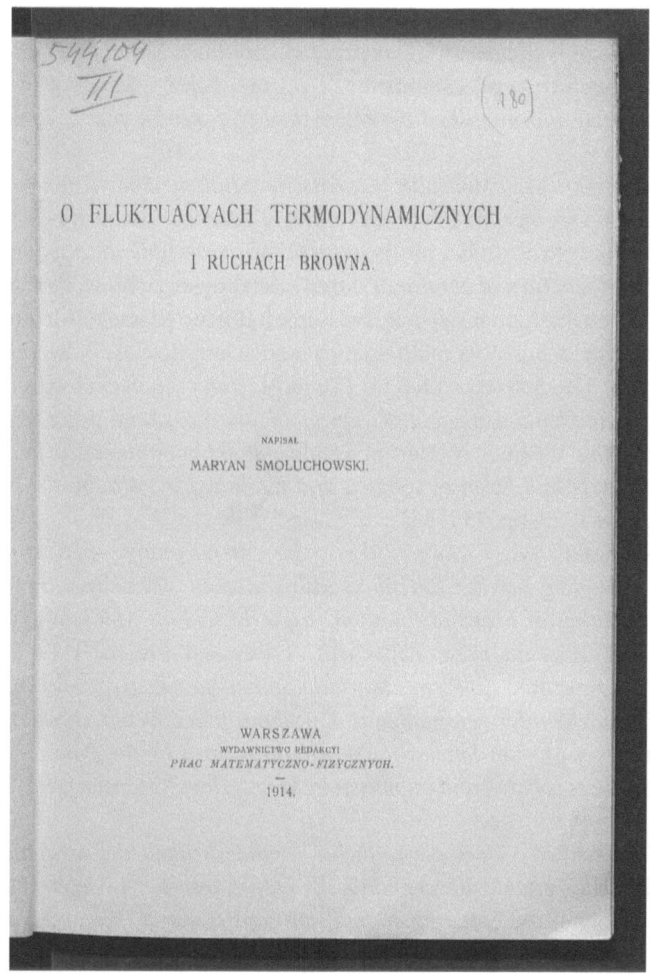

On thermodynamic fluctuations and Brownian motion

Smoluchowski M., *Über „durchschnittliche maximale Abweichung" bei Brown'scher Molekularbewegung und Brillouin's Diffusionsversuche* (*On 'average maximum deviation' in Brownian motion and Brillouin's diffusion experiments*) „Sitzungsberichte der Akademie der Wissenschaften mathematisch-naturwissenschaftliche Klasse" ("Meeting reports of the Academy of Sciences mathematical and scientific class"), Section IIa, No. 124, vol. 3–4, pp. 263–276;

reprinted in: *Pisma Mariana Smoluchowskiego z polecenia Polskiej Akademii Umiejętności zgromadzone i wydane przez Władysława Natansona* (*The Writings of Marian Smoluchowski commissioned by the Polish Academy of Arts and Sciences collated and published by Władysław Natanson*), vol. 2, Kraków 1927, pp. 422–434.

Smoluchowski M., *Molekulartheoretische Studien über Umkehr thermodynamisch irreversibler Vorgänge und über Wiederkehr abnorma- ler Zustände* (*Molecular theoretical studies on the reversal of irreversible thermodynamic processes and on the return of abnormal states*), „Sitzungsberichte der Akademie der Wissenschaften mathematisch-naturwissenschaftliche Klasse" ("Meeting reports of the Academy of Sciences mathematical and scientific class"), Section IIa, No. 124, H. 5, pp. 339–368; reprinted in: *Pisma Mariana Smoluchowskiego z polecenia Polskiej Akademii Umiejętności zgromadzone i wydane przez Władysława Natansona* (*The Writings of Marian Smoluchowski commissioned by the Polish Academy of Arts and Sciences collated and published by Władysław Natanson*), vol. 2, Kraków 1927, pp. 435–461.

Smoluchowski M., *Notiz über die Berechnung der Brown'schen Molekularbewegung bei der Ehrenhaft-Millikan'schen Versuchsanordnung* (*Note on the calculation of Brownian motion in the Ehrenhaft-Millikan experimental arrangement*) „Physikalische Zeitschrift" ("Physical Journal") 1915, vol. 16, pp. 318–321; reprinted in: *Pisma Mariana Smoluchowskiego z polecenia Polskiej Akademii Umiejętności zgromadzone i wydane przez Władysława Natansona* (*The Writings of Marian Smoluchowski commissioned by the Polish Academy of Arts and Sciences collated and published by Władysław Natanson*), vol. 2, Kraków 1927, pp. 477–485.

Smoluchowski M., *Über die zeitliche Veränderlichkeit der Gruppierung von Emulsionsteilchen und die Reversibilität der Diffusionserscheinungen* (*On the temporal variability of the grouping of emulsion particles and the reversibility of the diffusion phenomena*) „Physikalische Zeitschrift" ("Physical Journal") 1915, vol. 16, pp. 321–327.

Smoluchowski M., *Über gewisse Mängel in der Begründung des Entropiesatzes sowie der Boltzmann'schen Grundgleichung in der kinetischen Gastheorie* (*On certain deficiencies in the justification of the law of entropy and Boltzmann's basic equation in the kinetic theory of gases*), „Bulletin de l'Académie des Sciences de Cracovie", ("International Bulletin of the Academy of Sciences of Kraków"), Classe des Sciences Mathématiques et Naturelles (Class of Mathematical and Natural Sciences). Series A: Mathematical Sciences, June-July 1915, pp. 164–178; reprinted in: *Pisma Mariana Smoluchowskiego z polecenia Polskiej Akademii Umiejętności zgromadzone i wydane przez Władysława Natansona* (*The Writings of Marian Smoluchowski commissioned by the Polish Academy of Arts and Sciences collated and published by Władysław Natanson*), vol. 2, Kraków 1927, pp. 462-476.

Smoluchowski M., *Zur Theorie der Zustandsgleichungen* (*On the theory of the equations of state*) „Annalen der Physik" ("Annals in Physics") 1915, Edition 4, vol. 48, pp. 1098-1102; reprinted in: *Pisma Mariana Smoluchowskiego z polecenia Polskiej Akademii Umiejętności zgromadzone i wydane przez Władysława Natansona* (*The Writings of Marian Smoluchowski commissioned by the Polish Academy of Arts and Sciences collated and published by Władysław Natanson*), vol. 2, Kraków 1927, pp. 486-491.

Smoluchowski M., *Über Brownsche Molekularbewegung unter Einwirkung äusserer Kräfte und deren Zusammenhang mit der verallgemeinerten Diffusionsgleic hung* (*On Brownian molecular motion under the influence of external forces and its connection with the generalized diffusion equation*), „Annalen der Physik" ("Annals in Physics") 1915, Edition 4, vol. 48, pp. 1103-1112; reprinted in: *Pisma Mariana Smoluchowskiego z polecenia Polskiej Akademii Umiejętności zgromadzone i wydane przez Władysława Natansona* (*The Writings of Marian Smoluchowski commissioned by the Polish Academy of Arts and Sciences collated and published by Władysław Natanson*), vol. 2, Kraków 1927, pp. 492-502.

Smoluchowski M., *Studien über Kolloidstatistik und den Mechanismus der Diffusion* (*Studies on colloid statistics and the mechanism of diffusion*) „Kolloid Zeitschrift" ("Colloid Journal") 1916, No. XVIII, vol. 2, pp. 48-54; reprinted in: *Pisma Mariana Smoluchowskiego z polecenia Polskiej Akademii Umiejętności zgromadzone i wydane przez Władysława Natansona* (*The Writings of Marian Smoluchowski commissioned by the Polish Academy of Arts and Sciences collated and published by* Władysław Natanson), vol. 2, Kraków 1927, pp. 506-519.

Smoluchowski M., *Theoretische Bemerkungen über die Viskosität der Kolloide*, (*Theoretical remarks on the viscosity of colloids*) „Kolloid Zeitschrift" ("Colloid Journal") 1916, No. XVIII, vol. 5, pp. 190-195; reprinted in: *Pisma Mariana Smoluchowskiego z polecenia Polskiej Akademii Umiejętności zgromadzone i wydane przez Władysława Natansona* (*The Writings of Marian Smoluchowski commissioned by the Polish Academy of Arts and Sciences collated and published by* Władysław Natanson), vol. 2, Kraków 1927, pp. 520-529.

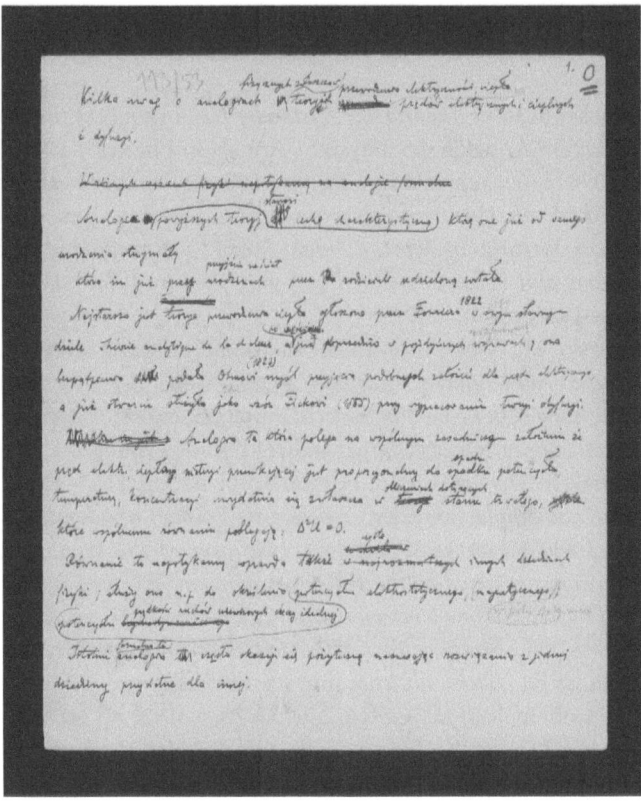

On physical analogies, especially in theories of electrical currents, heat and diffusion (handwritten)

Smoluchowski M., *Experimentelle Bestätigung der Rayleigh'schen Theorie des Himmelsblaus* (*Experimental confirmation of Rayleigh's theory of the blueness of the sky*), „Bulletin de l'Académie des Sciences de Cracovie", ("International Bulletin of the Academy of Sciences of Kraków"), Classe des Sciences Mathématiques et Naturelles (Class of Mathematical and Natural Sciences) Series A: mathematical Sciences, February–March–April 1916, s. 218–220; reprinted in: *Pisma Mariana Smoluchowskiego z polecenia Polskiej Akademii Umiejętności zgromadzone i wydane przez Władysława Natansona* (*The Writings of Marian Smoluchowski commissioned by the Polish Academy of Arts and Sciences collated and published by Władysław Natanson*), vol. 2, Kraków 1927, pp. 503–505.

Smoluchowski M., *Drei Vorträge über Diffusion, Brownsche Molekularbewegung und Koagulation von Kolloidteilchen* (Three lectures on diffusion, Brownian molecular motion and coagulation of colloids), „Physikalische Zeitschrift" ("Physical Journal") 1916, part 1: vol. 17, No. 22, pp. 557–571, part 2: vol. 17, No. 23, pp. 585– 599; reprinted in: *Pisma Mariana Smoluchowskiego z polecenia Polskiej Akademii Umiejętności zgromadzone i wydane przez Władysława Natansona* (The Writings of Marian Smoluchowski commissioned by the Polish Academy of Arts and Sciences collated and published by Władysław Natanson), vol. 2, Kraków 1927, pp. 585–599.

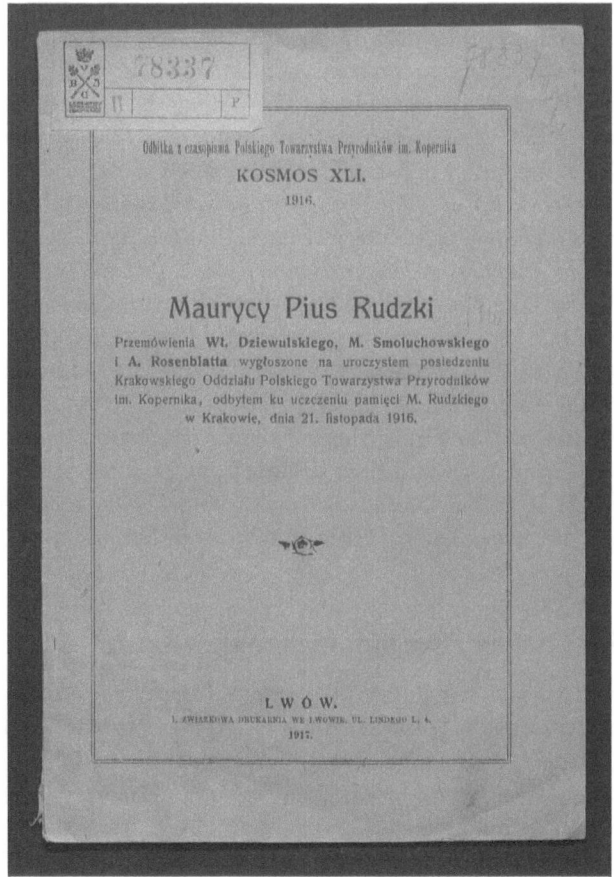

'*Kosmos*' ('Cosmos') – speech given at a meeting in memory of Maurycy Pius Rudzki (1862–1916)

Smoluchowski M., *Uwagi o pojęciu przypadku w zjawiskach fizycznych* (*Notes on the concept of chance in physical phenomena*), in: *Księga pamiątkowa ku czci Bolesława Orzechowicza* (Commemorative book in honour of Bolesaław Orzechowicz), vol. 2, Lviv 1916, pp. 445–458; reprinted in: *Pisma Mariana Smoluchowskiego z polecenia Polskiej Akademii Umiejętności zgromadzone i wydane przez Władysława Natansona* (*The Writings of Marian Smoluchowski commissioned by the Polish Academy of Arts and Sciences collated and published by Władysław Natanson*), vol. 3, Kraków 1928, pp. 74–86.

Smoluchowski M., *Versuch einer mathematischen Theorie der Koagula- tionskinetik kolloider Lösungen* (*Attempt at a mathematical theory of the coagulation kinetics of colloidal solutions*), „Zeitschrift für Physikalische Chemie" ("Journal of Physical Chemistry") 1917, vol. 92, pp. 129–168; reprinted in: *Pisma Mariana Smoluchowskiego z polecenia Polskiej Akademii Umiejętności zgromadzone i wydane przez Władysława Natansona* (*The Writings of Marian Smoluchowski commissioned by the Polish Academy of Arts and Sciences collated and published by Władysław Natanson*), vol. 2, Kraków 1927, pp. 595–639.

Smoluchowski M. (Ed.), *Fizyka* (Physics) in: *Poradnik dla samouków: wskazówki metodyczne dla studiujących poszczególne nauki* (*The Self- Study Handbook: methodological tips for studying particular sciences*), vol. 2. Warsaw 1917.

Smoluchowski M., *Znaczenie nauk ścisłych w wykształceniu ogólnym* (*The importance of the pure sciences in general education*) „Muzeum: czasopismo wydawane przez Towarzystwo Nauczycieli Szkół Wyższych" ("Museum: Magazine published by the Association of Higher School Teachers") 1917, vol. 32, book 2, pp. 286–294; reprinted in: *Pisma Mariana Smoluchowskiego z polecenia Polskiej Akademii Umiejętności zgromadzone i wydane przez Władysława Natansona* (*The Writings of Marian Smoluchowski commissioned by the Polish Academy of Arts and Sciences collated and published by Władysław Natanson*), vol. 3, Kraków 1928, pp. 124–131.

Publications issued after Smoluchowski's death

Smoluchowski M., *Maurycy Rudzki jako geofizyk* (*Maurycy Rudzki as a geophysicist*) „Kosmos: czasopismo Polskiego Towarzystwa Przyrodników im. Kopernika" ("Cosmos. Journal of the Polish Copernicus Society of Naturalists") 1917, vol. 41, pp. 105–119; reprinted in: *Pisma Mariana Smoluchowskiego z polecenia Polskiej Akademii Umiejętności zgromadzone i wydane przez Władysława Natansona* (*The Writings of Marian Smoluchowski commissioned by the Polish Academy of Arts and Sciences collated and published by Władysław Natanson*), vol. 3, Kraków 1928, pp. 111–123.

Smoluchowski M., *Karol Olszewski; ein Gelehrtenleben* (Karol Olszewski: a scholarly life), „Naturwissenschaften" ("Natural Sciences") 1917, No. 5, vol. 50, pp. 738-740; reprinted in: *Pisma Mariana Smoluchowskiego z polecenia Polskiej Akademii Umiejętności zgromadzone i wydane przez Władysława Natansona* (*The Writings of Marian Smoluchowski commissioned by the Polish Academy of Arts and Sciences collated and published by Władysław Natanson*), vol. 3, Kraków 1928, pp. 132-137.

Smoluchowski M., *Kobiety w naukach ścisłych* (*Women in the exact sciences*) „Rok Polski: czasopismo poświęcone zagadnieniom życia narodowego" ("Polish Year: magazine devoted to issues of national life") 1917, year 2, no. 12, pp. 7-24; reprinted in: *Pisma Mariana Smoluchowskiego z polecenia Polskiej Akademii Umiejętności zgromadzone i wydane przez Władysława Natansona* (*The Writings of Marian Smoluchowski commissioned by the Polish Academy of Arts and Sciences collated and published by Władysław Natanson*), vol. 3, Kraków 1928, pp. 138-152.

Smoluchowski M., *Grundrissder Koagulationskinetik kolloider Lösungen* (*Outline of the coagulation kinetics of colloidal solutions*) „Kolloid Zeitschrift" ("Colloid Journal") 1917, No. XXI, vol. 3, pp. 98-104; reprinted in: *Pisma Mariana Smoluchowskiego z polecenia Polskiej Akademii Umiejętności zgromadzone i wydane przez Władysława Natansona* (*The Writings of Marian Smoluchowski commissioned by the Polish Academy of Arts and Sciences collated and published by Władysław Natanson*) vol. 2, Kraków 1927, pp. 640-654.

Conclusion

The figure of Marian Smoluchowski has been neglected in Polish culture over the last hundred years, as manifested among other things in his systematic removal from common memory. This situation is difficult to accept for at least a few important reasons.

We are highly sensitised to all enrichment of Polish culture due to the successes of our countrymen. Over the course of the last hundred years, it has become obvious that Smoluchowski's scientific achievements turned out to be quite spectacular; over this time no other Polish physicist has appeared who would be in a position to replicate them in any way. A logical move would therefore be to make sure that he was duly appreciated in Polish science.

Throughout the whole century, the significance of Smoluchowskis works for world science have constantly grown. He introduced many solutions to physics which have become the basis for the development of whole branches of science. Examples of his achievements include equations of the theory of stochastic processes and the Smoluchowski continuity equation used in descriptions of sedimentation and coagulation processes, as well as a wide range of other equations and theories that have been described in detail in this book. This scientist has become one of the most cited Polish scholars in global scientific literature.

His legacy as a mountaineering pioneer is equally impressive. With his brother Tadeusz he is credited with numerous first ascents of peaks in the Dolomites and first traverses of previously impassable Alpine trails. It is due to them that Poland is counted among the forerunners of European mountaineering. Smoluchowski has considerable merit in the development of Polish mountaineering and was also at the forefront of Polish skiing.

It can safely be said that Marian Smoluchowski was a man of an extraordinary mind, righteous and worth following in every field. As Walery Goetal wrote:

> In (…) the whole of Smoluchowski's life, ethical value reigned supreme. An impeccable integrity of character, extraordinary nobility, simplicity, sincerity, an endearing modesty, created an image of him that was difficult to match. It was sufficient to see this person, to exchange a few words with him, to feel the power of his moral value, to feel carried away with it and uplifted by it. And this was not a verbal elevation, because no one ever heard moralistic words from Smoluchowski, but an exaltation by example, a living embodiment of the highest ethical ideals[505].

---

505   W. Goetel, *Ze wspomnień osobistych o Maryanie Smoluchowskim* (From personal recollections of Marian Smoluchowski), op. cit., p. 219.

He was also a European in the full meaning of the word, knew several languages and was passionate about painting and music. Even as a young man, immediately after his doctorate, he travelled with no problem from university to university, working with outstanding representatives of world physics.

In modern research, the Smoluchowski continuity equation is applied in the most diverse contexts, from theoretical work to industrial applications – for water purification, calculations to determine the formation of soot deposits in aircraft engines, the coagulation of milk, the formation of gel barriers, the growth of nanotubes, the aggregation of granulocytes and the adhesion of leukocytes. His discoveries can help in the application of the kinetics of diffusion-limited chemical reactions and in the study of a class of systems of stochastic differential equations describing diffusion phenomena. Valuable information can be found in Smoluchowski's work concerning problems of the simulation of potassium channels in cell membranes, fluid mechanics (the collision of droplets), protein electrophoresis, rotational diffusion, the effects of viscosity on electron transfer kinetics, analysis of cholera toxin interactions, the thermodynamics of information (numerical calculations), or Brownian dynamics. In all of these fields of science in which Smoluchowski's formulae and theories are used, they should be materials analysed at Polish universities in doctoral and post- doctoral work in chemistry and physics.

Smoluchowski made a huge contribution to the development of the philosophy of science and of natural philosophy. His philosophical thought was often ahead of its time and was a bold anticipation of modern solutions. He treated research of nature and the co-creation of science as a never- ending story that was written while it was being read, hence he never collected his philosophical views into a single compendium or philosophical system as such a system could never be complete, let alone true. Smoluchowski treated philosophy as a tool enabling a better understanding of the relationships occurring in nature. He believed that a proper trait of scientific work should be the search for and discovery of scientific truths, not burdened with philosophical limitations, and that philosophy should be important in that activity though merely a heuristically assistive tool.

This brief summary of the Polish scientist's achievements unfortunately does not translate into his position in Polish culture. In today's Poland, beyond a handful of scientists, hardly anyone has heard of this brilliant figure, not to mention the researcher's successes in the field of science. Even among practising physicists, not all know Smoluchowski's name and only a few are in a position to say anything about his scientific achievements. It is very telling that not a single school in Poland is named after him. There is no place where knowledge about him would be cultivated, although it should be shared far and wide among pupils and students.

It is without doubt an interesting fact that it was in the 1950s, when his name and work were exploited for political aims, that several publications about him appeared. Due to anniversaries during this period, three books devoted to him were also published. In addition to national cultural activities, local initiatives are very important. If we pay such great homage to Maria Skłodowska-Curie (who is often perceived and described as a French scientist), why has Marian Smoluchowski not gained a similar status over the last 30 years of a free Poland?

The centenary anniversary of his death passed almost unnoticed and, beyond a few scientific circles, nobody heard about it. To honour the anniversary, the Polish Physical Society dubbed 2017 the Year of Marian Smoluchowski. The scientist's memory was honoured by the Polish Senate. A sitting of the chamber was accompanied by a scientific conference and exhibition organised in the Senate building. An exhibition should also be mentioned called 'Under the guiding star of science. On the centenary of Marian Smoluchowski's death', which was jointly organised by staff of the Jagiellonian University Museum and the university's Faculty of Physics, Astronomy and Applied Computer Science. As can be seen, the celebrations were very limited and most of the events were able to organised thanks to the initiative of just a handful of people to whom Smoluchowski's memory was important. Let's hope that this inertia is finally broken, if only because Smoluchowski's scientific position continues to grow, and that the actions of people fascinated by the great Polish scholar at last start to bear the desired fruit.

Vernissage – powerful, eternal nature...

(...) when fatigue took on unusual proportions (...) he turned away from people, closed himself more deeply in his quite, generally secretive disposition and escaped to his faithful comforter and friend: nature. He did not need much here. It was enough to break free from the bonds of urban life for freedom, for a breath of fresh air, for the green of meadows or the white of snowy fields! There his soul was revived, it stretched out, in communing with nature it found peace and quiet. In a small pond in the meadows out of town, over which the first willow buds sprouted at the dawn of spring and the first living creatures spawned in the silent depths, or on solitary expeditions in a frail boat on the waves of the deep blue Adriatic, on a stroll to Sikornik or a carefree ramble in the hills around Kraków, where the gaze stretched far to the blue chains of the Carpathians and the toothy saw of the Tatras, or on the peaks of the Matterhorn or Jungfrau, after a hard day's tourist toil, in Park Stryjski in Lviv, remembering the charm of a copse of snow-covered trees, or hiking around the distant Norwegian fjords – everywhere he sought, saw, and felt the one eternal, unique Beauty of nature, untainted by human breath, and he relished it[506].

Smoluchowski's son Roman wrote in later years about the preserved water colours and sketches:

> Composition is generally treated as a subordinate factor compared to colour, which is undoubtedly the main subject of interest. (...) A significant share of the watercolours were done on the shores of seas or a lakes, or fjords, where there is always such a wealth and variety of colours, water, waves, skies and clouds. Sunsets and pre-storm moods are often repeated. And he stressed that in assessing these works it should be remembered that they were done lightly, by hand, as an interesting distraction with no pretensions to the art of painting[507].

---

506 W. Goetel, *Ze wspomnień osobistych o Maryanie Smoluchowskim* (From personal recollections of Marian Smoluchowski), op. cit., pp. 219–220.
507 See Z. Klemensiewicz, *Marian Smoluchowski*, op. cit., pp. 3–4 (citation after: M. Kluza, *Pod przewodnią gwiazdą nauki* (Under the guiding star of science), op. cit.).

Self-portrait[508], watercolour

---

508 Source: https://jbc.bj.uj.edu.pl/dlibra/publication/492587/edition/467084/con-tent (access: 12.04.2022).

The Alps 1, watercolour

The Alps 2, watercolour

The Alps 3, watercolour

The Alps 4, watercolour

The Alps 5, watercolour

Road, watercolour

England 1, watercolour

England 2, watercolour

England 3, watercolour

England 4, watercolour

England 5, watercolour

England 6, watercolour

England 7, watercolour

England 8, watercolour

Cliff, watercolour

Sea, watercolour

Fields, watercolour

River, watercolour

Tatras 1, watercolour

Tatras 2, watercolour

Tatras 3, watercolour

Tatras 4, watercolour

Tatras 5, watercolour

Tatras 6, watercolour

Tatras 7, watercolour

Tatras 8, watercolour

Italy 1, watercolour

Italy 2, watercolour

Italy 3, watercolour

Italy 4, watercolour

Italy 5, watercolour

Hills, watercolour

Acknowledgements

I wish to thank everyone who has contributed to the production of this book through their actions, valued help, and in some cases solid, hard work. I express my special gratitude to: Krystyna Bembennek, Professor Ordinarius Bogdan Cichocki, Dr. Piotr Chrząstowski, Dr Zofia Gołąb-Meyer, Prof. dr. hab. Paweł Horodecki, Dr. Maciej Kluza, Dr. hab. prof. KUL Marek Lechniak, Maciej Łyszczarz from the PAN and PAU Scientific Archive, Teresa Ossowska, Maria Pawłowska, Dr. hab. prof. UAM Tomasz Pospieszny, Dr. hab. prof. UJ Tomasz Pudłocki, Dr hab. prof. UJK Jacek Rodzeń, Prof. Dr. hab. Zenon Eugeniusz Roskal, Dr Ewa Szumilewicz and Dagmara Wachna.

Jestem pewny, że o wielu osobach zapomniałem, chciałbym jednak z tego miejsca serdecznie im wszystkim podziękować.

Jan Grzanka

List of illustrations

1. Smoluchowski's obituary in 'Nowa Reforma' magazine 1917, no 411, p. 2 (from the Smoluchowski family collections) 12
2. Tadeusz Godlewski on the death of Marian Smoluchowski (Jagiellonian Digital Library, https://jbc.bj.uj.edu.pl/dlibra/publication/400470#info) 12
3. Smoluchowski's demon (public domain) 18
4. Ernst Mach (1838–1916) (Wellcome Collection. Public Domain Mark) 22
5. Bolesław Józef Gawecki (1889–1984) (Jagiellonian Library, Rkp. BJ Przyb. 47/64) 24
6. Władysław Mieczysław Kozłowski (1858–1935) (public domain, Wikipedia) 26
7. Marian Smoluchowski during the Lviv years (from the Smoluchowski family collections) 34
8. The entrance to the former Cavendish Laboratory building (CC BY-SA 3.0) 40
9. Marian Smoluchowski during a stay at the University of Humboldt (picture taken in Berlin probably in 1897) (from the Smoluchowski family collections) 42
10. Announcement of the Polish Copernicus Society of Naturalists concerning public lectures (photographic collections of the Polish Copernicus Society of Naturalists) 43
11. Richard Zsigmondy (1865–1929) (public domain, Wikipedia) 49
12. Theodor Svedberg (1884–1971) (public domain, Wikipedia) 50
13. Jean Baptiste Perrin (1870–1942) (public domain, Wikipedia) 53
14. Ludwig Boltzmann (1844–1906) (public domain, Wikipedia) 61
15. Carl von Nägeli (1817–1891) (public domain, Wikipedia) 62
16. Wilhelm Ostwald (1853–1932) (public domain, https://www.britannica.com/biography/Wilhelm-Ostwald) 62
17. Brownian motion according to the observations of Perrin in 1908. (J. Perrin, *Les Atomes*, Paris 1913, p. 165) 64
18. Diagram of the formation of Brownian motion 66
19. A page from Einstein's paper *On the movement of small particles suspended in stationary liquids required by the molecular-kinetic theory of heat* (1905) 70

20. Extract from Smoluchowski's publication *An outline of a kinetic theory of Brownian motion and cloudy solutions* (1906) 71
21. Albert Einstein (1879–1955) (public domain) 72
22. A diagram by Prof. Bogdan Cichocki showing the different perspective in Einstein's and Smoluchowski's perception of Brownian motion (made by: B. Cichocki) 81
23. Einstein's letter to Smoluchowski 95
24. William Sutherland (1859–1911) at the age of 20. (public domain, Wikipedia) 96
25. Einstein's article on Smoluchowski („Naturwissenschaften" ("Natural Sciences"))1917, vol. 5, No. 50) 99
26. Mark Kac (1914–1984) 102
27. Wilhelm Ritter von Smolan-Smoluchowski (1831–1910), Marian Smoluchowski's father (from the Smoluchowski family collections) 106
28. Teofila Smoluchowska (Szczepanowska) (1847–1925), Marian Smoluchowski's mother (from the Smoluchowski family collections) 106
29. Marian's parents (from the Smoluchowski family collections) 107
30. Jan Smoluchowski – Marian Smoluchowski's grandfather, Wilhelm's father, Lord of Moszczenica (from the Smoluchowski family collections) 107
31. Wilhelm Smoluchowski (from the Smoluchowski family collections) 108
32. Teofila Smoluchowska (from the Smoluchowski family collections) ... 108
33. Antoni Popliński (1796–1868), Marian's great grandfather, father of Wanda Szczepanowska (from the Smoluchowski family collections) 109
34. Wanda Szczepanowska (Poplińska) (1829–1910), Marian's grandmother, mother of Teofila (from the Smoluchowski family collections) 109
35. Alleegasse in Vienna 110
36. Stanisław Szczepanowski (1846–1900), Marian's uncle, Teofila's brother (from the Smoluchowski family collections) 111
37. Stanisław Szczepanowski ski (from the Smoluchowski family collections) 111
38. Jan Władysław Szczepanowski (1813–1875), Marian's maternal grandfather 112
39. Five-year-old Marian and his older brother, Tadeusz (from the Smoluchowski family collections) 113

List of illustrations

40.	The Favorita Palace, designated by Empress Maria Theresa as the seat of the Collegium Therasianum (CC0, photo Gugerell)	114
41.	Marian Smoluchowski in his fourth year at school (from the Smoluchowski family collections)	115
42.	University of Vienna Physical Institute (1912) (University of Vienna)	116
43.	Jožef Stefan (1835–1893) (public domain, Wikipedia)	118
44.	A kustische Untersuchungen über die Elastizität weicher Körper – Marian Smoluchowski's doctoral dissertation (University of Vienna)	118
45.	Franz S. Exner (1849–1926) (University of Vienna)	119
46–47.	Zofia Baraniecka (both pictures taken at the same photographic studio, probably during one of Zofia's stays in Kraków before the wedding in 1901) (from the Smoluchowski family collections).	122
48–49.	Wedding photos of Zofia and Marian Smoluchowski (from the Smoluchowski family collections)	122
50.	Marian Smoluchowski with his sister-in-law, Jadwiga of the house of Braniecka (from the Smoluchowski family collections)	124
51–52.	Marian Smoluchowski with family in the Planty park in Kraków (around 1903) (from the family collections)	125
53.	Marian Smoluchowski with his wife Zofia and her sister, Jadwiga (from the Smoluchowski family collections)	127
54.	Marian Smoluchowski with his wife at Błonia Park in Kraków (from the Smoluchowski family collections)	128
55.	Marian Smoluchowski at lake Czarny Staw Pod Rysami (from the Smoluchowski family collections)	132
56.	Marian Smoluchowski on skis in the Tatra mountains (probably Kalatówki, December 27, 1913) (from the Smoluchowski family collections)	134
57.	On skis at Babia Góra (Polish Academy of Arts and Sciences)	134
58.	Fünffingerspitze in the Dolomites (2,918 MASL) (photo: Wolfgang Moroder, CC BY-SA 3.0)	139
59.	The Matterhorn (photo: chil, on Camptocamp.org Derivativework: Zacharie Grossen – Camptocamp.org, CC BY-SA 3.0)	141
60.	Marian Smoluchowski and family at Kałatówki (from the Smoluchowski family collections)	143

61. A sleigh from Zakopane to Morskie Oko (December 21, 1913) (from the Smoluchowski family collections) 143
62. Marian Smoluchowski with friends in the High Tatras (Mnich in the background) (from the Smoluchowski family collections) 144
63. Marian and Tadeusz Smoluchowski on Królowa Rówień in the Tatras (the route from Kuźnice to Hala Gąsienicowa) (photo: Adolf Guttenberg, Polish Academy of Arts and Sciences) 145
64. Above Zmarzły Staw in the Tatras (sitting: Marian Smoluchowski, behind him his nephew Bob Smoluchowski, standing probably Trzciński) (from the Smoluchowski family collections) 146
65. Pod Rysami (from the Smoluchowski family collections) 156
66. Marian Smoluchowski with Adolf Guttenberg on Królowa Rówień on the descent to Hala Gąsienicowa (from the Smoluchowski family collections) 157
67. Marian Smoluchowski in the Pięć Stawów valley (from the Smoluchowski family collections) 160
68. Walery Goetel (1889–1972) 161
69. Marian Smoluchowski and companions on the Thailhorn ridge (from the Smoluchowski family collections) 162
70. Marian Smoluchowski and wife on a trip to the mountains (from the Smoluchowski family collections) 162
71. Mieczysław Świerz (1891–1929) 163
72. Marian Smoluchowski on a trip with his wife and daughter (Library of the Jagiellonian University Faculty of Physics, Astronomy and Applied Computer Science) 163
73. Janusz Chmielowski (1878–1968) 164
74. Jorge Eduardo Hirsch (public domain, Wikipedia) 170
75. List of most cited pre-1930 author names (W. Marx, M. Cardona, *Blasts from the past*, "Physics World" 2004, vol. 17, no. 2) 172
76. Gabriel Lippmann (1845–1921) (CC BY-SA 4.0) 175
77. Hotel Orfila, where Smoluchowski lived in Paris (photo: WikimediaCommons / Mu, CC BY-SA 3.0), s. 226. 176
78. A plaque commemorating Smoluchowski's stay at Hotel Orfila (photo: WikimediaCommons / Mu, CC BY-SA 3.0) 176
79. The University of Glasgow (CC BY-SA 3.0) 177
80. Letter to Lord Kelvin 179
81. William Thomson, lord Kelvin (1824–1907) (public domain, Wikipedia) 180
82. Emil Gabriel Warburg (1846–1931) (public domain, Wikipedia) 182

List of illustrations

83. Marian Smoluchowski (from the Smoluchowski family collections) 185
84. Stanisław Ulam (1909–1984) (photo: Los Alamos National Laboratory) 188
85. Marian Smoluchowski after his doctorate (from the Smoluchowski family collections) 198
86. Portrait photo of Marian Smoluchowski – Lviv (from the Smoluchowski family collections) 201
87. Marian Smoluchowski – portrait photo taken at a photographic studio in Vienna (from the Smoluchowski family collections) 205
88. Marian Smoluchowski during the Kraków years (from the Smoluchowski family collections) 207
89. Karl Rajmund Popper (1902–1994) (public domain, Wikipedia) 209
90. Henri Poincaré (1854–1912) (public domain) 210
91. *The Writings of Marian Smoluchowski, commissioned by the Polish Academy of Arts and Sciences* 212
92. *The Writings of Marian Smoluchowski, commissioned by the Polish Academy of Arts and Sciences* – French version 213
93. Thomas Kuhn (1922–1996) (CC BY-SA 4.0) 231
94. Hilary Putnam (1926–2016) (CC BY-SA 2.5) 232
95. Ł. Storczak's article in 'Issues of Philosophy' magazine (Gdansk Library of the Polish Academy of Sciences) 243
96. Kazimierz Twardowski (1866–1938) (public domain, Wikipedia) 253
97. A letter from Kazimierz Twardowski 254
98. A telegram with congratulations on receiving the title of professor 255
99. In Lviv (photo: Bartosz Śmierciak) 256
100. Smoluchowski during the birthday celebrations of Prof. Viktor von Lang (Vienna 1908) (public domain) 257
101. Professors and students of the Faculty of Philosophy at the University of Lviv (June 1904) (V. Stefanyk, Lviv National Scientific Library of Ukraine) 258
102. August Witkowski (1854–1913) (public domain, Wikipedia) 259
103. The naming of Marian Smoluchowski as professor of experimental physics in Kraków (photo: Jagiellonian University archive) 260
104. Witkowski College (around 1912), ul. Gołębia 13 (photo: A. Pawlikowski) 261
105. Witkowski College, ul. Gołębia 13 (current view) (photo: M. Pawełek, Universitas) 262

106. The Collegium Physicum building, ul. Św. Anny 6, Kraków
(photo: M. Pawełek, Universitas) 263
107. Marian Smoluchowski and Tadeusz Zakrzewski (brothers-
in-law) at the exit of Wiślna Street (Olszewski College in the
background) (J. Zakrzewski photographic collections) 263
108. Marian Smoluchowski (right) in front of Collegium Novum;
to the left Witkowski College, currently the seat of the Faculty
of Physics of the Jagiellonian University (J. Zakrzewski
photographic collections) 265
109. The protocol of Prof. Marian Smoluchowski's election as rector
of the Jagiellonian University for the 1917–1918 academic year
(Jagiellonian University archive) 266
110. Posthumous recollection of Marian Smoluchowski in
'*Physikalische Zeitschrift*' (Physical Journal) of October 15, 1917 267
111. Article in the magazine '*Postępy w Naukach Fizycznych*'
(Advances in the Physical Sciences') (1917, Vol. 1, No. 1, p. 67) 268
112. Walery Goetel with his daughter, Wanda (1917) 271
113. Members of the Polish Copernicus Society of Naturalists at the
turn of the 19th and 20th centuries (photographic collections of
the Polish Copernicus Society of Naturalists) 272
114. An issue of '*Kosmos*' magazine with an article on Smoluchowski
(photographic collections of the Polish Copernicus Society of
Naturalists) 273
115. Election of the Board of the Tourist Section of the Tatra Society,
of which Smoluchowski was appointed Chairman (photographic
collections of the Polish Copernicus Society of Naturalists) 274
116. Not only skis but oars (Marian Smoluchowski at the middle oars) 276
117. Ernst Mach, as secretary of the Imperial Academy of Sciences,
informs Marian Smoluchowski of the sending to print of the
dissertation *Über den Temperatursprung bei Wärmeleitung in
Gasen* (On the temperature jump in heat conduction in gases)
with a request to make final amendments (1898) (manuscript
department of the Jagiellonian University) 278
118. Aldona (Donia) Smoluchowska (1902–1984) (from the
Smoluchowski family collections) 306
119. Donia and Romek Smoluchowski (1910) (from the
Smoluchowski family collections) 307
120. Zofia Smoluchowska, Romek, Teofila Smoluchowska and Donia
(Aldona) (from the Smoluchowski family collections) 307

List of illustrations

121. Aldona and her husband, Duncan H. Read, with their children (from the Smoluchowski family collections) 308
122. Zofia Smoluchowska with son Roman (from the Smoluchowski family collections) 308
123. Roman Smoluchowski (from the Smoluchowski family collections) 310
124. Roman Smoluchowski (public domain) 310
125. Zofia Smoluchowska with son Roman, his wife Louise, and grandson Peter (1952) (from the Smoluchowski family collections) 311
126. Zofia Smoluchowska (circa 1952) (from the Smoluchowski family collections) 312
127. Statue of Marian Smoluchowski, Wrocław University of Science and Technology Faculty of Chemistry (corner of Smoluchowski and Łukasiewicz streets) (Wrocław University of Science and Technology collections) 313
128. The Marian Smoluchowski Medal (public domain) 314
129. The Marian Smoluchowski-Emil Warburg Prize (public domain) 316
130. The Smoluchowski Award (public domain) 317
131. *The Self-Study Handbook* (public domain, Wikipedia) 319
132. Commemorative postcard 319
133. Marian Smoluchowski's grave (photo: Zygmunt Put, CC BY- SA 4.0) 320
134. *Maxwell's theory and the theory of electrons* (Library of the Faculty of Physics, Astronomy and Applied Computer Science, Jagiellonian University) 322
135. *Electricity and magnetism* (Library of the Faculty of Physics, Astronomy and Applied Computer Science, Jagiellonian University) 326
136. *Mechanics* (script from lectures) (Library of the Faculty of Physics, Astronomy and Applied Computer Science, Jagiellonian University) 327
137. *Reports on Polish papers in the field of physics for the years* 1901–1902 (collections of the Jagiellonian University Library), s. 412. 330
138. *Contribution to the theory of movement of viscous liquids; in particular of two-dimensional problems* (collections of the Jagiellonian University Library) 333
139. *Contribution to a theory of opalescence in gases in a critical state* (collections of the Jagiellonian University Library) 335

140. *A kinetic theory of gases* (collections of the Jagiellonian University Library) 337
141. *On a certain problem in the theory of elasticity and its relationship to the formation of fold mountains* (collections of the Jagiellonian University Library) 338
142. *A contribution to the theory of transpiration, diffusion and thermal conduction in rarefied gases* (collections of the Jagiellonian University Library) 341
143. *On the interaction of spheres moving in a viscous liquid* (collections of the Jagiellonian University Library) 344
144. *Contemporary atomic science* (collections of the Jagiellonian University Library) 346
145. *Number and size of molecules and atoms* (collections of the Jagiellonian University Library) 349
146. *On thermodynamic fluctuations and Brownian motion* (collections of the Jagiellonian University Library) 351
147. *On physical analogies, especially in theories of electrical currents, heat and diffusion* (handwritten) (collections of the Jagiellonian University Library) 354
148. *'Kosmos'* ('Cosmos') – speech given at a meeting in memory of Maurycy Pius Rudzki (1862–1916) (collections of the Jagiellonian University Library) 355
149. *Self-Portrait*, watercolour (from the Smoluchowski family collections) 364
150. *The Alps 1*, watercolour (from the Smoluchowski family collections) 365
151. *The Alps 2*, watercolour (from the Smoluchowski family collections) 366
152. *The Alps 3*, watercolour (from the Smoluchowski family collections) 366
153. *The Alps 4*, watercolour (from the Smoluchowski family collections) 367
154. *The Alps 5*, watercolour (from the Smoluchowski family collections) 367
155. *Road*, watercolour (from the Smoluchowski family collections) 368
156. *England 1*, watercolour (from the Smoluchowski family collections) 368
157. *England 2*, watercolour (from the Smoluchowski family collections) 369

158.	*England 3*, watercolour (from the Smoluchowski family collections)	369
159.	*England 4*, watercolour (from the Smoluchowski family collections)	370
160.	*England 5*, watercolour (from the Smoluchowski family collections)	370
161.	*England 6*, watercolour (from the Smoluchowski family collections)	371
162.	*England 7*, watercolour (from the Smoluchowski family collections)	371
163.	*England 8*, watercolour (from the Smoluchowski family collections)	372
164.	*Cliff*, watercolour (from the Smoluchowski family collections)	372
165.	*Sea*, watercolour (from the Smoluchowski family collections)	373
166.	*Fields*, watercolour (from the Smoluchowski family collections)	373
167.	*River*, watercolour (from the Smoluchowski family collections)	374
168.	*Tatras 1*, watercolour (from the Smoluchowski family collections)	374
169.	*Tatras 2*, watercolour (from the Smoluchowski family collections)	375
170.	*Tatras 3*, watercolour (from the Smoluchowski family collections)	376
171.	*Tatras 4*, watercolour (from the Smoluchowski family collections)	376
172.	*Tatras 5*, watercolour (from the Smoluchowski family collections)	377
173.	*Tatras 6*, watercolour (from the Smoluchowski family collections)	377
174.	*Tatras 7*, watercolour (from the Smoluchowski family collections).	378
175.	*Tatras 8*, watercolour (from the Smoluchowski family collections)	379
176.	*Italy 1*, watercolour (from the Smoluchowski family collections)	380
177.	*Italy 2*, watercolour (from the Smoluchowski family collections)	380
178.	*Italy 3*, watercolour (from the Smoluchowski family collections)	381
179.	*Italy 4*, watercolour (from the Smoluchowski family collections)	381
180.	*Italy 5*, watercolour (from the Smoluchowski family collections)	382
181.	*Hills*, watercolour (from the Smoluchowski family collections)	382

Bibliography

Amsterdamski S., Augustynek Z., Mejbaum W., *Prawo, konieczność, prawdopodobieństwo*, (*Laws, Necessity and Probability*), Spółdzielnia Wydawniczo-Handlowa "Książka i Wiedza", Warsaw 1964.

Aristotle, *Dzieła wszystkie. Fizyka* (*The complete works. Physics*) trans. K. Leśniak, Państwowe Wydawnictwo Naukowe, Warsaw 1990. Auerbach F., *Geschichtstafeln der Physik* (*History of Physics*) Leipzig 1910. Bafia S., *Fizyka w Quaestiones super octo libros "Physicorum" Aristotelis John of Głogów*, Scriptum, Kraków 2013.

Bernstein J., *Einstein and the Existence of Atoms*, "American Journal of Physics" 2006, vol. 74, no. 10.

Białobrzeski C., *Podstawy poznawcze fizyki świata atomowego* (*Cognitive foundations of the physics of the atomic world*), Państwowe Wydawnictwo Naukowe, Warsaw 1984.

Bohr N., *On the Constitution of Atoms and Molecules*, "Philosophical Magazine and Journal of Science" 1913, vol. 26, no. 1, pp. 1–25.

Bouckaert L.P., Smoluchowski R., Wigner E., *Theory of Brillouin Zones and Symmetry Properties of Wave Functions*, "Crystals. Physical Review" 1936, no. 50.

Casimir H., *Haphazard Reality: Half a Century of Science*, Harper Books, New York 1984.

Chandrasekhar S., *Stochastic Problems in Physics, and Astronomy*, "Reviews of Modern Physics" 1943, vol. 15, no. 1.

Chramow J.A., *Biografia fiziki* (*A biography of physics*) Kyiv 1983. Cichocki B., *Nagroda Nobla* (*The Nobel Prize*) "Delta" 1997, no 12. Comte A., *Metoda pozytywna w szesnastu wykładach* (*The positive method in sixteen lectures*), trans. W. Wojciechowska, Państwowe Wydawnictwo Naukowe, Warsaw 1961.

Copleston F., *Historia filozofii* (*A History of Philosophy*) vol. 8, *Bentham to Russell*, trans. B. Chwedeńczuk, Instytut Wydawniczy PAX, Warsaw 2006. Corbalán F., *Poskromienie przypadku. Teoria prawdopodobieństwa*, (*The taming of chance. A theory of probability*), trans. K. Rejmer, Buka Books Sławomir Chojnacki, Warsaw 2012.

Crombie A.C., *Nauka średniowieczna i początki nauki nowożytnej* (*Medieval and Early Modern Science*) vol. 1, *Nauka w średniowieczu w okresie V–XIII w.* (*Science in the Middle Ages: V-XIII Centuries.*), tłum. S. Łypacewicz, Instytut Wydawniczy PAX, Warsaw 1960.

Dawes G.W., *Belief Is Not the Issue: A Defence of Inference to the Best Explanation*, "Ratio. An International Journal of Analytic Philosophy" 2013, vol. 26, no. 1.

Dąmbska I., *O poglądach metanaukowych Władysława Natansona i Mariana Smoluchowskiego*, (*On the metascientific views of Władysław Natanson and Marian Smoluchowski*) "Zagadnienia Naukoznawstwa" 1979, vol. XV, book. 1 (57).

Dryden G., Vos J., *Rewolucja w uczeniu (The Learning Revolution)* trans. B. Jóźwiak, Zysk i S-ka, Poznań 2000.

Duplantier B., *Brownian Motion, "Diverse and Undulating"*, in: *Einstein, 1905–2005 Poincaré Seminar 2005*, eds. T. Damour, O. Darrigol, B. Duplantier, V. Rivasseau, "Progress in Mathematical Physics" book series (PMP, vol. 47), Birkhäuser Basel, Basel 2006.

Eftekhari A., *Ludwig Boltzmann (1844–1906)*, https://pdfs.semanticscholar.org/5c96/924ab515da7ebb6cb7601ec916099b03aed0.pdf.

Einstein A., *5 prac, które zmieniły oblicze fizyki (Five papers that changed the face of physics)*, trans. P. Amsterdamski, University of Warsaw Press, Warsaw 2005.

Einstein A., *Ist die Trägheit eines Körpersvon seinem Energieinhalt abhängig?(- Does the inertia of a body depend upon its energy content?)* "Annalen der Physik" ("Annals in Physics") 1905, No. 18.

Einstein A., *Marian v. Smoluchowski*, "Die Naturwissenschaften" ("The Natural Sciences") 1917, vol. 50.

Einstein A., *Über einen die Erzeugung und Verwandlung des Lichtes betreffendenheuristischen Gesichtspunkt (On a heuristic point of view about the creation and conversion of light)* "Annalen der Physik" ("Annals in Physics") 1905, No. 17, pp. 132–148.

Einstein A., *Über die von der molekular kinetischen Theorie der Wärmegeforderte Bewegung von in ruhenden Flüssigkeiten suspendierten Teilchen (On the movement of small particles suspended in stationary liquids required by the molecular-kinetic theory of heat)*, "Annalen der Physik" ("Annals in Physics") 1905, vol. 322, No. 8.

Einstein A., *Wspomnienie o Smoluchowskim (Recollections of Smoluchowski)*, „Problemy" ("Problems") 1972, no 8 (317).

Einstein A., *Zur Theorie der Brownschen Bewegung (On the theory of Brownian motion)*, „Annalen der Physik" ("Annals in Physics") 1906, vol. 324, no. 2.

Einstein A., *Zur Elektrodynamik bewegter Körper (On the electrodynamics of moving bodies)* „Annalen der Physik" ("Annals in Physics") 1905, vol. 17.

Einstein A., Besso M., *Correspondence 1903–1955*, transl., notes and introduction P. Speziali, Hermann, Paris 1979.

Fuliński A., *Współczesne zastosowania równań Smoluchowskiego* (Contemporary applications of the Smoluchowski equations) in: *Marian Smoluchowski – od teorii atomistycznej do fizyki współczesnej* (*Marian Smoluchowski – from atomic theory to modern physics*), ed. A. Strzałkowski, Polish Academy of Arts and Sciences, Kraków 2003.

Gałecki A., *Badania M. Smoluchowskiego w dziedzinie układów mikroskopijnych* (*M. Smoluchowski's research in the field of microscopic systems*) speech given at a ceremonial sitting of the Kraków Branch of the Polish Copernicus Society of Naturalists dedicated to honouring the memory of Prof. Marian Smoluchowski, December 11, 1917, „Kosmos" ("Cosmos") 1917, yearbook XII, books. 5–12.

Gawecki B.J., *Zagadnienie przyczynowości w fizyce* (*The problem of causality in physics*), Instytut Wydawniczy „Pax", Warsaw 1969.

Godlewski T., *Marian Smoluchowski. Jego życie i działalność naukowa* (*Marian Smoluchowski. His life and scientific activity*) „Wiadomości Matematyczne" ("Mathematical News") 1919, vol. 23, books 1–3.

Goetel W., *Marian Smoluchowski – człowiek gór* (*Marian Smoluchowski –man of the mountains*)„Wierchy" ("Peaks") 1953, vol. XXII.

Goetel W., *Zewspomnień osobistych o Maryanie Smoluchowskim* (*From personal recollections of Marian Smoluchowski*) „Kosmos" ("Cosmos") 1917, no 5–12.

Gołąb-Meyer Z., *Poglądy Mariana Smoluchowskiego na nauczanie fizyki z perspektywy stulecia*, w: *Marian Smoluchowski – od teorii atomi-stycznej do fizyki współczesnej* (Marian Smoluchowski's views on the teaching of physics from a century's perspective), in *Marian Smoluchowski – od teorii atomistycznej do fizyki współczesnej* (*Marian Smoluchowski – from atomic theory to modern physics*) ed. A. Strzałkowski, Polish Academy of Arts and Sciences, Kraków 2003.

Grotowski K.A., *Marian Smoluchowski – taternik i narciarz* (Marian Smoluchowski – mountaineer and skier) „PAUza Akademicka. Tygodnik PAU" (Academic PAUza, the Weekly of the Polish Academy of Arts and Sciences) 2017, no 380–381.

Grotowski K.A., *Marian Smoluchowski – taternik i narciarz* (Marian Smoluchowski – mountaineer and skier) w: *Marian Smoluchowski – od teorii atomistycznej do fizyki współczesnej* (*Marian Smoluchowski – from atomic theory to modern physics*) ed. A. Strzałkowski, Polish Academy of Arts and Sciences, Kraków 2003.

Hawking S.W., *Krótka historia czasu. Od wielkiego wybuchu do czarnych Dziur* (*A brief history of time: From the big bang to black holes*) trans. P. Amsterdamski, Alfa, Warsaw 1990.

Hawking S.W., *Krótkie odpowiedzi na wielkie pytania* (*Brief answers to the big questions*) trans. M. Krośniak, Wydawnictwo Zysk i S-ka, Warsaw 2019.

Hawking S.W., *Wszechświat w skorupce orzecha* (*The universe in a nutshell*) trans. P. Amsterdamski, Wydawnictwo Zysk i S-ka, Poznań 2018.

Hajduk Z., *Filozofia przyrody. Filozofia przyrodoznawstwa. Metakosmologia* (*The philosophy of nature. The philosophy of the natural sciences. Metacosmology*) Towarzystwo Naukowe KUL, Lublin 2007.

Heisenberg W., *Część i całość. Rozmowy o fizyce atomu* (*The part and the whole. Talks in the vicinity of atomic physics*) trans. K. Napiórkowski, Państwowy Instytut Wydawniczy, Warsaw 1987.

Heller M., *Czas i przyczynowość* (*Time and causality*) Towarzystwo Naukowe Katolickiego Uniwersytetu Lubelskiego (Scientific Society of the Catholic University of Lublin), Lublin 2002.

Herschel J.F.W., *Wstęp do badań przyrodniczych* (*Preliminary discourse on the study of natural philosophy*), trans. T. Pawłowski, Państwowe Wydawnictwo Naukowe, Warsaw 1955.

Herbut J., *Pragmatyzm* (*Pragmatism*) in: *Powszechna encyklopedia filozofii* (*The Universal Encyclopedia of Philosophy*) vol. VIII, Polskie Towarzystwo Tomasza z Akwinu (The Polish Society of Thomas Aquinas), Lublin 2007.

Hegel G.W.F., *Zasady filozofii prawa* (*Elements of the philosophy of right*) trans. A. Landman, Państwowe Wydawnictwo Naukowe, Warsaw 1969.

Home R.W., *Sutherland, William (1859–1911)*, in: *Australian Dictionary of Biography*, vol. 12, http://adb.anu.edu.au/biography/sutherland-william-8719. Hossenfelder S., *Zagubione w matematyce. Fizyka w pułapce piękna* (*Lost in math: How beauty leads physics astray*), trans. T. Miller, Copernicus Center, Kraków 2019.

Isaacson W., *Einstein. Jego życie, jego wszechświat* (*Einstein: His life and universe*) trans. J. Skowroński, Wydawnictwo W.A.B., Warsaw 2014.

James W., *Pragmatyzm. Nowa nazwa kilku starych metod myślenia. Popularne wykłady z filozofii* (*Pragmatism, a new name for some old ways of thinking; popular lectures on philosophy*) trans. M. Filipczuk, Zielona Sowa, Warsaw 2004.

Juszkiewicz A.P., *Historia matematyki* (*The history of mathematics*) vol. 3, PWN, Warsaw 1977.

Kac M., *Marian Smoluchowski and the Evolution of Statistical Physics*, in: S. Chandrasekhar, M. Kac, R. Smoluchowski, *Polish Men of Science. Marian Smoluchowski. His Life and Scientific Work*, ed. R.S. Ingarden, Wydawnictwo Naukowe PWN, Warsaw 1986.

Kac M., *Zagadki losu* (*Enigmas of chance*), trans. K. i H. Lipszycowie, Polska Fundacja Upowszechniania Nauki (Polish Foundation for Science Advancement), Warsaw 1997.

Kampen N.G. van, *Procesy stochastyczne w fizyce i chemii* (*Stochastic processes in physics and chemistry*) ed. Ł.A. Turski, trans. M. Dudyński, M. Ekiel-Jeżewska, D. Śledziewska-Błocka, Wydawnictwo Naukowe PWN, Warsaw 1990.

Kapczyński M., *Indeks Hirscha – zastosowanie oraz metody obliczania* (*The Hirsch Index – application and calculation methods*) Thomson Reuters Scientific, July 2, 2012, http://biblioteka.ans.pila.pl/download/dVKic4 GDFuImwDIHxbdnNwNHQ2KBkbfWzLhs_L3Y1KQs5J3JbMGIRYBRea R8qPWwwYhxuJRkzMDI6Gw8sJDAqEzhmFFYjexFqBmY1SHZzZThq JWNAXDYxND4cIFw8GjIeInEnCjl8HGsOZxkHPyVsNX8fLhgSNjwqOhk 6Yno1LBlzKWoJJGQXYAR2GQM-czljMGxjFh8xOHJhVSNvdjg/indeks_h_ zastosowanie_metody_obliczania.pdf.

Kelly T., *Hume, Norton, and Induction without Rules*, "Philosophy of Science" 2010, vol. 77, no. 5.

Klemensiewicz Z., *Marian Smoluchowski*, „Taternik. Organ Sekcji Tury- stycznej Polskiego Towarzystwa Tatrzańskiego"("Mountaineer. A unit of the Tourist Section of the Polish Tatra Society" Kraków 1915–1921 (photocopy with no number).

Kluza M., *Pod przewodnią gwiazdą nauki: Marian Smoluchowski w stulecie śmierci* (*Under the guiding star of science: Marian Smoluchowski on the centenary of his death*) https://jbc.bj.uj.edu.pl/dlibra/publication/492587/ edition/ 467084/content.

Kołakowski L., *Główne nurty marksizmu* (*The Main Currents of Marxism*), vol. 3, Wydawnictwo Naukowe PWN, Warsaw 2009. Kosmulski M., *Potencjał ζ i równanie Smoluchowskiego* (*Zeta potential and the Smoluchowski equation*) „PAUza Akademicka. Tygodnik PAU" (Academic PAUza, the Weekly of the Polish Academy of Arts and Sciences) 2017, no 380–381.

Kotowa B., *Scjentyzm jako światopogląd nauki* (*Scientism as a worldview of science*) „Nowa Krytyka" ("New Criticism") 2004, no 16.

Kozłowski W.M., *Przyczynowość jako podstawowe pojęcie przyrodoznawstwa* (*Causality as a fundamental concept of the natural sciences*), E. Wende i S-ka, Warsaw 1906.

Kostro L., *Alberta Einsteina koncepcja nowego eteru: jej historia, sens fizyczny i uwarunkowania filozoficzne* (*Albert Einstein's conception of the new ether: its history, physical meaning and philosophical conditions*), Scientia, Gdańsk 1999.

Krajewski W., *Światopogląd Mariana Smoluchowskiego* (*Marian Smoluchowski's worldview*) Wydawnictwo Naukowe PWN, Warsaw 1956.

Krajewski W., *Wielki fizyk i filozof materialista (w 80-lecie urodzin Mariana Smoluchowskiego)* (*The great physicist and materialist philosopher (on the 80th anniversary of Marian Smoluchowsi's birth*) „Trybuna Ludu" ("The People's Tribune"), 1952, no 148.

Kuhn T., *Przewrót kopernikański. Astronomia planetarna wdziejach myśli* (*The Copernican revolution: Planetary astronomy in the development of western thought*), trans. S. Amsterdamski, Państwowe Wydawnictwo Naukowe, Warsaw 1966.

Kumaniecki K., *Zbiór najważniejszych przepisów uniwersyteckich* (*A collection of the most important university regulations*), Kraków 1913.

Lemons D.S., Gythiel A., *Paul Langevin's 1908 paper "On the Theory of Brownian Motion"*, "American Journal of Physics" 1997, vol. 65, no. 11, November.

Loria S., *Marian Smoluchowski i jego dzieło (1872–1917)* (*Marian Smoluchowski and his works (1872–1917)* „Postępy Fizyki" ("Progress of Physics") 1953, vol. 4, book 1, pp. 5–38.

Litwinowicz-Droździel M., *Indukcje i przepływy. Michael Faraday – mikrostudium o romantycznej nauce* (*Inductions and flows. Michael Faraday – a microstudy of romantic science*) „Wiek XIX. Rocznik Towarzystwa Literackiego im. Adama Mickiewicza" ("The XIX Century. Yearbook of Adam Mickiewicz Literary Society") 2015, vol. VIII (L).

Loria S., *Marian Smoluchowski i jego dzieło (1872–1917)* (*Marian Smoluchowski and his work (1872–1917)* „Postępy Fizyki" ("Progress of Physics") 1953, vol. 4, book 1

Maiocchi R., *The Case of Brownian Motion*, "The British Journal for the History of Science" 1990, vol. 23, no. 3.

Marian Smoluchowski (1872–1917). Fizyk, taternik, romantyk nauki (*Marian Smoluchowski (1872–1917). Physicist, mountaineer, romantic of science*), catalogue of the Collegium Maius temporary exhibition, May 17–July 14, 2002.

Marshall S.J., *Shaping the University of the Future*, University of Wellington, New Zealand 2018.

Maryniarczyk A., *Tomizm* (*Thomism*) in: *Powszechna encyklopedia filozofii* (*The universal encyclopedia of philosophy*) vol. IX, Polskie Towarzystwo Tomasza z Akwinu (The Polish Society of Thomas Aquinas), Lublin 2008.

Marx W., Cardona M., *Blasts from the Past*, "Physics World" 2004, vol. 17, no. 2.

Maślanka J., *Początki narciarstwa i taternictwa polskiego* (*The beginnings of Polish skiing and mountaineering*) Archiwum Nauki PAN i PAU(PAN and PAU Scientific Archive), K III 36.

Maślanka J., *Zaranie polskiego alpinizmu* (*The dawn of Polish mountaineering*) Archiwum Nauki PAN i PAU (PAN and PAU Scientific Archive), K III-36, typescript. T. II 103.

Maxwell J.C., *On Faraday's Lines of Force*, "Transactions of the Cambridge Philosophical Society" 1864, vol. X, part I, no. III.

Maxwell J.C., *Theory of Heat*, trans. E. Szumilewicz, on the service: science20. com, https://pl.wikipedia.org/wiki/Demon_Maxwella. Metallmann J., *Zagadnienie przypadku* (*The issue of chance*) „Przegląd Współczesny" (" Modern Review"), 1933, year XII, vol. XLIV, pp. 85–95.

Michelson A.A., *Speech at the University of Chicago*, in: A.K. Wróblewski, *Historia fizyki: od czasów najdawniejszych do współczesności* (*The history of physics: from antiquity to modernity*), Wydawnictwo Naukowe PWN, Warsaw 2006.

Miecznik J.B., *U źródeł geologicznych zainteresowań Mariana Smoluchowskiego* (*The sources of Marian Smoluchowski's geological interests*), „Przegląd Geologiczny" ("Geological Review"), 2013, vol. 61, no 5.

Nicholas of Cusa, *O oświeconej niewiedzy* (*On learned ignorance*), trans. I. Kania, Znak, Kraków 1997.

Morales J.M.R., *Kościół i nauka. Konflikt czy współpraca?*(*The Church and science. Conflict or cooperation?*), trans. S. Jędrusiak, Wydawnictwo WAM, Kraków 2003.

Nernst W., *Stöchiometrie Verwandtschaftslehre* (*Stoichiometry relationship theory*), „Zeitschrift für Physikalische Chemie" ("Journal of Physical Chemistry"), 1888, vol. 2.

Newton's Dream, ed. M.S. Stayer, Queen's University Press, Montreal 1988.

Niezgoda A., *Philip Zimbardo: Młodzi mężczyźni wycofują się z życia społecznego* (*Philip Zimbardo: young men withdraw from social life*), Focus.pl, https://www.focus.pl/artykul/wylogowani-z-zycia.

Norton J.D., *A Material Theory of Induction*, "Philosophy of Science" 2003, vol. 70, no. 4.

Palczewski A., *Marian Smoluchowski – alpinista* (*Marian Smoluchowski – mountaineer*), „Delta", December 1997, http://www.deltami.edu.pl/temat/roznosci/historia i filozo- fia/2017/08/17/Marian Smoluchowski-alpinista/.

Perrin J., *Les Atomes* Librairie Felix Alcan, Paris 1913; English edition: *Atoms*, D. van Nostrand Company, New York 1916.

Planck M., *Über irreversible Strahlungsvorgänge* (*On irreversible radiation processes*), „Annalen der Physik" ("Annals in Physics) 1900, vol. 306, No. 1.

Planck M., *Vom Relativen zum Absoluten* (*From the Relative to the Absolute*), lecture given in Munich n December 1 1924, in: M. Planck, *Jedność fizycznego obrazu świata. Wybór pism filozoficznych* (*The unity of the physical worldpicture. A selection of philosophical writings*), trans. R. i S. Kernerowie, Książka i Wiedza, Warsaw 1970.

Polak P., *Byłem pana przeciwnikiem [profesorze Einstein]…* (*I was your opponent [Professor Einstein]…*), Copernicus Center Press, Kraków 2012.

Polak P., *Koncepcja przypadku w pismach Mariana Smoluchowskiego* (*The concept of chance in Marian Smoluchowski's writings*), in: *Krakowska filozofia przyrody w okresie międzywojennym* (*The Krakovian philosophy of nature in the inter-war period*), vol. 3, *Smoluchowski – Natanson – others*, ed. M. Heller, J. Mączka, Ośrodek Badań Interdyscyplinarnych przy Wydziale Filozoficznym Papieskiej Akademii Teologicznej – Wydawnictwo Diecezji Tarnowskiej Biblos, Kraków–Tarnów 2007.

Portell Bueso X., *SUSY Searches at the Tevatron and the LHC*, presentation given at the "Physics in Collision" international symposium, Vancouver, Canada, August–September 2011.

Protocol of Bolesław Gawecki's rigour in physics of 27.01.1914 in the presence of Władysław Natanson and Marian Smoluchowski, Card 23 Jagiellonski University Archive.

Putnam H., *Czym jest realizm?* (*What is "realism"?*), trans. P. Zeidler, "Colloquia Communia" 1991, no 1 –3.

Putnam H., *Philosophical Papers*, vol. 2, *Mind, Language and Reality*, Cambridge University Press, Cambridge 1975.

Roskal Z.E., *Mariana Smoluchowskiego ujęcie zasady przyczynowości w badaniach ruchów Browna* (*Marian Smoluchowski's approach to the principle of causality in research on Brownian motion*) „Zagadnienia Filozoficzne w Nauce" ("Philosophical Problems in Science") 2017, vol. 62.

Roszkowska E., *Alpejska działalność Mariana Smoluchowskiego* (*Marian Smoluchowski's alpine activity*) The Bronisław Czech University of Physical Education in Krakow „Folia Turistica" 2012, no 26, http://www.folia-turistica.pl/attachments/article/402/FT_26_2012.pdf.

Sławianowski J.J., *Przyczynowość w mechanice kwantowej* (*Causality in quantum mechanics*) Wydawnictwo „Wiedza Powszechna", Warsaw 1969.

Słomski W., *Władysław Krajewski*, in: *Polska filozofia powojenna* (*Post-war Polish philosophy*), ed. W. Mackiewicz, Agencja Wydawnicza Witmark, Warsaw 2001.

Smith A., *Badania nad naturą i przyczynami bogactwa narodów* (*An inquiry into the nature and causes of the wealth of nations*), vol. II, trans. A. Prejbisz, Państwowe Wydawnictwo Naukowe, Warsaw 1954. Smoluchowski M., *Dwie książki z dziedziny „filozofii przyrody"* (*Two books from the field of "natural philosophy"*) „Ateneum Polskie" 1909, vol. IV.

Smoluchowski M., *Dzisiejszy stan teorii atomistycznej* (*The current state of atomic theory*) lecture given on teachers' courses in Lviv, March 12, 1913, printed in "Kosmos" (Cosmos) magazine, 1918, vol. 38, pp. 355–373, reprinted in *Pisma Mariana Smoluchowskiego z polecenia Polskiej Akademii Umiejętności zgromadzone i wydane przez Władysława Natansona* (*The writings of Marian

Smoluchowski commissioned by the Polish Academy of Arts and Sciences, collated and published by Władysław Natanson), vol. 3, Jagiellonian University Press, Kraków 1928

Smoluchowski M., *Ewolucjateorii atomistycznej* (*The evolution of atomic theory*), in: *Pisma Mariana Smoluchowskiego z polecenia Polskiej Akademii Umiejętności* zgromadzone i wydaneprzez Władysława Natansona (*The writings of Marian Smoluchowski commissioned by the Polish Academy of Arts and Sciences, collated and published by Władysław Natanson*), vol. 3, Jagiellonian University Press, Kraków 1928

Smoluchowski M., *Kierunki i zagadnienia fizyki dzisiejszej* (*Themes and issues in today's physics*) in: *Pisma Mariana Smoluchowskiego z polecenia Polskiej Akademii Umiejętności zgromadzone i wydane przez Władysława Natansona* (*The writings of Marian Smoluchowski commissioned by the Polish Academy of Arts and Sciences, collated and published by Władysław Natanson*), vol. 3, Jagiellonian University Press, Kraków 1928

Smoluchowski M., *Kilka uwag o analogiach fizycznych, zwłaszcza w teoriach prądówelektrycznych, prądówcieplnych i zjawiska dyfuzji* (*Several observations on physical analogies, especially in theories of electrical currents, heat currents and diffusion phenomena*) „Wiadomości Matematyczne" ("Mathematical News") 1918, vol. XXII; reprinted in: *Pisma Mariana Smoluchowskiego z polecenia Polskiej Akademii Umiejętności zgromadzone i wydaneprzez Władysława Natansona* (*The writings of Marian Smoluchowski commissioned by the Polish Academy of Arts and Sciences, collated and published by Władysław Natanson*), vol. 3, Jagiellonian University Press, Kraków 1928.

Smoluchowski M., *Kobiety w naukach ścisłych* (*Women in the exact sciences*), lecture given at the Scientific-Literary Society of Lviv in 1912, reprinted in: *Pisma Mariana Smoluchowskiego z polecenia Polskiej Akademii Umiejętności zgromadzone i wydaneprzez Władysława Natansona* (*The writings of Marian Smoluchowski commissioned by the Polish Academy of Arts and Sciences, collated and published by Władysław Natanson*), vol. 3, Jagiellonian University Press, Kraków 1928.

Smoluchowski M., *List do S. Michalskiego w sprawie „Poradnika dla samouków"* (*Letter to S. Michalski on the matter of "The Self-Study Handbook"*) Jagiellonian Library, signature BJ Rkp. 9412, k. 86.

Smoluchowski M., *Maurycy Rudzki jako geofizyk* (*Maurycy Rudzki as a geophysicist*), speech given on November 21, 1916 at a sitting of the Kraków Branch of the Polish Copernicus Society of Naturalists held in memory of M.P. Rudzki, „Kosmos" ("Cosmos") 1916, vol. XLI. Smoluchowski M., *Mihailecul (1926 m) and Farcaul (1961 m)*, „Taternik. Organ Sekcji Turystycznej Polskiego Towarzystwa Tatrzańskiego" ("Mountaineer. A unit of the Tourist Section of the Polish Tatra Society") 1913, no 6.

Smoluchowski M., *O fluktuacjach termodynamicznych i ruchach Browna* (*On thermodynamic fluctuations and Brownian motion*), „Prace Matematyczno-Fizyczne" ("Mathematical-Physical papers") 1924, vol. 2525, reprinted in: *Pisma Mariana Smoluchowskiego z polecenia Polskiej Akademii Umiejętności zgromadzone i wydaneprzez Władysława Natansona* (*The writings of Marian Smoluchowski commissioned by the Polish Academy of Arts and Sciences, collated and published by Władysław Natanson*), vol. 2, Jagiellonian University Press, Kraków 1927.

Smoluchowski M., *On irregularities in the distribution of gas molecules and their influence on entropy and the equation of state* (*Über Unregelmässigkeiten in der Verteilung von Gasmolekülen und deren Einfluss auf Entropie und Zustandsgleichung*), trans. B.J. Gawecki, Państwowe Wydawnictwo Naukowe, Warsaw 1956, first edition: *Über Unregelmäßigkeiten in der Verteilung von Gasmolekülen und deren Einfluß auf Entropie und Zustandsgleichung*, in: *Festschrift Ludwig Boltzmann gewidmet zum sechzigsten Geburtstage*, (*Commemorative book dedicated to Ludwig Boltzmann on the occasion of his sixtieth birthday*), Leipzig 1904.

Smoluchowski M., *O nowszych postępach na polu kinetycznych teorii materii*, (*On newer advances in the field of kinetic theories of matter*) w: *Pisma Mariana Smoluchowskiego z polecenia Polskiej Akademii* Umiejętności *zgromadzone i wydane przez Władysława Natansona* (*The writings of Marian Smoluchowski commissioned by the Polish Academy of Arts and Sciences, collated and published by Władysław Natanson*), vol. 1, Jagiellonian University Press, Kraków 1921.

Smoluchowski M., *O pojęciu przypadku i pochodzeniu praw fizyki opartych na prawdopodobieństwie* (*On the concept of chance and the origin of the laws of probability in physics*), „Wiadomości Matematyczne" ("Mathematical News")1923, vol. 27, book 2, reprinted in: *Pisma Mariana Smoluchowskiego z polecenia Polskiej Akademii Umiejętności zgromadzone i wydaneprzez Władysława Natansona* (*The writings of Marian Smoluchowski commissioned by the Polish Academy of Arts and Sciences, collated and published by Władysław Natanson*), vol. 3, Jagiellonian University Press· Kraków 1928.

Smoluchowski M., *Organizacja i działalność zakładów fizycznych* (*The organisation and activity of physics facilities*) „Nauka Polska, jej potrzeby, organizacja i rozwój"("Polish Science: requirements, organisation and development") Yearbook of the Dr Józef Mianowski Aid Fund for People working in the scientific field), vol. I, Warsaw 1918.

Smoluchowski M., *Poradnik dla samouków: wskazówki metodyczne dla studiującychposzczególnenauki.Fizyka,Geofizyka, Meteorologia* (*Self-study handbook: methodological tips for students of specific sciences. Physics, Geophysics, Meteorology*) vol. I i II, Wydawnictwo A. Heflicha i St. Michalskiego, Warsaw 1917.

Smoluchowski M., *Teoria kinetyczna opalescencji w stanie krytycznym oraz innych zjawisk pokrewnych* (*Molecular-kinetic theory of the opalescence of gases in the critical state and a few related phenomena*), dissertations of the Faculty of Mathematics and Natural Sciences of the Polish Academy of Arts and Sciences, issued by the Academy of Art and Sciences, Kraków 1907, vol. XLVII, Series A.

Smoluchowski M., *Über den Begriff des Zufalls und den Ursprung der Wahrscheinlichkeitsgesetze in der Physik* (*On the concept of chance and the origin of the laws of probability in physics*), „Naturwissenschaften" ("Natural Sciences") 1918, vol. 6 (17).

Smoluchowski M., *Über Unregelmäßigkeiten in der Verteilung von Gasmolekülen und deren Einfluß auf Entropie und Zustandsgleichung* (*On irregularities in the distribution of gas molecules and their influence on entropy and the equation of state*), in: *Festschrift Ludwig Boltzmann gewidmet zum sechzigsten Geburtstage* (*Commemorative book dedicated to Ludwig Boltzmann on the occasion of his sixtieth birthday*), Leipzig 1904.

Smoluchowski M., *Uwagi o pojęciu przypadku w zjawiskach fizycznych* (*Notes on the concept of chance in physical phenomena*) in: *Księga Pamiątkowa ku czci Bolesława Orzechowicza* (*Memorial book in honour of Bolesław Orzechowicz*) Towarzystwo dla Popierania Nauki Polskiej, Lwów (Society for the support of Polish Science, Lviv) 1916, reprinted in: *Pisma Mariana Smoluchowskiego z polecenia* Polskiej Akademii Umiejętności *zgromadzone i wydaneprzez Władysława Natansona* (*The writings of Marian Smoluchowski commissioned by the Polish Academy of Arts and Sciences, collated and published by Władysław Natanson*), Jagiellonian University Press, Kraków 1928.

Smoluchowski M., *Uwagi o roli przypadku we fizyce* (*Notes on the role of chance in physics*) Towarzystwo Filozoficzne w Krakowie (Philosophical Society of Kraków), lecture delivered on March 1, 1917, hand-written manuscript, Jagiellonian Library, signature 9398 IV, k. 3; reprinted in: „Zagadnienia Filozoficzne w Nauce" ("Philosophical Problems in Science") 2017, no 62 (special edition).

Smoluchowski M., *Vortrag im Philosophischen Seminar 1893/1894* (*Lecture during philosophical seminar 1893/1894*) University of Vienna, Vienna (1893/1894), ed. M. Dziekan, „Zagadnienia Filozoficzne w Nauce" ("Philosophical Problems in Science") 2017, vol. LXII.

Smoluchowski M., *Wycieczki górskie w Szkocji* (*Highland trips in Scotland*) „Taternik. Organ Sekcji Turystycznej Polskiego Towarzystwa Tatrzańskiego" ("Mountaineer. A unit of the Tourist Section of the Polish Tatra Society") years 1915–1921, Kraków 1921.

Smoluchowski M., *Zarys kinetycznej teorii ruchów Browna i roztworów mętnych* (*An outline of a kinetic theory of Brownian motion and cloudy solutions*) „Rozprawy Wydziału Matematyczno- Przyrodniczego Akademii Umiejętności"("Dissertations of the Faculty of Mathematics and Natural Sciences of the Polish Academy of Arts and Sciences"), vol. 6, section A, Kraków 1906; reprinted in: *Pisma Mariana Smoluchowskiego z polecenia Polskiej Akademii Umiejętności zgromadzone i wydaneprzez Władysława Natansona* (The writings of *Marian Smoluchowski commissioned by the Polish Academy of Arts and Sciences, collated and published by Władysław Natanson*), vol. 1, Jagielloniam University Press, Kraków 1924.

Smoluchowski M., *Znaczenie nauk ścisłych w wykształceniu ogólnym* (*The importance of the exact sciences in general education*), speech given during the congress of members of the Society of Higher School Teachers, May 27, 1917, reprinted in „Muzeum" ("Museum") magazine 1917, vol. 32, pp. 286–294, and also in: *Pisma Mariana Smoluchowskiego z polecenia Polskiej Akademii Umiejętności zgromadzone i wydaneprzez Władysława Natansona* (*The writings of Marian Smoluchowski commissioned by the Polish Academy of Arts and Sciences, collated and* published by Władysław Natanson*)*, vol. 3, Jagiellonian University Press, Kraków 1928.

Smoluchowski M., *Zur kinetischen Theorie der Brownschen Molekularbewegung und der Suspensionen* (*On the kinetic theory of Brownian motion and suspensions*) „Annalen der Physik" ("Annals in Physics") 1906, vol. 21.

Smoluchowski R., *Anisotropy of the Electronic Work Function of Metals*, "Physical Review" 1941, no. 60.

Smullyan R., *Na zawsze nierozstrzygnięte. Zagadkowy przewodnik po twierdzeniach Gödla* (*Forever undecided: A puzzle guide to Gödel*), trans. J. Pogonowski, Książka i Wiedza, Warsaw 2007.

Söderbaum I.G., speech given by the secretary of the Royal Swedish Academy of Sciences during the ceremony to award Theodor Svedberg the Nobel Prize, May 19, 1927.

Stawarz M., *Punkt wyjścia filozoficznych rozważań Mariana Smoluchowskiego na temat przypadku i prawdopodobieństwa* (*The starting point of Marian Smoluchowski's philosophical deliberations on the subject of chance and probability*) „Semina Scientiarum" 2008, no 7, pp. 82–95.

Stawarz M., *Rekonstrukcja i krytyczna analiza poglądów filozoficznych Mariana Smoluchowskiego* (*Reconstruction and critical analysis of Marian Smoluchowski's philospohical views*) doctoral thesis, promoter: Dr hab. Paweł Polak, Kraków 2016.

Storczak Ł.I., *Diskussija o prirodie fiziczeskogo znania* (Discussion on the origin of physical knowledge), „Woprosy Fiłosofii" ("Problems of Philosophy") 1948, no. 1.

Strzałkowski A., *Wstęp (Introduction)*in: *Marian Smoluchowski (1872– 1917). Fizyk, taternik, romantyk nauki (Marian Smoluchowski (1872– 1917). Physicists, mountaineer, romantic f science)*, catalogue of the Collegium Maius temporary exhibition, May 17–July 14, 2002.

Superstrings: A Theory of Everything?, eds. P. Davies, J. Brown, Cambridge University Press, Cambridge 1988.

Sutherland W., *A Dynamical Theory of Diffusion for Non-Electrolytes and the Molecular Mass of Albumin*, "Philosophical Magazine and Journal of Science" 1905, series 6, vol. 9.

Sutherland W., *Causes of Osmotic Pressure and of the Simplicity of the Laws of Dilute Solutions*, "The London, Edinburgh, and Dublin Philosophical Magazine and Journal of Science" 1897, series 5, vol. 44.

Sutherland W., *Ionization, Ionic Velocities, and Atomic Sizes*, "The London, Edinburgh, and Dublin Philosophical Magazine and Journal of Science" 1902, series 6, vol. 4.

Svedberg T., Inouye K., *Eine neue Methodezur Prüfungder Gültigkeitdes Boyle-Gay-Lussacschen Gesetzesfürkolloide Lösungen (A new method to test the validity of the Boyle-Gay-Lussac law for colloidal solutions)* „Zeitschrift für Physikalische Chemie-Stöchiometrie und Verwandtschaftslehre" ("Journal of physical chemistry stoichiometry and relationship theory" 1911, vol. 77.

Szczepanowski S.W., *Zarys życia i prac Stanisława Prus Szczepanowskiego (Outline of Stanisław Prus Szczepanowski's life and works)* Wrocław 2020, http://www.rp-gospodarna.pl/Szczepanowski_2.pdf.

Szumilewicz I., *Koncepcja przyczynowości u Macha w świecie współczesnego determinizmu (Mach's concept of causality in the world of modern determinism)* „Studia Filozoficzne" ("Philosophical Studies") 1959, no 4.

Szumilewicz I., *Poincaré*, Wiedza Powszechna, Warsaw 1978. Średniawa B., *Rola współpracy Mariana Smoluchowskiego i Teodora Svedberga w prowadzonych w pierwszych latach XX wieku badaniach ruchów Browna i fluktuacji* (The role of the cooperation between Marian Smoluchowski and Theodor Svedberg in the study of Brownian motion conducted in th first years of the 20th century) „Postępy Fizyki" ("Advances in Physics") 1991, vol. 42, book 4.

Świerz M., *W dwudziestopięciolecie Sekcji Turystycznej Polskiego Towarzystwa Tatrzańskiego 1903–1928 (In a quarter-century of the Tourist Section of the Polish Tatra Society1903–1928)*, „Taternik. Organ Sekcji Turystycznej

Polskiego Towarzystwa Tatrzańskiego" ("Mountaineer. A unit of the Tourist Section of the Polish Tatra Society") 1928, no 4–6.

Taleb N.N., *Czarny łabędź* (*The Black Swan*), trans. O. Siara, Kurhaus Publishing, Warsaw 2016.

Tarski A., *Pisma logiczno-filozoficzne. Prawda* (*Logical philosophical writings. Truth*), vol. I, trans. J. Zygmunt, Wydawnictwo Naukowe PWN, Warsaw 1995

Teske A., *Marian Smoluchowski*, „Fizyka i Chemia: czasopismo dla nauczycieli" ("Physics and Chemistry: a magazine for teachers") 1951, no 3 (17).

Teske A., *Marian Smoluchowski: życie i twórczość* (*Marian Smoluchowski: life and works*), Państwowe Wydawnictwo Naukowe, Warszawa 1955.

The Collected Papers of Albert Einstein, vol. 2, *The Swiss Years: Writings, 1900–1909*, ed. J. Stachel, trans. A. Beck, Princeton University Press, Princeton 1989.

The Philosopher's Tree. Michael Faraday's Life and Work in His Own Words, ed. P. Day, CRC Press, London 1999.

Thomson J.J., *Recollections and Reflections*, G. Bell, London 1936.

Turek J., *Materializm* (*Materialism*) w: *Powszechna encyklopedia filozofii* (*The Universal encyclopedia of philosophy*) vol. VI, Polskie Towarzystwo Tomasza z Akwinu (The Polish Society of Thomas Aquinas), Lublin 2005. *The Collected Papers of Albert Einstein. Vol. 2: The Swiss Years: Writings, 1900–1909*, eds. J. Stachel, D.C. Cassidy, J. Renn, R. Schulmann, Princeton University Press, Princeton, New Jersey 1989.

Ulam S.M., *Marian Smoluchowski and the Theory of Probabilities in Physics*, Los Alamos, New Mexico 1956.

Ulam S.M., *Przygody matematyka* (Adventures of a mathematician), trans. A. Górnicka, Prószyński i S-ka, Warsaw 1996.

Wittgenstein L., *Tractatus logico-philosophicus*, trans. B. Wolniewicz, Wydawnictwo Naukowe PWN, Warszawa 2004.

Wróblewski A., *Marian Smoluchowski: Polak, który stworzył nową gałąź fizyki* (*Marian Smoluchowski: The Pole who created a new branch of physics*) Interia – historia, June 20, 2017, https://historia.interia.pl/aktualnosci/news-marian-smoluchowski- polak-ktory-stworzyl--nowa-galaz-fizyki,nId,2407763.

Wszołek S., *Esencjalizm transcendentalny K.R. Poppera* (*The transcendental essentialism of K.R. Popper*), „Zagadnienia Filozoficzne w Nauce" ("Philosophical Problems in Science") 2002, no 31.

Wyka E., *Marian Smoluchowski we wspomnieniach bliskich i przyjaciół* (*Marian Smoluchowski in the recollections of friends and relatives*), „Zwoje" ("Scrolls") 2003, no 2 (35).

Zakrzewski J., *Dzieci Mariana Smoluchowskiego* (*Marian Smoluchowski's children*), „Pauza Akademicka" 2017, no 380–381. Zsigmondy R., *Properties of Colloids*, Nobel Lecture, Chemistry 1922–1941, Singapore, New Jersey, London, Hong Kong 1999.

Zurek K., *High Energy Theory, Particle Astrophysics and Early Universe Cosmology*, http://www.kzurek.theory.caltech.edu/.

Życiński J., *Język i metoda* (*Language and method*). Znak, Kraków 1983.

Index of Names

Alembert, Jean Le Rond d' 203
Alicki, Robert 85
Allan, James D. 318
Altarelli, Guido 221
Alton, Giovanni Battista 137
Amsterdamski, Piotr 67, 68, 228, 396, 397, 398
Amsterdamski, Stefan 33, 230, 395, 400
Archimedes, z Syrakuz 193
Arkani-Hamed, Nima 302
Arystarch z Samos 197
Aristoteles 193–199, 283, 395
Auerbach, Felix 281, 285, 286, 395
Augustynek, Zdzisław 33, 395
Avenarius, Richard 204, 205

Bacon, Francis 195, 196, 197, 199
Bacon, Roger 195
Bafia, Stanisław 197, 395
Banach, Stefan 9, 48
Barabasz, Stanisław 157
Baraniecki, Marian 122, 305
Barnaś, Józef 315
Beatty, Alfred Chester 178
Becquerel, Antoine Henri 281, 297
Benedykt XVI 277
Benesch, Friedrich 153
Berg, Alban 120
Berkeley, George 315
Bernard, Giorgio 147, 148
Bernstein, Jeremy 298, 395
Berthelot, Marcellin Pierre 282
Bertram, H. (alpinista) 137, 152
Besso, Michele 76, 79, 80, 103, 396
Bettega, Michel 147, 148
Białas, Andrzej 315, 316

Białobrzeski, Czesław 234, 395
Bilczewski, Józef 276, 277
Bill, Robert 257
Binn, M. (alpinista) 137, 152
Birkhoff, George David 187
Bobrzyński, Michał 276
Bohr, Niels 248, 284, 301, 395
Boltzmann, Ludwig 21, 51, 58, 60, 61, 68, 73, 74, 77, 83, 87, 115, 173, 187, 262, 264, 283, 286, 293, 294, 295, 296, 329, 352, 396
Borra, Jean-Pascal 318
Born, Max 171, 224
Bouckaert, L.P. (imię nieustalone) 309, 395
Bouty, Edmond Marie 174
Boyle, Robert 231
Brahms, Johannes 121
Brentano, Franz 115
Bridgman, Percy Williams 68
Brown, Julian 284, 407
Brown, Robert ruchy Browna 58, 59, 75, 287
Bruckner, Anton 121
Brzozowski, Stanisław 110
Buras, Andrzej 316
Bunsen, Robert Wilhelm 290
Bunz, Helmut 317
Burtscher, Heinz 317

Cardano, Girolamo 15
Cardona, Manuel 171, 172
Carriere, Moritz 116
Casimir, Hendrik 282, 395
Cassidy, David C. 67, 408
Cassirer, Ernst 211
Carathéodory, Constantin 301

Chałubiński, Tytus 159
Chandrasekhar, Subrahmanyan 37, 54, 68, 187, 313, 314, 395, 398
Chmielowski, Janusz 145, 159, 160
Chramow, Jurij 281, 285, 395
Chwedeńczuk, Bohdan 395
Cichocki, Bogdan 50, 81, 383, 395
Clausius, Rudolf 174, 181, 287, 293
Cline, Douglas 315
Cohen, Hermann 211
Comte, August 202, 203, 204, 283, 395
Copleston, Frederick 202, 395
Corbalán, Fernando 16, 17, 395
Crombie, Alistair Cameron 194, 198, 395
Crookes, William 288
Curie, Piotr 281, 284, 297

Dahlgren, Erik Wilhelm 298
Dalton, John 57, 63
Damour, Thibault 54, 396
Danysz, Marian 314
Da-Ren, Chen 318
Darrigol, Olivier 54, 396
Darwin, Karol 202, 231, 234, 250, 277
Davies, Paul 284, 407
Davy, Humphry Bartholomew 288
Dawes, Gregory W. 224, 229, 230, 396
Day, Peter 288, 408
Dąmbska, Izydora 203, 396
Debye, Peter 171
Dietl, Tomasz 315
Dirac, Paul 221
Dorn, Friedrich Ernst 281
Dryden, Gordon 283, 396
Du Bois-Reymond, Emil 222
Dudyński, Marek 167, 399
Duhem, Pierre 209

Duplantier, Bertrand 54, 72, 73, 79, 80, 82, 96, 101, 102, 103, 396
Dybowski, Benedykt 277
Dziekan, Małgorzata 35, 117, 405

Eberly, Joseph Henry 314
Edwards, D.A. 318
Eftekhari, Ali 295, 396
Einstein, Albert 37, 46, 50, 52, 53, 54, 55, 58, 63, 67, 68, 69, 70, 72, 73, 74, 75, 76, 77, 78, 79, 80, 81, 82, 83, 84, 85, 86, 87, 88, 89, 90, 91, 92, 93, 94, 95, 96, 97, 98, 99, 100, 101, 102, 103, 171, 172, 188, 208, 224, 227, 231, 235, 277, 284, 285, 286, 291, 292, 297, 298, 299, 300, 304, 395, 396, 398, 399, 401, 408
Ehrenfest, Tatiana 269
Ekiel-Jeżewska, Maria 167, 399
Engels, Fryderyk 241, 247
Exner, Franz Serafin 59, 61, 79, 92

Faraday, Michael 284, 288, 289, 290, 291, 292, 293, 400
Fermat, Pierre de 15
Filipczuk, Michał 398
Filippov, Andrey V. 317
Flagan, Richard C. 317
Ford, Ian J. 318
Florow, Georgij Nikołajewicz 314
Franck, César 121
Freud, Sigmund 120
Friedman, Aleksandr 238
Fulde, Peter 316
Fuliński, Andrzej 167, 168, 186, 397

Gadacz, Tadeusz 286
Galileusz (wł. Galileo Galilei) 15, 196, 197, 198, 199, 228, 283, 292, 297, 303

Index of Names

Gałązka, Robert R. 315
Gałecki, Antoni 191, 397
Ganán-Calvo, Alfonso M. 318
Gradoń, Leon 317
Gawecki, Bolesław J. 22, 23, 24, 25, 26, 28, 29, 37, 75, 397, 402, 404
Gibbs, Josiah Willard 72, 73, 187
Gierula, Jerzy 314
Gleichen-Rußwurm, Wilhelm Friedrich von 59
Goetel, Walery 45, 105, 121, 131, 184, 190, 192, 249, 270, 271, 274, 275, 359, 363, 397
Gödel, Kurt 229
Godlewski, Tadeusz 41, 42, 46, 47, 74, 75, 126, 174, 181, 183, 189, 191, 192, 397
Gołąb-Meyer, Zofia 46, 269, 383, 397
Gombrowicz, Witold 9
González, Loscertales Ignacio 318
Gosiewski, Władysław 23, 32
Gombaud, Antoine (Kawaler de Méré) 15
Gouy, Louis Georges 63
Gozzini, Adriano 314
Górnicka, Agnieszka 37, 408
Graham, Thomas 74
Grinshpun, Sergey 317
Grotowski, Jerzy 126
Grotowski, Kazimierz A. 126, 129, 136, 138, 140, 397
Gythiel, Anthony 90, 400

Haake, Fritz 316
Hajduk, Zygmunt 304, 398
Harman, Gilbert 229
Hasenöhrl, Friedrich 262, 264
Hawking, Stephen W. 18, 67, 227, 228, 238, 397, 398

Hegel, Georg Wilhelm Friedrich 241, 245, 247, 285, 286, 398
Heinrich, Władysław 23, 24
Heisenberg, Werner 33, 221, 224, 248, 398
Heller, Michał 18, 215, 398, 402
Helmholtz, Hermann von 63, 287, 288, 289, 294
Henri, Victor 78
Herbut, Józef 226, 398
Hermite, Charles 174
Herschel, John Frederick William 193, 195, 196, 398
Hertz, Heinrich 291, 296
Hirsch, Jorge E. 170, 171
Hobson, Ernest William 257
Hoerlin, Hermann 166
Höfler, Alois 115, 262, 264
Hoff, Jacobus van 't 81, 82, 87, 88, 103
Hofmann, Werner 316
Hofmannsthal, Hugo von 120
Hogan, Christopher J., Jr. 318
Home, Roderick W. 95, 398
Hossenfelder, Sabine 220, 221, 302, 303, 304, 398
Hrynkiewicz, Andrzej 315
Hume, David 22, 210, 237, 399

Ingarden, Roman S. 37, 398
Ingen-Housz, Jan 58
Inouye, Katsuji 51, 407
Isaacson, Walter 284, 398

Jabłoński, Aleksander 314
James, William 206, 226, 227
Jayasinghe, Suwan 318
Jędrusiak, Szymon 197, 401
Jolly, Philipp von 282

Joule, James Prescott 177, 287, 293
Jóźwiak, Bożena 396
Juszkiewicz, Adolf Pawłowicz 16, 398

Kac, Mark 9, 36, 37, 54, 73, 101, 186, 187, 398
Kalberer, Markus 318
Kallmus, Dora Philippine (Madame D'Ora lub Madame d'Ora) 120
Kálmán Emmerich 120
Kampen, Nicolaas Godfried van 167, 399
Kania, Ireneusz 401
Kant, Immanuel 203, 210, 211
Kapczyński, Marcin 170, 399
Karłowicz, Mieczysław 121
Kerminen, Veli-Matti 318
Karpowicz, Stanisław 165
Kelly, Thomas 237, 399
Kepler, Johannes 231, 34, 235
Kerner, Ryszard 282, 401
Kerner, Samuel 282, 401
Kielich, Stanisław 315
Kirchhoff, Gustav 174, 183, 284, 290, 291
Klemensiewicz, Zygmunt 11, 121, 130, 132, 133, 141, 145, 146, 147, 154, 251, 363, 399
Klimt, Gustav 120
Kluza, Maciej 313, 321, 363, 383, 399
Kokoschka, Oskar 120
Kołakowski, Leszek 245, 246, 399
Kopernik, Mikołaj 325, 329, 339, 342, 348, 356, 400
Kordys, Roman 145
Kosmulski, Marek 169, 399
Kossowicz, Tadeusz 141, 154
Kostro, Ludwik 224, 399
Kotowa, Barbara 204, 399

Kozłowski, Władysław Mieczysław 23, 25, 26, 27, 28, 29, 399
Krajewski, Władysław 242, 244, 245, 246, 247, 248, 277, 399, 402
Krośniak, Marek 18, 398
Królikowski, Wojciech 315
Kruis, Einar 318
Kuhn, Thomas 193, 194, 230, 400
Kulczyński, Władysław jun. 145
Kulmala, Markku 318
Kumaniecki, Kazimierz 24, 400
Kurlbaum, Ferdinand 281
Kürten, Andreas 318

Landman, Adam 398
Langevin, Paul 89, 90, 95, 400
Langmuir, Irving 171
Laplace, Pierre Simon de 16, 17, 18
Laue, Max von 301
Lavoisier, Antoine 287, 291
Lehár, Franz 120
Lehmann, Otto 60
Lemańska, Anna 33, 34
Lemons, Don S. 90, 400
Lenk, Robert 135, 150, 151
Lenin, Włodzimierz 241, 245
Leo XIII 213
Le Roy, Édouard 210
Leśniak, Kazimierz 395
Lippmann, Gabriel 174, 322
Lipszyc, Henryk 398
Lipszyc, Katarzyna 398
Litwinowicz-Droździel, Małgorzata 288. 289. 290. 400
Lorenz, Hans 135, 137, 140, 150, 151, 152, 153, 154
Lorentz, Hendrik 296, 297
Loria, Stanisław 37, 68, 77, 80, 81, 181, 182, 184, 185, 190, 400
Lukács, György 245

Index of Names

Lukierski, Jerzy 315
Łazarewicz, Ginzburg Witalij 314
Łoś, Jan 24
Łypacewicz, Stanisław 194, 395

Mackiewicz, Witold 245, 402
Mach, Ernst 21, 22, 23, 24, 26, 27, 28, 29, 31, 34, 58, 60, 69, 204, 219, 278, 295
Mädler, Lutz 318
Mahler, Gustav 120
Maiocchi, Roberto 75, 76, 78, 400
Maria, Teresa Habsburg 113
Marshall, Stephen James 284, 400
Markov, Andriej Andriejewicz 32, 167, 242
Marx, Karl 241, 247
Marx, Werner 171, 172, 400
Maryniarczyk, Andrzej 214, 400
Maślanka, Jerzy 112, 131, 145, 400
Maxwell, James Clerk 17, 65, 73, 85, 93, 187, 281, 283, 284, 289, 291, 292, 293, 294, 296, 297, 302, 400, 401
Mayer, Julius Robert von 63
Mączka, Janusz 18, 402
Mehlberg, Henryk 10
Meinong, Alexius 19
Meitner, Lise 269
Mejbaum, Wacław 33, 395
Mendelsohn, Bartholdy Felix 121
Merton, Robert 101
Merz, Walter 135, 139, 150, 151, 152, 153
Metallmann, Joachim 23, 36, 401
Michelson, Albert A. 282, 284, 285, 401
Mięsowicz, Marian 314
Miller, Tomasz 398
Miłosz, Czesław 9
Mises, Ludwig von 120
Misiewicz, Jan 315

Mlodinow, Leonard 228
Möller, Wolfgang 317
Morales, José María Riaza 197, 401
Moroziewicz, Józef 192
Moser, Kaspar 139
Moser, Koloman 120
Mottelson, Ben R. 313, 314
Musil, Robert 120

Nafe, O. (alpinista) 137, 150, 152
Nägeli, Carl Wilhelm von 64, 65
Napiórkowski, Kazimierz 303, 398
Nasibulin, Albert G. 318
Natanson, Władysław 14, 18, 23, 25, 47, 64, 203, 321, 322, 323, 324, 325, 326, 327, 328, 329, 330, 331, 332, 334, 336, 337, 338, 339, 340, 342, 343, 344, 345, 346, 347, 348, 349, 350, 352, 353, 354, 355, 356, 357, 396, 402, 403, 404, 405, 406
Natorp, Paul 211
Needham, John Turberville 59
Niessner, Reinhard 317
Nernst, Walther 81, 103, 301, 401
Norman-Neruda, Ludwig 138
Neurath, Otto 120
Newton, Isaac 68, 219, 222, 227, 228, 231, 234, 235, 283, 284, 291, 292, 299, 401
Nicolussi, Matteo 133, 147, 148
Niezgoda, Agnieszka 269, 401
Norton, John D. 220, 236, 237, 238, 303, 399, 401

Ockham, William 197
O'Dowd, Colin 318
Olbrich, Joseph Maria 120
Oleś, Andrzej M. 316
Olszewski, Karol 185, 357
Opęchowski, Władysław 314
Ostwald, Wilhelm 60, 69, 80, 295

Paczyński, Bohdan 315
Palczewski, Andrzej 130, 401
Pascal, Blaise 15, 16, 318
Pawłowski, Tadeusz 193, 398
Pearson, Gerald Leondus 314
Peirce, Charles Sanders 206, 226, 229
Penrose, Roger 68, 292, 299
Perrin, Jean Baptiste 49, 50, 52, 53, 64, 67, 70, 72, 78, 94, 100, 300, 401
Peverone, Giovanni Francesco 15
Piekara, Arkadiusz Henryk 314
Pius X 213
Planck, Max 167, 171, 281, 282, 286, 291, 295, 296, 298, 304, 348, 401
Pniewski, Jerzy 314
Pogonowski, Jerzy 229, 406
Poincaré, Jules Henri 54, 174, 209, 210, 211, 222, 232, 235, 283, 396, 407
Polak, Paweł 18, 46, 135, 277, 401, 402, 406, 408
Popliński, Antoni 110
Popper, Karl Raimund 208, 209, 225, 408
Portell, Bueso Xavier 221, 402
Pratsinis, Sotiris E. 317
Prentki, Jacek 315
Priestly, Joseph 287
Prus, Bolesław (właśc. Aleksander Głowacki) 110, 407
Pokorski, Stefan 315
Pomorski, Krzysztof 315
Prather, Kimberly A. 318
Prejbisz, Antoni 238, 402
Ptolemeusz 196, 197, 234, 235
Pui, David Y.H. 317
Purtscheller, Ludwig 165
Putnam, Hilary 232, 233, 402

Read, Duncan Hicks 305

Redlich, Krzysztof 316
Rejmer, Krzysztof 16, 395
Renn, Jürgen 67, 408
Riemann, Bernhard 238
Riggs, Louise 311
Riipinen, Ilona 318
Rivasseau, Vincent 54, 396
Röntgen, Wilhelm Conrad 297, 322, 323
Roskal, Zenon E. 203, 383, 402
Roszkowska, Ewa 135, 137, 140, 142, 147, 402
Rubczyński, Witold 24
Rubens, Heinrich 281, 298
Rubinowicz, Wojciech 314
Rudzki, Maurycy 184, 356, 403
Rutherford, Ernest 257, 301
Rzewuski, Jan 314

Schleiermacher, August 181
Schmidt-Ott, Andreas 317
Schönberg, Arnold 120
Schrödinger, Erwin 224
Schubert, Franz 121
Schulmann, Robert 67, 408
Schumann, Robert 121
Schuster, Oscar 139, 153
Shugar, David 315
Siara, Olga 408
Sienkiewicz, Henryk 110
Simonyi, Károly 67
Shapiro, Michael 317
Skłodowska-Curie, Maria 9, 169, 171, 269, 281, 297, 361
Skowroński, Jarosław 398
Sławianowski, Jan Jerzy 19, 402
Słomski, Wojciech 245, 402
Smith, Adam 238, 402
Smoluchowska, Aldona (Donia) 306, 307
Smoluchowska Irena

Index of Names

Smoluchowska, Teofila 106, 108, 110, 147, 148, 149, 150, 151, 152, 153, 307
Smoluchowska, Zofia (z domu Baraniecka) 191, 307, 308, 311, 312
Smoluchowski, Marian 9, 10, 11, 13, 14, 17, 18, 19, 21, 22, 23, 24, 25, 29, 30, 31, 32, 33, 34, 35, 36, 37, 39, 40, 41, 42, 43, 44, 45, 46, 47, 48, 49, 50, 51, 52, 53, 54, 55, 57, 58, 59, 60, 61, 62, 63, 64, 65, 66, 67, 68, 69, 71, 72, 73, 74, 75, 76, 77, 78, 79, 80, 81, 83, 84, 85, 86, 87, 88, 89, 90, 91, 92, 93, 94, 95, 96, 97, 98, 99, 100, 101, 102, 103, 105, 106, 112, 113, 114, 115, 116, 117, 119, 120, 121, 122, 124, 126, 127, 129, 130, 131, 132, 133, 135, 136, 137, 138, 139, 140, 142, 144, 145, 146, 147, 154, 155, 156, 157, 158, 159, 160, 161, 162, 163, 164, 165, 166, 167, 168, 169, 170, 172, 173, 174, 177, 178, 180, 181, 182, 183, 184, 185, 186, 187, 188, 189, 190, 191, 192, 193, 195, 196, 197, 199, 201, 202, 203, 204, 205, 206, 207, 208, 209, 210, 211, 212, 214, 215, 216, 217, 218, 219, 220, 221, 222, 223, 225, 226, 227, 228, 229, 230, 233, 235, 239, 241, 242, 244, 245, 246, 247, 248, 249, 250, 251, 253, 255, 257, 258, 261, 262, 263, 264, 265, 266, 267, 268, 269, 270, 271, 273, 275, 276, 277, 278, 279, 283, 286, 299, 300, 301, 302, 305, 313, 315, 316, 318, 320, 321, 322, 323, 324, 325, 326, 327, 328, 329, 330, 331, 332, 333, 334, 335, 336, 337, 338, 339, 340, 341, 342, 343, 344, 345, 346, 347, 348, 349, 350, 351, 352, 353, 354, 355, 356, 357, 359, 360, 361, 363, 395, 396, 397, 398, 399, 400, 401, 402, 403, 404, 405, 406, 407, 408, 409
Smoluchowski, Peter 311
Smoluchowski, Roman 37, 308, 309, 395, 398
Smoluchowski Tadeusz 10
Smoluchowski Wilhelm Ritter von 50, 108
Smullyan, Raymond 229, 406
Sobolewski, Andrzej L. 316
Sommerfeld, Arnold 37, 46
Sosnowski, Leonard 314
Sosnowski, Ryszard 315
Soto, Domingo de 197
Söderbaum, Henrik Gustaf 52, 67, 406
Speziali, Pierre 79, 80, 396
Stachel, John 67, 68, 74, 77, 83, 93, 299, 408
Stalin, Józef 246
Stark, Wendelin J. 318
Stawarz, Małgorzata 35, 277, 406
Stayer, Marcia Sweet 222, 401
Stefan, Jožef 117, 293
Steinhaus, Hugo 9, 48
Storczak, Ł. (imię nieustalone) 36, 242, 244, 407
Stern, Otto 301
Strutt, John William 172
Strzałkowski, Adam 46, 48, 261, 397, 407
Sutherland, William 68, 72, 73, 74, 79, 80, 81, 82, 91, 95, 96, 97, 103, 300, 398, 407
Svedberg, Theodor 49, 50, 51, 52, 54, 63, 67, 94, 100, 406, 407
Szczepanowski, Stanisław
Szczepanowski, Stanisław Wiktor 110, 407
Szumilewicz, Irena 17, 210, 219, 407
Szumilewicz Ewa 383, 401
Szymanski, Vladek 317

Szymański, Zdzisław 315
Szymczak, Henryk 315
Śledziewska-Błocka, Danuta 167, 399
Średniawa, Bronisław 52, 407
Świątecki, Władysław 315
Świerz, Mieczysław 145, 146, 159, 164, 166, 407

Taleb, Nassim Nicholas 282, 286, 295, 408
Tarski, Alfred 9, 228, 229, 408
Teleki, Alexandra 318
Teske, Armin 47, 114, 119, 120, 121, 124, 142, 173, 178, 180, 191, 244, 249, 255, 258, 265, 266, 276, 408
Thomson, Joseph John 257, 297, 408
Thomson, William lord Kelvin 177, 292, 293
Tomasz z Akwinu 398, 400, 408
Trautman, Andrzej 314
Trzebiatowski, Włodzimierz 314
Turek, Józef 241, 408
Turski, Łukasz A. 167, 399
Twardowski, Kazimierz 9, 48, 253

Ulam, Stanisław M. 9, 37, 39, 48, 120, 166, 177, 178, 183, 186, 187, 188, 192, 257, 302, 408

Villard, Paul Ulrich 281
Vinci, Leonardo da 196
Vos, Jeanette 283, 396

Wagner, Paul E. 317
Wagner, Otto 120
Wagner, Richard 121
Warburg, Emil 180, 315, 316
Warczak, Andrzej 316
Weber, Alfred P. 318
Webern, Anton 120
Weinberg, Steven 309

Wessely, Victor 135, 137, 140, 151, 152, 153, 154
Weisskopf, Victor Frederick 314
Wien, Wilhelm 296, 298, 324
Wigner, Eugene 309, 395
Winkelmann, Adolf 181
Witkowski, August 185, 259, 308
Wittgenstein, Ludwig 120, 251, 408
Whymper, Edward 140
Wojciechowska, Wanda 202, 395
Wolfendale, Arnold Whittaker 315
Wolniewicz, Bogusław 408
Wolska, Benigna 114
Wolszczan, Aleksander 315
Wöste, Ludger 316
Wróblewski, Andrzej Kajetan 285, 301, 315, 401
Wróblewski, Artur 46, 135, 408
Wszołek, Stanisław 209, 408
Wyka, Ewa 191, 408

Yao, Maosheng 318

Zakrzewski, Jakub 305, 309, 409
Zakrzewski, Janusz 305, 309, 316, 409
Zakrzewski, Wincenty 145
Zalewski, Kacper 315
Zawadzki, Włodzimierz 315
Zeeman, Pieter 282, 296
Zeidler, Paweł 402
Zermelo, Ferdinand 60
Zimbardo, Philip 269, 401
Zsigmondy, Emil 130, 131
Zsigmondy, Richard A. 49, 50, 54, 165, 186, 409
Zurek, Kathryn 304, 409
Zygmunt, Jan 228, 408
Żłobicki, Władysław 45
Żuławski, Jerzy 159, 160
Żurek, Wojciech 315
Życiński, Józef 229, 409

Studies in Social Sciences, Philosophy and History of Ideas

Edited by Andrzej Rychard

Vol.	1	Józef Niżnik: Twentieth Century Wars in European Memory. 2013.
Vol.	2	Szymon Wróbel: Deferring the Self. 2013.
Vol.	3	Cain Elliott: Fire Backstage. Philip Rieff and the Monastery of Culture. 2013.
Vol.	4	Seweryn Blandzi: Platon und das Problem der Letztbegründung der Metaphysik. Eine historische Einführung. 2014.
Vol.	5	Maria Gołębiewska / Andrzej Leder/Paul Zawadzki (éds.): L'homme démocratique. Perspectives de recherche. 2014.
Vol.	6	Zeynep Talay-Turner: Philosophy, Literature, and the Dissolution of the Subject. Nietzsche, Musil, Atay. 2014.
Vol.	7	Saidbek Goziev: Mahalla – Traditional Institution in Tajikistan and Civil Society in the West. 2015.
Vol.	8	Andrzej Rychard / Gabriel Motzkin (eds.): The Legacy of Polish Solidarity. Social Activism, Regime Collapse, and the Building of a New Society. 2015.
Vol.	9	Wojciech Klimczyk / Agata Świerzowska (eds.): Music and Genocide. 2015.
Vol.	10	Paweł B. Sztabiński / Henryk Domański / Franciszek Sztabiński (eds.): Hopes and Anxieties in Europe. Six Waves of the European Social Survey. 2015.
Vol.	11	Gavin Rae: Privatising Capital. The Commodification of Poland´s Welfare State. 2015.
Vol.	12	Adriana Mica / Jan Winczorek / Rafał Wiśniewski (eds.): Sociologies of Formality and Informality. 2015.
Vol.	13	Henryk Domański: The Polish Middle Class. Translated by Patrycja Poniatowska. 2015.
Vol.	14	Henryk Domański: Prestige. Translated by Patrycja Poniatowska. 2015.
Vol.	15	Cezary Wodziński: Heidegger and the Problem of Evil. Translated into English by Agata Bielik-Robson and Patrick Trompiz. 2016.
Vol.	16	Maria Gołębiewska (ed.): Cultural Normativity. Between Philosophical Apriority and Social Practices. 2017.
Vol.	17	Anita Williams: Psychology and Formalisation. Phenomenology, Ethnomethodology and Statistics. 2017.
Vol.	18	Mikołaj Pawlak: Tying Micro and Macro. 2018.
Vol.	19	Franciszek Sztabiński / Henryk Domański / Paweł B. Sztabiński (eds.): New Uncertainties and Anxieties in Europe. Seven Waves of the European Social Survey. 2018.
Vol.	20	Adriana Mica / Katarzyna M. Wyrzykowska / Rafał Wiśniewski / Iwona Zielińska (eds.): Sociology of the Invisible Hand. 2018.

Studies in Philosophy, Culture and Contemporary Society

Edited by Bogusław Paź

Vol.	21	Jan Felicjan Terelak: Psychology of the Operator of Technical Devices. 2019.
Vol.	22	Dorota Maria Leszczyna: Del idealismo al realismo crítico. La política como realización en José Ortega y Gasset. 2019.

Vol. 23 Zbigniew Drozdowicz: La république des savants. Sans révérence. Traduit du polonais par Catherine Popczyk. 2019.

Vol. 24 Andrzej Waśkiewicz: The Idea of Political Representation and Its Paradoxes. Translated from Polish by Agnieszka Waśkiewicz and Marilyn Burton. 2019.

Vol. 25 Ilona Błocian / Dmitry Prokudin (eds.): Imagination – Art, Science and Social World. 2019.

Vol. 26 Zbigniew Drozdowicz: Faces of the Enlightenment. Philosophical sketches. 2019.

Vol. 27 Włodzimierz Piątkowski: From Medicine to Sociology. Health and Illness in Magdalena Sokołowska's Research Conceptions. 2020.

Vol. 28 Roman Witold Ingarden: Die Mitschriften von den Vorlesungen Martin Heideggers über die Phänomenologische Interpretation von Kants *Kritik der reinen Vernunft* (Wintersemester 1927/28). Aus dem Manuskript abgeschrieben und das Vorwort verfasst haben: Radosław Kuliniak und Mariusz Pandura. 2020.

Vol. 29 Krzysztof Wielecki / Klaudia Śledzińska (eds.): The Relational Theory of Society. Archerian Studies vol. 2. 2020.

Vol. 30 Zenon Gajdzica / Robin McWilliam / Miloň Potměšil / Guo Ling: Inclusive Education of Learners with Disability – The Theory versus Reality. 2020.

Vol. 31 Jan Felicjan Terelak: Antarctic Winter-Over Syndrome. Narrative Perspective. 2021.

Vol. 32 Nuria Sánchez Madrid / Julia Muñoz Velasco / José Luis Villacañas Berlanga (eds.): El ethos del republicanismo cosmopolita. Perspectivas euroamericanas sobre Kant. 2021.

Vol. 33 Zbigniew Drozdowicz: Academic Culture. Traditions and the Present Days. 2021.

Vol. 34 Jan Felicjan Terelak: Antarctic Isolation as a Mars Habitat Analogue. A Psychological Perspective. 2021.

Vol. 35 Aleksandra Horowska (ed.): The Labyrinths of Leibniz's Philosophy. 2022.

Vol. 36 Bogusław Szuba / Tomasz Drewniak (eds.): Beauty in Architecture. Harmony of Place. 2022.

Vol. 37 Monika Bukowska / Krzysztof Wielecki (eds.): The Transformations of Contemporary Culture and Their Social Consequences. Archerian Studies Vol. 3. 2023.

Vol. 38 Kinga Anna Gajda (ed.): (Non)Commemoration of the Heritage in Eastern Europe. 2024.

Vol. 39 Jan Grzanka: The Forgotten Genius of Physics, a work on Marian Smoluchowski. The Story of Marian Smoluchowski. 2025.

www.peterlang.com

www.ingramcontent.com/pod-product-compliance
Ingram Content Group UK Ltd.
Pitfield, Milton Keynes, MK11 3LW, UK
UKHW041924210426
5322IPUK00002B/54